© Yoichiro Kawai

About the Author

YOSSEF BODANSKY, author of *The Secret History of the Iraq War* and the #1 *New York Times* bestseller *Bin Laden: The Man Who Declared War on America*, was the director of the Congressional Task Force on Terrorism and Unconventional Warfare for sixteen years (1988–2004). He is also the longtime director of research at the International Strategic Studies Association. The author of ten books, he has written frequently for numerous periodicals, including *Global Affairs*, *Jane's Defence Weekly*, *Defense and Foreign Affairs: Strategic Policy*, and *BusinessWeek*. A member of the Prague Society for International Cooperation, he is a former senior consultant for the U.S. Department of Defense and the U.S. Department of State. Bodansky was also a visiting scholar in the Johns Hopkins University's School of Advanced International Studies. He divides his time between the Middle East and the Washington, D.C., area.

Also by Yossef Bodansky

The Secret History of the Iraq War

The High Cost of Peace: How Washington's Middle East Policy Left America Vulnerable to Terrorism

Bin Laden: The Man Who Declared War on America

Islamic Anti-Semitism as a Political Instrument

Some Call It Peace: Waiting for the War in the Balkans

Offensive in the Balkans: Potential for a Wider War as a Result of Foreign Intervention in Bosnia-Herzegovina

Terror: The Inside Story of the Terrorist Conspiracy in America

Target America and the West: Terrorism Today

CHECHEN JIHAD

AL QAEDA'S TRAINING GROUND AND THE NEXT WAVE OF TERROR

YOSSEF BODANSKY

HARPER

NEW YORK • LONDON • TORONTO • SYDNEY

HARPER

A hardcover edition of this book was published in 2007 by HarperCollins Publishers.

HarperCollins books may be purchased for educational, business, or sales promotional use. For information please write: Special Markets Department, HarperCollins Publishers, 10 East 53rd Street, New York, NY 10022.

FIRST HARPER PAPERBACK EDITION PUBLISHED 2008.

Designed by William Ruoto

Library of Congress Cataloging-in-Publication Data is available upon request.

ISBN 978-0-06-142977-4

09 10 11 12 13 NMSG/RRD 10 9 8 7 6 5 4 3 2 1

When will blood cease
to flow in the mountains?
When sugar-canes grow
in the snows.

—A CAUCASIAN PROVERB

CONTENTS

Introduction ix

1 Why Should We Care? 1
2 Legacy and Roots 7
3 Exploiting the Collapse of the Soviet Union 20
4 First Steps toward Jihad 32
5 The Ascent of Islamist-Jihadist Chechnya 44
6 Interlude 56
7 Establishing a Terrorist State 66
8 Escalating and Waiting 75
9 The Slide to Crisis 87
10 The Slide to War 106
11 The Second Chechen War 117
12 War Again 131
13 The Internationalization of the Jihad 139
14 A "New" War 153
15 Money Matters 167
16 The Routine of War 179
17 Chechnya and the Palestinian Problem 190
18 The Chechen Jihad after 9/11 199
19 Chechenization in Afghanistan 210
20 Moscow Strikes Back in Chechnya 223
21 Strikes in Moscow and in Western Europe 236
22 The Black Widows 257

23 The Road to Beslan **271**
24 Self-Devouring. **295**
25 Going International **305**
26 In the Theater of Global Jihad **323**
27 Pacification in Chechnya,
Eruption in the Caucasus **334**
28 Center Stage in Global Jihad **354**
29 End Game **377**

Postscript **389**
A Note on Sources and Methods **420**
Acknowledgments **427**
Index **430**

INTRODUCTION

ON JULY 15, 2006, THE LEADERS OF THE WORLD'S MOST POWERFUL governments arrived in St. Petersburg to discuss international economics, politics, and issues of security at the annual G-8 Summit. Russian president Vladimir Putin hosted the meeting, and the guest list included President George W. Bush, Prime Minister Tony Blair, Chancellor Angela Merkel of Germany, Prime Minister Stephen Harper of Canada, President Jacques Chirac of France, Prime Minister Romano Prodi of Italy, and Prime Minister Junichiro Koizumi of Japan.

Someone else wanted to crash the meeting. Shamil Basayev, also known by his nom de guerre, Amir Abdallah Shamil Abu-Idris, was planning to make this a G-8 meeting never to be forgotten.

Basayev, a tall, long-bearded forty-one-year-old, was a hero to some. But to most he was the nefarious chief of the Islamist-Jihadists in Chechnya, one of the most feared terrorists in the world. He had masterminded the 2002 siege of the Dubrovka Theater in Moscow (which led to the deaths of 129 hostages and 42 terrorists) and the 2004 seige of a secondary school in Beslan, Russia (where 344 civilians were killed, 186 of them children). Basayev was capable of planning elaborate attacks, and he viewed the deaths of innocent people not as collateral damage but as political justice.

For Basayev and his fellow Islamist-Jihadist leaders in Chechnya, the G-8 summit was an irresistible target. A spectacular strike deep into the heart of Russia was perhaps his last chance to refocus world attention on the lingering struggle in Chechnya and generate international pressure on the Kremlin to make concessions to the Chechens.

Aware of the logistical challenges such a mission would involve, the Chechen jihadist leaders consulted with the supreme Islamist-Jihadist leadership, including members of the inner circle of Osama bin Laden and Ayman al-Zawahiri, in both Tehran and the mountains of the Afghanistan-Pakistan border area. The al Qaeda leaders promised them every possible assistance, deeming the proposed operation so important that they instructed the leading Arab volunteer commanders then in Chechnya to participate in the strike personally alongside their Chechen counterparts.

In the second half of May, the Islamist-Jihadist leadership in Chechnya evolved two contingency plans: best-case and worst-case scenarios. Their preferred plan involved sending a strike team of martyrdom seekers—suicide terrorists—to St. Petersburg to conduct a spectacular two-phase strike at the time of the summit. In the first phase, they would shower the summit's compound with improvised rockets—called *shaytan-truba* ("devil-tubes") by the Russian security services—in order to scare, if not kill, the world leaders. The second phase called for the terrorist team to withdraw into a crowded public place (a school, hospital, or shopping mall) in the center of St. Petersburg, take as many hostages as possible, and commit to a protracted standoff. The plan specifically required the terrorists to commit martyrdom at the end—causing massive carnage among the civilian hostages. The Islamist-Jihadist leadership in Chechnya stressed the importance of hostage taking because such operations take time to resolve. With the international media gathered in St. Petersburg for the summit, the operation would attract unprecedented international attention, shaming President Putin and his fellow antijihadist guests—particularly President George W. Bush and Prime Minister Tony Blair.

The alternative plan addressed the possibility that the strike team might fail to breach the security cordons around St. Petersburg or that they might arouse suspicion before they reached the city. Under such circumstances, the terrorist team was to proceed to the nearest small town and conduct a localized version of the plan. They would first rocket the center of the town in order to attract attention and sow confusion. Then the terrorists would seize a school or a hospital full of hostages, barricade themselves, and maintain a standoff for as long as possible. At a given time, they would provoke a clash with the security forces that would culminate in the mass murder of the hostages and the martyrdom of the terrorists. Throughout the standoff, the terrorists were instructed, they should demand access to foreign officials as well as both Russian and foreign media, ostensibly to be used as mediators but actually to serve as instruments for transmitting the terrorists' message to the world.

Regardless of whether the terrorists reached St. Petersburg, both the Chechnya-based jihadist leadership and the supreme al Qaeda leadership were certain that such an audacious strike would provoke massive retaliation by the Russian military and security forces. The leaders hoped that such a Russian strike would antagonize the civilian population of Chechnya, and the northern Caucasus as a whole, enough to scuttle the trend of pacification and normalization that had dampened the Islamists' momentum in the region in recent years. The leaders looked forward to the spectacle of casting the northern Caucasus back into the flames of jihad.

In the first half of June, the region's most prominent Islamist-Jihadist commander, the native Chechen terrorist Shamil Basayev conducted a series of meetings with his senior commanders in Chechnya and southern Russia to begin implementing the strike. The most important of these was a "war council" convened in Krasnodar Krai, southern Russia, with a group of *amirs* (senior commanders) based in the krais of Adygeya, Karachayevo-Cherkessia, Krasnodar, and Stavropol. In that meeting, Basayev outlined his plans and demanded that the amirs start preparing the logistical and transportation infrastructure required to smuggle the terrorist team deep into the heart of Russia—with luck, all the way to St. Petersburg.

During this planning phase, however, the supreme al Qaeda leaders warned the Chechen jihadists that U.S. and British intelligence personnel—the latter feared most—would be taking extra precautions to prevent any viable threat to the Western leaders. The Islamist-Jihadists would have to launch a major diversion in order to attract the attention of the intelligence services—both Russian and Western—and perhaps cause friction between them. The diversion should be conducted outside Russia and the Caucasus, the supreme leaders dictated, but the Chechen leadership must seem to be directly involved in, if not responsible for, the operation. After some deliberation, the leaders settled on a strike in Iraq, where trusted jihadist resources in the Baghdad area could be activated and directed on a moment's notice.

On June 8, Basayev started laying groundwork to create the impression that the Chechens were responsible for the upcoming decoy operation, even though his forces would actually have nothing to do with it. He convened "an urgent commanders' meeting" in Chechnya, knowing full well that Russian intelligence would learn about the contents soon afterward. Basayev also made sure that a report on the meeting would be transmitted to his allies in the Middle East through channels likely to be intercepted by Western intelligence services. In the meeting, Basayev announced that "Chechen special services" would now be used "abroad to arrest and kill international terrorists and war criminals [i.e., the Russians] who are responsible for the genocide of the Chechen people and have been sentenced to death by the Shariah court." Basayev concluded his announcement with a dramatic statement: "We attach great significance to the punishment of war criminals and international terrorists responsible for the genocide of the Chechen people and violence against the Muslims of the Caucasus and the whole of Russia. Thank God, our forces and capabilities are growing. We have been watching some odious figures. God willing, none of the criminals will be able to escape due punishment. Retribution will not be limited to any time frame."

On June 8, Basayev offered his stamp of approval for an operation that was already unfolding. Back on June 3, a group of masked gunmen attacked an SUV belonging to the Russian embassy in the upscale Baghdad neighborhood of Mansour, some four hundred yards from the embassy compound. It was a swift and professional operation. Within a few minutes the terrorists had killed a security officer named Vitaly Titov and seized four other diplomats: Fyodor Zaytsev, Rinat Aglyulin,

Anatoly Smirnov, and Oleg Fedosseyev, blindfolding and tying them up before pushing them into waiting vehicles and speeding away. No one intervened to help the Russians; nearby residents later insisted they saw and heard nothing. Though the attack bore all the hallmarks of a jihadist operation, no one claimed responsibility or made any demand in connection with the seizure. Meanwhile, U.S. intelligence sources in Baghdad were quick to leak that "all but one [of the Russian captives were] considered to be staff members of Russia's secret services."

It wasn't until June 19, ten days after Basayev spoke to his commanders and a week after the report of the meeting was sent to the Middle East, that the first communiqué emerged concerning the kidnapping of the Russian diplomats. It came from the Mujahedin Shura Council, an umbrella group comprising the eight key Islamist-Jihadist organizations in Iraq: al Qaeda in the Land of Two Rivers, the Victorious Army Group, the Army of al-Sunnah wal-Jamaa, the Jamaa al-Murabiteen, the Ansar al-Tawhid Brigades, the Islamic Jihad Brigades, the Strangers Brigades, and the Horrors Brigades. The Shura's charter declares that these organizations are collaborating to "confront the unbelievers gathering from different sides" and "to defend Islam" against all enemies.

The communiqué from the Mujahedin Shura Council acknowledged responsibility for the operation. "Allah has enabled the lions of monotheism to seize four Russian diplomats in Iraq and kill a fifth," it said. The message also issued an ultimatum for Moscow: "The Council's Shariah court has decided to give the Russian government forty-eight hours to meet the following demands or bear the consequences: the immediate withdrawal from Chechnya, and the release of all our brothers and sisters held in Russian jails." In carrying out this operation, the council stressed, it was putting the all-Islamic jihadist cause ahead of the interests of Iraq. "We know in advance that there will be appeals . . . to release those [Russian hostages] under the pretext that Russia took a clear stand in rejecting the [U.S.-led] war on Iraq," the communiqué acknowledged. However, Moscow had already forfeited all goodwill by "sending its diplomats to Iraq in support of the Crusader project led by America."

Two days later, with Moscow refusing to consider surrendering to the terrorists' demands, the council issued a follow-up statement: "The Shariah court of the Mujahedin Shura Council in Iraq decided to apply the rule of God against them and ruled that they should be killed." The communiqué added that the execution of the Russians would "serve as a lesson . . . to those who would still defy the mujahedin and dare to set foot in the proud Land of Two Rivers." The next day, June 22, the Mujahedin Shura Council announced that the verdict had been carried out, contending that the Russians' deaths proved only that Russia "put no value on the lives of its citizens."

On June 25, the council posted a ninety-second video clip on an Islamist website showing the execution of the Russian diplomats by beheading and shooting. An accompanying statement justified the verdict in all-Islamic terms. "Allah's verdict has been carried out on the Russian diplomats . . . in revenge for the torture, killing, and expulsion of our brothers and sisters by the infidel Russian government," the statement said. The message demanded that Russia withdraw all troops from Chechnya and release all "Muslim hostages" in Russian hands. Since Moscow had not met the council's demands, the execution should be considered as a symbolic revenge. "We present the implementation of Allah's rule against the Russian diplomats to comfort the believers," the statement explained. The execution "is also in revenge for our brothers and sisters and the torture, killing, and displacement they have suffered at the hands of the infidel Russian government." After careful study, Moscow confirmed the authenticity of the videotape a couple of days later.

On July 10, with the G-8 strike about to unfold, Basayev formally reacted to the kidnapping and execution of the Russian diplomats in Baghdad. A posting on the rebel website Kavkaz-Tsentr reported that Basayev had sent a telegram to the commanders of the Mujahedin Shura Council. "Mujahedin of the Caucasus express huge gratitude to those who have carried out the elimination of the Russian diplomats the spies in Iraq," Basayev's telegram read. "Allahu Akbar!"

As the terrorists had expected, throughout this period Russian intelligence had focused first on locating and saving the diplomats, and ultimately on finding those responsible for their execution. Putin made this a high priority. U.S. security authorities in Baghdad declined to accommodate Russian efforts, however, and a flurry of intense high-level exchanges followed between the United States and Russia, both in Baghdad and between Washington and Moscow. Still, other intelligence services with long experience fighting jihadist terrorism rallied to help their Russian counterparts, cooperating closely with Russia in various operational fields.

The Islamist-Jihadist and Iranian intelligence networks, which have thoroughly penetrated the Iraqi governing entities, enabled the jihadist leadership to follow the crisis; they learned of the Americans' reluctance to help the Russians, but missed all other relationships the Russians were augmenting.

By the time this decoy operation had run its course, a number of senior commanders from the Islamist-Jihadist network were making final preparations for the trip north, slated to participate personally in what was expected to be a suicide mission at St. Petersburg. The Chechen jihad had been in decline for some time, and the leadership was intent on effecting a dramatic shake-up. Among them were Basayev himself, Ali Musaevich Taziyev (whose nom de guerre was Magas), one of the leaders of the Beslan school siege of 2004, and Abu-Hafs al-Urduni (born Youssuf Amirat, the senior Arab jihadist commander in the Caucasus).

As they prepared for the St. Petersburg operation, however, the Chechen jihadist leaders were confronted by the realities of their declining jihad, in the form of dwindling stockpiles of weapons, ammunition, high explosives, and, most important, sophisticated fuses and other sabotage systems. Once the G-8 strike was triggered, Abu-Hafs al-Urduni personally asked al Qaeda's supreme leadership to help supply highly specialized equipment and high explosives from overseas resources. It was an unusual request, particularly coming directly from Abu-Hafs, and it reflected both the urgent priority of the operation and the growing shortages in Chechnya and the northern Caucasus. Placing similar importance on the operation, the al Qaeda leaders agreed quickly.

Yet it was this request that would ultimately doom the operation.

The Chechnya-based jihadist leadership, it seems, was unaware of how deeply it had been penetrated by Russian intelligence services. At first, a spy in the jihadists' ranks alerted Moscow to their active preparations for the fall-back option: "On the eve of the G-8 Summit, the agent reported [that] gunmen were planning to seize a major center of population in the North Caucasus." When a Western intelligence service informed Moscow at around the same time that an emissary of Abu-Hafs, and other representatives of Chechen Jihadist leaders, had met with senior al Qaeda operatives to discuss large weapons and equipment shipments for a major operation, Russian intelligence connected the two reports and instructed their agents to investigate.

One of the sources was an agent of the Russian Federal Security Service (FSB) who had infiltrated the Islamist-Jihadist supply system, which helped facilitate the flow of people and materiel in and out of Russia through a particular country. Meanwhile, at Moscow's request, the Western intelligence service increased its monitoring of the local Islamist-Jihadist network and their scheme to ship the requested materiel to Chechnya. It was to be a highly professional undertaking: The foreign network would deliver the goods to a third party in a neighboring country. From there, the locally based Chechen network would take control of the goods, pack them in a truck, smuggle them across the Russian border, and deliver the truck to a concealed storage site in Ingushetiya, just west of Chechnya. Once both networks were prepared to withdraw safely and conceal their tracks, Abu-Hafs and his colleagues would be notified by al Qaeda from overseas where to pick up the hidden goods—leaving no trail of connection between the Chechen terrorist networks and their supporting elements overseas.

And so it was . . . almost.

The Western intelligence service succeeded in tracking the shipment for Abu-Hafs all the way to the neighboring country. The deep-cover FSB agent was already there, waiting for the shipment. Moscow had supplied him with a small gadget that included an electronic beacon, a remote-controlled detonator, and a powerful mini-bomb, and instructed him to attach the gadget to the KamAZ-brand truck the network was to use.

When the truck crossed the border, it was picked up by Russian intelligence. The Russians followed it to a safe site near the village of Ekazhevo in Nazranovskiy District, Ingushetiya. To ensure operational safety, the initial monitoring was done from afar using a small, unmanned Pchela (Bee) spy plane. As they waited for the truck to be picked up and delivered to Abu-Hafs, Basayev, and the terrorist elite, FSB special forces selected and organized ambush sites on the road leading to the storage site.

On the night of July 9–10, several individuals arrived in a few small cars to pick up the truck. A couple of people got into the truck's cabin, and the vehicles drove back in a small convoy, unaware of the Pchela following them from a safe distance. The ambush was waiting just outside Ekazhevo. When the convoy reached the ambush point, the remote-control detonator was activated, and the truck exploded with such power that the small cars around it were also destroyed. Numerous flechette rods were found in the cabin area, demonstrating the agent's success in aiming the directed-explosion charge at the cabin to ensure the driver's death. Indeed, the explosion was so powerful that most of the mujahedin simply evaporated. Subsequent investigation revealed that a dozen mujahedin were killed, including Shamil Basayev (Moscow's public enemy number one), Magas (wanted for his role in Beslan), local commander Tarkhan Gaziyev, and Isa Kushtov, a special assistant who handled weapons acquisition for Chechen separatist "president" Dokka Umarov. Kushtov's presence proved that Umarov had known and approved of the planned operation.

The next morning, the head of the FSB met with President Putin to report on the success of the operation. The operation, he told Putin in the public part of the event, "resulted in the elimination of Shamil Basayev and a number of bandits who had prepared and carried out acts of terrorism in Ingushetiya. This entire operation became possible thanks to the fact that operational positions had been created abroad. Primarily in those countries where weapons were collected and then supplied to us in Russia to commit acts of terrorism."

Beyond killing Basayev and several veteran senior terrorists, the FSB strike was a severe blow to the Islamist-Jihadist movement in the northern Caucasus—threatening to derail a center of jihad that had been a potent source of international threat for years. Yet the strike left one of the jihad's most powerful leaders, Abu-Hafs al-Urduni, at large, along with several other expert terrorists still in Chechnya or the northern Caucasus, leaving open the probability of future major strikes on Russia in the name of the jihad. Indeed, the remaining rebel forces responded to the FSB counterstrike with a defiant vow to continue the jihad. Movladi Udugov, the head of the rebel Chechen National Information Service, declared that Basayev's killing "will not change anything." Udugov's statement reflected the overall grim state of the Chechen jihad: "Leaders come and go, commanders get changed, those who get tired leave, and those who are strong in their belief will never relent," Udugov said. And he contended that "the Jihad against Russia will continue."

Meanwhile, the Russian security services intensified their hunt of the remaining jihadist commanders. By late summer, most commanders had left Chechnya and gone underground in neighboring republics. Abu-Hafs hid in southern Dagestan, not far from the border with Azerbaijan. But he was losing control over his remaining forces in Chechnya, so he agreed to come to Khasavyurt, near the Chechen border, to meet with emissaries from Chechnya in order to discuss the resumption of fighting in the coming spring. Russian intelligence discovered the plans and waited for Abu-Hafs to settle in what he thought to be a safe house. On November 26, Russian special forces surrounded the house, and, after a lengthy firefight, stormed and killed all five muijahedin inside—including Abu-Hafs.

It took the jihadist leadership three days to acknowledge that "Amir Abu-Hafs, the Deputy Military Amir of Chechen Mujahedin [was] martyred"—although they insisted that only two fighters died with him. Abu-Hafs's successor—Commander Muhannad, a Sudanese national—is of lower rank and posture in the jihadist movement. His ascent is indicative of jihadists' inability to deploy and sustain a senior figure in the northern Caucasus.

Thus, by all evidence the destruction of that single truck marked not just a momentary blow but a major downturn for the grassroots jihad in the northern Caucasus, which the supreme leaders had hoped to galvanize with their planned G-8 strike. It marked the collapse of one of the jihadist movement's most audacious ambitions: to take over a national liberation struggle and convert it into an instrument for the transformation of society. The result was a profound setback for the global jihad, one that came about not so much because of the Kremlin's impressive antiterrorism efforts but because the jihadists overplayed their hand, failing to grasp the native population's commitments to their own heritage and traditions.

The story of the Islamist-Jihadists' quest to seize the strategic ground of Chechnya and the Caucasus is a unique and critical case study for anyone seeking to understand the means and goals of the worldwide Islamist jihad. And its failure—if failure it be—contains lessons that may be instructive as the secular world confronts continued Islamist-Jihadist surges in places such as Iraq and Afghanistan.

CHECHEN JIHAD

Chapter 1

WHY SHOULD WE CARE?

TO MOST AMERICANS, THE "WAR ON TERRORISM"—THE POPULAR euphemism for the series of U.S.-led military campaigns all over the world—had a distinct starting point in the spectacular terrorist strikes against the World Trade Center in New York and the Pentagon in Washington, D.C., on September 11, 2001.

In reality, however, these strikes were milestones in what was already an ongoing global war. The Islamists' quest to dominate Islam and control the Muslim world, as well as to cordon off the Islamic world from Westernized modernity until it could be taken over by the Islamists through a fateful jihad, has been unfolding in various degrees of intensity since Napoleon set foot on Egyptian soil in the late eighteenth century. At present, the primary "front" of the Islamist jihad is the Hub of Islam—the Middle East along with South and Central Asia—where the jihadist movement is trying to confront Western modernity while preserving the Islamic sociopolitical character of society. Rather than adapt to the ethos of the information age and globalization, the jihadists see controlling and dominating the West as their only salvation.

With the collapse of the Soviet Union, the jihadists resolved to pursue three historic axes of advance into lands within reach of the Hub of Islam, lands that have been claimed by Islam since its ascent: the Caucasus (the historic avenue into the heart of Russia and Eastern Europe), the Balkans (the historic road to Western Europe), and Kashmir (the entrée into the Indian subcontinent). By the mid-1990s, the Islamists were already escalating their jihad in each of these regions. During the same period, the United States was involved in conflicts in the Balkans, the Caucasus, and Afghanistan/Pakistan—a series of entanglements that were problematic for U.S. interests at the time and quite counterproductive, in retrospect. In each of those regions, Washington was pursuing near-term political interests while disregarding historic and global megatrends. So the idea that Washington "discovered" the jihadist menace on 9/11, and has since been leading the global campaign to defeat and reverse the phenomenon, simply ignores the crucial role played by key regional powers who have battled Islamist-Jihadism for more than a decade.

Critical, in these years, was Russia's role in combating Islamist terrorism on the Caucasus jihadist front, in a drawn-out conflict commonly referred to as the "war in Chechnya." For the Russians, the importance of containing the jihad in the Caucasus went beyond their desire to control this small republic, with a population of slightly over a million and a land mass smaller than the state of Vermont. The rebellion in Chechnya may have begun as an indigenous nationalist movement, but it was soon co-opted by the international Islamist movement as an element of its global jihad. By the turn of the twenty-first century, the jihadists were well on their way to transforming the Caucasus into a springboard for strikes into Russia and Europe, and a site of sociopolitical transformation that threatened to affect the entire Hub of Islam and beyond.

In order to comprehend the process that has become known as "Chechenization," its role in the Islamist-Jihadist movement, and what it tells us about the potential vulnerabilities of jihadist activity elsewhere in the world, it is critical to examine the course of the Islamist jihad as it has played out in the Caucasus in the last decade.

Chechenization is a relatively new concept, still whispered about by experts on Islamist-Jihadist terrorism and attacked by Western politicians, mainly American, who are loath to face the reality of the conflicts their countries are mired in—or to acknowledge Russia's preeminent role in the worldwide war on terrorism.

Chechenization refers to the profound transformation of a predominantly Muslim society from its traditional, largely pre-Islamic structure to one dominated by Islamist-Jihadist elements that historically have been alien to that society. Chechenization involves not only the Arabization of that society's value system, social structure, and way of life, but a near-complete abandonment of a society's own cultural heritage in favor of subservience to pan-Islamic jihadist causes, even if those causes are detrimental to the self-interest of that society.

The process of Chechenization—which is now arguably at play in significant parts of Iraq, the Palestinian Authority, and Indonesia, as well as several Muslim communities in both Central Asia and the Balkans—was named for the jihadist campaign in Chechnya in the mid-1990s. There, the national liberation struggle of a secularized Muslim population, inspired by a rich historical legacy of quests for self-determination, was taken over from within by the Islamist-Jihadists—transforming the liberation struggle into a regional anti-Russian terrorist jihad, at the expense of the Chechens' own self-interest. The process—which included the intentional destruction of Chechnya's own socioeconomic infrastructure and the forfeiting of Chechnya's ability to benefit from agreements with Moscow—could not have been accomplished without lavish funding from charities based in Saudi Arabia and several Persian Gulf states. Since the mid-1990s, the radicalization and transformation of Muslim societies from within has bred and nourished the waves of the Islamist-Jihadist terrorists, which not only kill their own kin but also strike out at the heart of the West.

Today, the Chechenization of other regional conflicts, subversions, and insurgencies around the world is fast becoming the key to al Qaeda's rapid expansion and further consolidation, despite the U.S.-led war on terrorism. The Islamist-Jihadists see Chechenization as the profound transformation of a "jihad front" (to use their own term) from a besieged community on the defense to a springboard for the expansion of their fateful onslaught on Western civilization. The first cycle of Chechenization saw the jihadists' struggle for the heart of Asia and the Caucasus cross a major milestone. The Islamists were no longer intent merely on consolidating their hold over the Muslim states of South and Central Asia, or on "liberating" traditionally contested territories such as Russia's northern Caucasus, Indian Kashmir, and the state of Israel. Rather, the Islamist-Jihadists launched an offensive into Russian territory aimed to transform the very shape of Eurasia. Given the strategic and economic potential of the Caucasus and Central Asia, it was the sponsoring states of this Islamist-Jihadist upsurge—not the peoples of the Caucasus—who would reap the primary benefits of this strategic upheaval.

Meanwhile, the integration of the Chechen jihad into the global Islamist-Jihadist movement made both native Chechen and Chechen-trained expert operatives and terrorists available to participating in other jihad fronts all over the world. (Foreign mujahedin who volunteered, trained, and fought in Chechnya and the northern Caucasus came to be referred to as "Chechen mujahedin," an umbrella term that comprised all mujahedin native to the region—not just ethnic Chechens but Dagestanis, Avars, Ingushets, and so on—as well as Chechens and other Caucasians from Central Asia, and Circassians from Jordan, Saudi Arabia, and other Middle East nations.) Despite their relatively small numbers, the Chechen mujahedin came to play an increasingly important role in the global jihad. Many of the Chechens had extensive military knowledge and expertise gleaned during their service in the Soviet and Russian military, including service in Afghanistan. Starting in the mid-1990s, the Chechens established an elaborate training system in which such Chechen veterans, as well as highly experienced Ukrainian and Balt mercenaries/volunteers, trained the Chechen mujahedin in sabotage, communications, military and combat engineering, logistics, intelligence work, information technology, weapons of mass destruction, and the like, offering a level of expertise that exceeded what was available in other Islamist-Jihadist training programs in South Asia and the Middle East. Many of the Chechen trainees, too, were professional fighters—disciplined and responsible, having a combination of skills, expertise, and character. Today, these trainees are the most sought-after "force multipliers" (to use the U.S. military term) in the Islamist movement—their unique fingerprints increasingly noted with each new phase in the jihad.

Another overlooked aspect of Chechenization was its growing impact and influence over the movement known as pan-Turkism. Originally a nationalistic trend with only vaguely Muslim overtones, pan-Turkism was co-opted as an instrument for the spread of jihadism in the Turkic lands from the Balkans to Xinjiang, China. In the early 1990s, the revival of nationalistic pan-Turkism in Turkey helped to destabilize the Caucasus, aggravating the Chechen revolt of 1994 by providing international recognition and support to the rebels. Pan-Turkism has grown in recent years in Turkey, supported by the military elite as a counterideology to Islamism; militant pan-Turkism is extremely popular with a military nostalgic for the glory of long-ago wars in the Caucasus and the days of the Ottoman Empire. Even though the Turkish military elite is essentially anti-Islamist, this pan-Turkic nostalgia has undergirded Ankara's commitment to the Chechen revolt, and it endures despite the Islamist-Jihadist nature of Chechenization. A steady stream of Turkish volunteers—most of whom are highly trained military veterans—continues to fill jihadist ranks today, even as Ankara ignores the Islamist support for Chechnya and the drug trade that funds the jihad.

The growth of pan-Turkism also had major political ramifications for the West, and especially for the United States. Washington has long considered Turkey an ally, a status that survived even the crisis over access to Iraq that began in 2003. At first, this traditional alliance led the West to extend some support to the Chechen rebellion. Even today, this legacy provides a reluctant Bush administration with a fig-leaf excuse for not confronting the Chechen threat head-on, despite its immersion in a war on terrorism that has been aggravated by the spread of Chechenization.

Chechenization involves mobilizing a country, or a region, against the West, in large part by conditioning its local society to commit to the spread of jihad. For Russia, the Chechenization of its war in the Caucasus meant unleashing waves of jihadist terrorism at the heart of Russia—a terrorist campaign that has already taken hundreds of innocent lives and is far from over. For the United States and the West, the Bush administration's mounting contretemps with Putin's Russia has prevented the United States from understanding Chechenization or benefiting from the vast intelligence data and operational knowledge accumulated by the Russians. The increasing casualties in the American quagmires in Iraq and Afghanistan testify to the self-inflicted ramifications of official Washington's adamant refusal to confront, let alone learn from, the realities of the Caucasus and Moscow's war on terrorism—at a time when the United States could hardly afford to ignore the experience of the rare nation that had battled the Islamists to a draw.

In this context, a somber word about the ongoing U.S. war on terrorism. Among other things, the unfolding global jihad should be understood as the latest, and most significant and intense, phase in the profound struggle *within* Islam over its interaction with Western modernity, and over its own future. Since the march of Western modernity is unlikely to be reversed or stopped, the war is unlikely to end until a genuine reformer—a Martin Luther—rises from the ranks of Islam to lead its followers into the twenty-first century. The United States may be an object of wrath for the Islamist-Jihadist movement, but it does not have a side in this conflict within Islam itself. That is why the United States could lose this war—if the Islamist-Jihadist movement triumphs—but cannot win it. Victory is solely in the hands of the Muslim world. Until then, it is in the interest of the West to understand, and combat, the strategy represented by Chechenization, in order to help the sane and responsible elements within Islam triumph over adversity while bringing modernity to their world.

Chapter 2

LEGACY AND ROOTS

THE WESTERN UNDERSTANDING OF THE SITUATION IN CHECHNYA, and the Caucasus as a whole, has been distorted and complicated for decades by a legacy of romanticized accounts of Russia's involvement with the Caucasus. In the West, the violence in Chechnya and the Caucasus is commonly associated—by the media and most governments alike—with the legacy of Imam Shamil and his anti-Russia rebellion in the mid–nineteenth century. Shamil was an Avar, not even Chechen, but Western eyes still consider the myths surrounding his rebellion as a yardstick for assessing the contemporary struggle in Chechnya. The enduring impact of accounts such as Lesley Blanch's masterful 1960 epic history, *The Sabres of Paradise,* influenced generations of Western would-be experts and analysts, offering an alluring interpretation of Shamil's struggle against Russia's advance into the Caucasus. Indeed, the modern Chechen jihadist Shamil Basayev made an intense effort to associate himself with Imam Shamil—even taking his name—thus conflating the jihad of the mid–nineteenth century with that of the early twenty-first.

In order to comprehend the real war that has transformed Chechnya in the last two decades, it is important to understand the actual legacy of Imam Shamil (also spelled Shamyl or Schamyl in older books) and his nineteenth-century struggle against Russia.

Imam Shamil (1797–1871), a Muslim and a powerful national military leader, commanded a Caucasian guerrilla force for twenty-five years, from 1834 to 1859. In *The Sabres of Paradise,* Blanch, the romantic chronicler of the Caucasus wars, stressed the centrality of Shamil's personality and leadership to the revolt in the mountains of the Caucasus:

> To Shamyl, who straddled these mountains like a legendary giant, they were his birthright, his kingdom. From their shadows he first unfurled his black standard. In the name of *Ghazavat!* Holy War! he wielded the dissenting mountain tribes into the implacable army of fanatics whose private feuds were submerged in their common hatred of the Infidel invaders. For twenty-five years, he dominated both land and people. For twenty-five years, the Caucasians accepted lives of bleak abnegation and hardship for Shamyl. His Murids revolved around his dark presence with the slow set circling of planetary forces. In life and death his word was law. All of them were vowed to resist Russia to the death.

Despite such monumental efforts, however, the impact of Shamil's revolt would have been marginal at best if not for larger political and social dynamics at play during his times. Indeed, the strategic impact of Shamil's revolt peaked during the Crimean War (1853–1856) because the revolt in the Caucasus affected Russia's ability to wage war against both Turkey and Great Britain. The Caucasus theater was a minor front in the Crimean War, one that had only limited influence on the greater conflict. Yet the story of Shamil left an indelible mark: Even today Moscow sees the Caucasus as the archetypal hotbed of dangerous local revolt, and as a reminder that Turkish forces might one day support a local uprising in the Caucasus. Modern Russian and Soviet military studies stress that the most important legacy of the Russian military operations in the Caucasus during the Crimean War was the effect that fighting the Shamil uprising had on Russian military performance against the Turkish army.

Indeed, Shamil's impact on Russian strategic decisions transcended his actual military capabilities. In the early 1850s, when fighting between Russia and Turkey erupted in the Caucasian theater, Shamil so intensified his activities that by the fall of 1853 the Russians were hard-pressed to commit sufficient troops to the Turkish front because so many were required to hold off Shamil. By mid-1855, Shamil was engaging some two hundred thousand Russian troops, who were needed at the Crimean front.

Yet Shamil's greatest influence on Russian military operations was a direct outcome of his regional popularity as a Muslim leader. The Caucasus theater of the 1850s was basically identical to that of the 1828–1829 Russo-Turkish War. Despite the Russians' successful experience in fast-maneuver attack in the area—and despite the inferiority and insufficiency of the Turkish fortifications in Kars—Russian general N. N. Muraviev chose to engage in conventional, protracted siege warfare. It was a cautious strategy, born of fear that a failed Russian attack, or any setback for Muraviev's side, would encourage both the Turks and Shamil's rebels, emboldening the Chechens to intensify their attacks on the Russians' already strained lines of communication on the few mountain roads.

Since 1850, when Shamil had emerged as a significant threat, the Russian command had been determined to contain and eventually defeat him. Although Shamil's heartland was Dagestan, the bulk of his troops were Chechens, and the Russians attempted to destroy Shamil with a series of devastating assaults on Chechen villages, destroying the uprising's economic base and Shamil's greatest source of popular support and manpower. The Russians followed these raids with a wide-scale settlement effort, seizing the most fertile lands and forcing the Chechens to plead for compromise. Fearing a loss of his Chechen support, Shamil launched a series of harsh punitive raids against his former allies, who in turn sought and received Russian protection. By 1852, most Chechen citizens were settled under Russian protection—yet the mountains remained beyond Russian control.

When Turkey declared war on Russia in 1853, both sides realized the potential significance of Shamil. The Turks encouraged him to intensify his struggle under a unified banner of Islam, and the British rushed weapons and ammunition to his forces. Despite this support, Shamil's irregulars were still no match for the experienced, skilled Russian troops, who had superior firepower and an advanced grasp of tactical warfare. Still, the Russians overreacted to the potential threat Shamil posed. Fearing that the now-pacified Chechens might rejoin their past leader, the Russians launched a series of preemptive raids, destroying Chechen villages and literally driving their inhabitants into the mountains. In the fall of 1853, Shamil tried to revive the rebellion, capitalizing on the Turkish advance and scarcity of Russian forces. He gathered a large force and launched a major attack in the direction of Tbilisi. However, his undisciplined troops eagerly attacked the Georgian villages, looting and burning them. The locals organized their own irregular bands, which immediately started instigating skirmishes against Shamil, eventually forcing him to stop and reorganize his troops. As he fought off the spontaneous Georgian resistance, the Russians took the time they needed to regroup for battle.

By now, Imam Shamil was the most powerful leader in the Caucasus. Yet the actual size of his forces was relatively small, and their lack of training, supplies, and competent command echelons made his forces far less valuable than their size suggested. Contemporary analysts such as Charles Duncan, in his 1855 book, *A Campaign with the Turks in Asia*, offered uncomplimentary assessments of Shamil's forces. Even in 1854, at the height of the revolt, Duncan noted that "the indisciplined and badly armed rabble that comprises the followers of Schamyl, though invincible in their mountain fastnesses, are utterly harmless in the plains of Georgia. . . . A single Russian dragoon regiment, backed by a troop of horse artillery, would suffice to rout any force that Schamyl could bring into the plains." Duncan felt that even Shamil recognized his own limitations: "Nobody is better aware of this than the chieftain himself, and he has displayed consummate wisdom in never having committed himself in any similar expedition." Indeed, by the spring of 1854, ferocious Russian raids drove the Chechens back into Russian control, where they would benefit from a regular supply of food and land, as well as freedom from taxes. In the late fall, the Russians drove the remaining Shamil loyalists into the mountains and starvation under the harsh winter conditions. Now desperate, Shamil tried to reverse the trend by launching a series of vengeance raids. Though he matched the Russians' skill at destroying and burning villages, Shamil was unable to provide food and land for resettlement. The Russians could.

By the spring of 1855, the starving Shamil loyalists who had survived the harsh winter started coming down from the mountains to seek Russian help and protection. That summer, the Russians sent their cavalry in to protect farmers harvesting their crops in remote mountain villages and resupplied food to villages whose fields were burnt by Shamil. In the meantime, the winds in the Caucasus theater were shifting dramatically in Russia's favor. In November 1855, General Muraviev captured Kars, and Shamil's hopes of teaming up with Turkish forces collapsed. The combined impact of the fall of Kars and the Russians' relentless operations throughout Chechnya and Dagestan convinced the local population that the Russians were still the dominant power and that Shamil could not stop them.

Contemporary observers stressed the importance of Russia's highly mobile Cossack cavalry detachments in destroying Shamil's forces. Because the Muslim chieftains were reluctant to commit forces to a direct confrontation with the Russians, the Russians had to seek out the rebel forces and launch deep raids on them. Since their first encounters with Muslim uprisings, the Russians had realized that intelligence was the key to interdicting the flexible, elusive rebel forces: They paid special attention to cutting off the rebels' lines of communication and to destroying Shamil's popular and economic support. The Cossacks' raiding parties, which gathered intelligence from agents they had placed in Shamil's camps, were the most effective Russian forces in the fight against Shamil.

One of the most crucial lessons the Russians had learned in their long struggle against the Caucasian Muslims, and especially from Shamil's uprising, was that these mobile forces could outperform their opponents by drawing on such timely intelligence, which they often used to intercept and destroy Shamil's raiding forces. Mackie J. Milton stressed the importance of these tactics in his 1856 book, *Life of Schamyl and Narrative*:

> Sometimes the Circassians dash through between forts without stopping to attack them, and suffering perhaps, somewhat from crossfire, gain the country beyond the line, where they find more abundant spoils and no resistance. But on their return, they are sure to encounter the Cossacks drawn up at the ford, or some other point

> convenient for disputing the passage to an enemy encumbered with booty. These Russian hirelings, however, the freemen of the mountains despise, and with superior horses ride them down. Only when the espionage which is maintained among all the tribes on the border—for everywhere there are souls which can be bought for gold—succeeds in procuring for their enemies information of any incursion before it takes place, is the foray rendered unsuccessful and the troop cut off.

By the late 1850s, the Russians were winning, having exacted repeated—albeit localized—defeats of Shamil's key forces, while carefully eroding his popular support. Ultimately, Shamil's growing reliance on the hardcore Murids (who were oblivious to the civilian population's capitulation to advancing Russian forces), rather than on volunteers from local tribes, contributed to his downfall. As John F. Baddeley stressed in his 1908 book, *The Russian Conquest of the Caucasus,* "Shamil had lost heavily in the field, and his best lieutenants had been killed off one by one or had gone over to the enemy; while whole districts, wearied of constant warfare, had gladly submitted to the Russians as soon as they saw that the latter were in a position to protect them against the wrath of the fanatical Murids."

The struggle against Shamil in the Crimean War taught the Russians lessons that would influence their approach to the Muslim population for more than a century. They followed a simple and efficient "pacification" policy: Force the Muslim population to submit to Russian rule or face physical destruction. As J. S. Curtiss wrote in his 1979 book, *Russia's Crimean War*: "Although the treatment of the Chechens by the Russians was cruel, the Russians had no choice but to continue fighting against an elusive foe or subdue the population and gain control of the territory. They believed the latter course was the only safe one to follow, especially since it cost them relatively few losses in manpower. Hence they presented the Chechens with three options: to be ruined or destroyed by the Russian raids; to flee into the wilderness to try to escape their punitive measures; or to submit and break their ties with Shamil." Gradually, the majority elected the third option.

The Russians finally subdued the insurgency in the Caucasus in

1859, when Shamil surrendered to the tsar and went into exile at the heart of Russia. Shamil's surrender on August 25 was a characteristically pragmatic act: He simply recognized the futility of continued struggle against the vastly superior and determined forces of the Russian empire. Indeed, unlike the portrait drawn by his contemporary self-anointed successors, Shamil was fundamentally a pragmatic traditionalist leader—far more grounded in the sociopolitical roots of his heritage than in Islamic traditions. As Israeli historian Moshe Gammer wrote in his 1994 milestone book, *Muslim Resistance to the Tsar: Shamil and the Conquest of Chechnia and Daghestan*, "Shamil was a born leader, commander, diplomat and politician. He repeatedly outmanoeuvred the Russians in battles, intrigues and negotiations. Contrary to Russian propaganda, he was far from extremism or blind fanaticism. He tended to use force sparingly and tried to come to an accommodation with both his internal rivals and the Russians. Furthermore, while engaged continually in these battles, intrigues and negotiations, he managed to unite a multiplicity of tribes and forge them into a unified state."

Ultimately, it was the inherent character of the peoples of the northern Caucasus—their unwavering commitment to their own distinctive tribal and ethnic self-identities and their fierce resistance to unification under any ideological-theological banner—that was Shamil's undoing. For the various clans that comprised Shamil's constituency, the only real hope of rebuffing Russian domination would have been to commit to a unified group effort. But the people would not sacrifice their distinctive self-identities on the altar of independence from Russia. "Shamil had failed because success was impossible," Baddeley concluded. "He had to contend from the beginning not only against Russia but against a far worse foe—internal dissension; and from the nature of the circumstances he could overcome neither the one nor the other." Russian occupation, harsh as it was, permitted the people of the Caucasus to preserve the one thing that independence under Murid rule would not: their distinct self-identities. It was a profound lesson—one that Shamil's self-anointed successors have yet to learn.

* * *

The most profound change in the sociopolitical posture of the peoples of the northern Caucasus in the aftermath of Shamil's surrender was the spread of the Qadiriya Sufi brotherhoods *(tariqat)*. Until then, the Naqshbandiya Sufi brotherhoods had been the dominant Islamic order in the northern Caucasus. In principle, the Naqshbandiya was a diffuse, nonhierarchical order, focused on preserving Islamic practices among individuals and families without challenging the political or social status quo. The Qadiriya was exactly the opposite. Founded in Baghdad by Abd al-Qadir al-Ghilani around 1166, the Qadiriya Sufi Tariqat had spread widely but sparsely, following the key commercial routes of the Muslim world. It was a radical, clandestine order, preoccupied with communities rather than individuals.

The Qadiriya Sufi Tariqat was brought to the North Caucasus in the 1850s by a Kumyk named Kunta Haji Kishiev. Its strict social order and well-defined centralized hierarchy appealed to a population devastated by the long war and its attendant social upheaval. Throughout the second half of the nineteenth century, a localized adaptation of the order spread among the Chechens and Ingush, playing an important role in the Islamicization and recovery of the population. The distinct Qadiriya of the Caucasus, known as the Kunta Haji Tariqat, incorporated the region's strong tribal and family traditions, evolving into a clandestine web of loyalties tied together by rigid communal and individual discipline, as well as personal devotion to hereditary sheikhs. The Kunta Haji Tariqat strengthened the peoples of the North Caucasus at the very moment when they were most vulnerable to Russian pressure: The tariqat operated largely under the radar, which served it well during the Soviet era and the First Chechen War; some elements of the Kunta Haji Tariqat even evolved into more or less closed societies, which the Russian and Soviet security services proved powerless to penetrate.

After Shamil's surrender, the peoples of the Caucasus would resist Russian "liberation" for half a century; their insurrection was put down only through harsh repression. The population appeared to have been subdued until World War I broke out, when locally drafted Russian Muslim soldiers refused to go to the front and fight their brethren Turks. The Turkish invasion of the southern Caucasus, under the banner of pan-Turkic solidarity and liberation, reverberated in the northern Caucasus and created great anticipation for the arrival of the Turkish armies. Yet British military intervention in Azerbaijan—mounted to preserve their control over oil resources—blocked the Turks' advance northward. The Muslim population of the northern Caucasus remained hostile to Russia through the collapse of the empire and renewed their struggle soon after the Bolshevik revolution of 1917. Their resistance peaked between the summers of 1920 and 1921, just as civil war was winding down in the region. Sporadic fighting continued at least until mid-1925, when Soviet intelligence finally caught and executed the last of the local sheikhs who inspired and legitimized the regional ghazavat (holy war).

After the revolution, the Soviets pledged to afford the people of the Caucasus equality and moderate self-rule, but their promises were met with disbelief. The Soviets continued their harsh oppression of the Muslim opposition throughout these years, but local forces offered continued resistance well into the early 1940s, with distinct uprisings in 1924, 1928, and 1936.

By now, nearly a century after the collapse of Shamil's revolt, his legacy and his descendants remained a significant factor in the politics of the Caucasus. This lesson was driven home to Moscow by the efforts of Shamil's grandson, Said Shamil, to organize the mountain tribes of the North Caucasus during World War II.

The story began in November 1939, when the U.S.S.R. was still allied with the Nazis under the 1939 Molotov-Ribbentrop Treaty. Concerned that the Azerbaijan oil fields could fall into Axis hands, Winston Churchill prioritized the formulation of Operation Pike, a plan to seize and/or destroy the fields. By April 1940, the RAF had worked up detailed contingency plans for bombing oil installations throughout the Caucasus, including sites at Baku, Batum, and Grozny. But another key component of Operation Pike was a plan to instigate an anti-Soviet popular revolt in the northern Caucasus in order to tie up Soviet forces and prevent them from holding onto the Baku area. To help stir the pot, the British and the French decided to rejuvenate the Shamil legacy.

In 1940, Said Shamil visited the French commander in chief in the

Middle East, General Weygand. "Said Shamil represented the interests of the fighting mountain tribes in North Caucasus, and the purpose of his visit was to effect liaison with the army of Weygand, which formed the potential nucleus of an expeditionary force to the Caucasus," explained Nikolai Tolstoy in his 1981 book, *Stalin's Secret War*. The French and British forces were too small to mount such a mission alone, and both sides hoped Turkey would join the venture, providing large forces but also inspiring the local population to rally around Said Shamil. Churchill himself made repeated efforts to convince Turkey to join the Allies' war effort and repeatedly urged the nation to exercise its pan-Turkic responsibilities, particularly in the Caucasus.

But Said Shamil's talks with the French did not keep him from working the other side of the street. When Germany attacked the Soviet Union on June 22, 1941, Shamil sought and received extensive help from Germany. In 1942, Germany launched a major offensive into the Caucasus, surging toward Azerbaijan in hopes of seizing oil resources it considered vital to sustaining the war. The move would spark intense fighting in the rugged and harsh Caucasus Mountains that would continue until late 1943. Throughout this campaign, Germany made repeated efforts to incite the local population into rebelling against the Soviet Union and supporting the German drive. In April 1942, Berlin invited some forty emigré leaders of Soviet Muslim "nations" for discussions in Berlin. One of the prime guests was Said Shamil, who was promised independence of a sort for the North Caucasus after the German victory. The Germans lavished such special status because General Reinhard Gehlen considered the "Caucasians and tribes of Turkish origin" especially well suited for anti-Soviet covert operations.

In their plan to absorb the Caucasus, the Germans also had the help of Hajj Amin al-Husseini, the former mufti of Jerusalem, who provided Islamic propaganda for the German drive. On August 30, 1942, Said Shamil was nominated to be Hajj Amin's representative with "the Muslim tribes of the northern Caucasus," as well as the German army's liaison officer with the tribal leadership. At first the Germans distributed only Amin's propaganda, but it didn't take long for them to realize that Shamil's influence was their true secret weapon. Soon Said Shamil was issuing a series of calls for ghazavat, in the name and tradition of his grandfather's anti-Russian struggle.

The combination of Islamist incitement and traditional nationalist propaganda proved effective: The local population cooperated with the Germans even as they were being pushed out of the Caucasus by a series of Soviet offensives. In 1943, the Abwehr launched a series of covert operations in which special forces teams were inserted deep into Soviet-controlled territory for purposes of sabotage and inciting the local population. One such mission was Operation Shamil, conducted by a battalion of Soviet POWs and local volunteers recruited from the Caucasus. They distinguished themselves fighting for some months in guerrilla operations in the Caucasus Mountains, repeatedly sabotaging local oil installations.

The Caucasian revolt against Soviet rule continued for months, bolstered by the cooperation of German forces. The rebel forces inflicted heavy casualties, tying down sizeable Soviet forces in a series of ambushes and sporadic irregular warfare. The Soviets were unable to suppress the revolt until mid-1944—months after the German withdrawal from the area. The Soviets' brutal crackdown culminated in a mass exile of the local population, particularly the Chechens, into the heart of Central Asia. There are discrepancies in the official Soviet figures, but roughly half a million civilians were evicted from the area in spring 1944. Two years later, in June 1946, the U.S.S.R. formally abolished the Chechen-Ingush Autonomous Soviet Socialist Republic (A.S.S.R.) because of the population's cooperation with the Germans. Although the A.S.S.R. was reinstated in 1957 as part of the de-Stalinization effort, and some Chechens were permitted to return to the Caucasus, by then between seventy and one hundred thousand exiled civilians had already died during the harsh "resettlement" in Central Asia.

Moscow has not forgotten the Said Shamil episode.

In the postwar era, the Soviets tried to portray Shamil's revolt as a part of the popular antitsarist struggle that preceded the Communist revolution of 1917. But the effort backfired, serving merely to introduce younger generations to Shamil and his legacy. In his 1982 *Islam in USSR*, E. G. Filimonov stressed the importance of the Shamil legacy in subverting the youth of the Soviet Caucasus:

> The supposed accomplishments of muridism [i.e., Sufi Islam] for the mountaineers were in every possible way advertised and extolled by the leaders of the religious communities. They lay at the basis of the ideological maneuvering of the defenders of muridism under present-day conditions and are the grounds on which the religious-nationalistic vestiges survive. In endeavoring to represent the murid communities as a national form of life, they play on the interests of the people, particularly the youth, in their past, and endeavor to insinuate in the believers and nonbelievers the idea of a supposed progressive role of muridism in the life of the waynahs, which ostensibly provides a "class harmony" among the Chechens and Ingush.
>
> Under the influence of such propaganda, among certain young Chechens and Ingush, there are ideas current that Islam played a progressive role in the anti-feudal and anti-colonial struggle of the mountaineers under the leadership of Shamil. [This had the effect of emphasizing] not the socio-class essence of this mountaineer movement, but rather its religious-nationalistic form.

By the late twentieth century, as they faced yet another Muslim revolt—this time in Afghanistan—Moscow had begun to reexamine the essence and legacy of Shamil's uprising. Indeed, during the mid-1980s, Soviet military analysts were reexamining the lessons of the Caucasus War for insight into contemporary conflicts. The seemingly popular uprising in the North Caucasus, they now concluded, had actually been the work of militant feudal leaders who had manipulated the region's poor and oppressed population. To a degree, they claimed, the uprising itself was an internal affair among the Muslim population—a grass-roots reaction to the religious leaders' efforts to extend their power and impose restrictive economic rules in the mountains. This tension was then manipulated by external forces, namely Turkey, which was committed to preserving the old repressive ways for its own political benefit.

This manipulation had led to an escalation of fighting, especially under Shamil, and consequently to the Russian annexation of the area, to save the population from Muslim despotism.

This analysis was captured in official Soviet histories such as the 1983 edition of the Soviet *Military Encyclopedic Dictionary*, which presented the Caucasus Wars as a historical process, an "important military-political mission" that "intensified as a result of a mountainous movement created in Chechnya and Dagestan under the banner of *gazavat* . . . led by Gazi-Magomed [and later] by Shamil" and "ended with the joining of the Caucasus to Russia, which was of progressive nature [because] it freed its peoples from enslavement by backward oriental despotism (Iran, Turkey) and assisted their social-economic, political and cultural development, their preparedness to the revolutionary struggle, along with the Russian people, against Tsarism."

By the mid-1980s, however, this vision of heroic Muslim revolt had taken on a life of its own, feeding into a revival of nationalist sentiment throughout the Soviet Union. For most of the population, the facts of the Shamil revolt had long since faded into obscurity; instead, it became fodder for the mythmaking efforts of interested bodies with economic and political agendas. The mythical legacy of Shamil was seized upon as a means of arousing separatist militant aspirations and hostility toward not just Soviet authorities but the Russian people.

Thus, the table was set for the Chechnya crisis of the 1990s.

Chapter 3

EXPLOITING THE COLLAPSE OF THE SOVIET UNION

BY THE LATE 1980S, THE SUFI STRAIN OF ISLAM HAD BECOME A manifestation of nationalism and self-identity for much of the Caucasian population. Yet the inhabitants of the Caucasus—Muslims, Chechens and other peoples—rarely followed strictly traditional, let alone fundamentalist, Islam. Through prolonged exposure to Russian culture (from the time of the Russian empire through the Soviet occupation), the Caucasians had undergone a near-complete Russification. Most Chechens spoke Russian even among themselves, and few of them knew how to pray.

Moscow was confident in the stability of the North Caucasus, in part because the local economic infrastructure was strongly interdependent with the rest of the U.S.S.R. In 1991, fifteen million tons of petroleum produced by other Soviet regions were refined in Chechnya and Ingushetiya. By then, the Grozny-area refineries furnished ninety percent of the aviation fuel for the entire Soviet Union—a strategic asset the Chechen leadership should have known Moscow could not afford to give up. Moreover, this economic buildup, combined with the expulsion of the mid-1940s, had changed the demographics of Soviet Chechnya. In 1989, of more than fifty thousand workers and engineers working in the Combinat Grozneft petroleum plant, only a few hundred were ethnically Chechen. The rest were predominantly Russian and Ukrainian. During the great economic development of the North Caucasus in 1960–1980, no effort had been made to educate or train the local population to assume a greater role in these industries. Consequently, during these two decades, roughly one hundred thousand Chechens—now known as the Chabachniki—were driven out of their native Republic in search of work as seasonal farmhands or construction workers elsewhere in the U.S.S.R.

Indeed, the original leader of the Chechen revolt of the 1990s, Major General Dzhokar Dudayev, rose through the ranks of the Soviet Air Force, even serving for a time as commander of a bomber regiment equipped with nuclear weapons—a very sensitive assignment that would be entrusted only to a highly assimilated commander like Dudayev, who was thoroughly Russified and had even married a Russian woman. When Dudayev broke with Moscow, it was a reaction to the violent suppression of nationalist movements in the Baltic states, not the situation in the Caucasus. Indeed, when the first wave of the Chechen revolt began in the early 1990s, it too had a distinct nationalist character and was an integral part of the disintegration of the U.S.S.R. Another future Chechen leader who returned from the Baltics at that time was Colonel Aslan Maskhadov, who (unlike Dudayev) did participate in the January 1991 suppression of an insurrection by Lithuanian separatists. Thereafter, Maskhadov was convinced by Dudayev to return to help seize power in the Chechen revolt in Grozny. It was personal ambition rather than disillusionment with the U.S.S.R.—by then already collapsing—that prompted Maskhadov to leave the Soviet armed forces.

The Chechnya-Dagestan crisis that continues today began in 1991, as the U.S.S.R. was disintegrating, and was driven by motivations that were common among the various dependencies of the former Soviet Union, personal ambition among them. Sensing an opportunity to seize control, Dudayev, Maskhadov, and a host of senior officers and officials returned to Chechnya from postings throughout the U.S.S.R. and Eastern Europe in this period. These men, all relatively junior in rank and position within the Soviet system, saw a chance to ascend to power with little or no resistance in their native republic.

When the state of Chechnya unilaterally declared its independence on September 6, 1991, the move went virtually ignored in Moscow. Dzhokar Dudayev was elected president of the self-proclaimed

Chechen Republic on a nationalist separatist platform on October 27, and on November 1 he reiterated the validity of Chechnya's declared independence from Russia. Yet Dudayev was reluctant to take irreversible steps toward separation from Russia, and Boris Yeltsin refused to take the Dudayev Mutiny seriously. The resulting standoff put Chechnya in a kind of suspended legal status—an opportunity quickly exploited by the rising "businessmen" of Russia (now commonly known as the "oligarchs"), who began moving money and goods in and out of Russia via Chechnya to avoid taxation. These business interests also poured money into the fledgling Chechen leadership. Since some of these oligarchs were also supporters and financiers of the Yeltsin political machine, however, the Kremlin had little incentive to close this financial loophole. For the next eighteen months Moscow and Grozny maintained a tense coexistence, conducting cycles of inconclusive negotiations as tension and personal animosity grew between Yeltsin and Dudayev.

By the time Dudayev finally decided to cut Grozny's umbilical cord with Moscow in 1993, his popular and political support was fading. In April, citing "emergency conditions," he disbanded the Chechen parliament by decree and established unitary rule. Even in the chaotic posture of Russia's early days, as noted historian Robert Service observed in his 2002 book *Russia: Experiment with a People,* Dudayev's rule was "a disgrace to minimal standards of political decency." It's little wonder that this move was resisted by wide segments of the local population—both Chechen and Slav (Russian, Ukrainian, and so on)—who preferred to continue formal relations with Russia. Tensions escalated, and by June a series of skirmishes had transformed into a full-blown war between Dudayev's supporters and pro-Russian elements, aided by special and covert operations by Russian forces. By the summer, there was heavy street fighting in Chechnya's capital of Grozny, though Moscow largely remained in the background. Indeed, for the next eighteen months the Russians would provide clandestine support to both sides in Chechnya, maintaining close contact with Dudayev through military channels, including old and trusted military friends. Ultimately, Moscow recognized Dudayev's popular support and accepted the legitimacy of his government.

Starting in mid-1994, the Kremlin made repeated efforts to negotiate a settlement with Dudayev. But these efforts collapsed in the fall, a victim of Dudayev's hard-line posture (which helped him shore up popular support), and the gradual escalation of military clashes between Russian security forces and Chechen separatist forces. As the military buildup continued in the Caucasus, both sides began prodding the lines and trying to gauge the limits of the other side.

A major escalation in the crisis took place in early September 1994, when "Chechen forces"—in actuality, a few symbolic Chechen units loyal to Moscow, backed heavily by Russian troops—captured a few strategically located villages on the approaches to Grozny. Dudayev recognized this as Moscow's attempt to strengthen its military position in case Yeltsin decided to break the deadlock before the winter. Fighting then escalated throughout the Grozny area, with Dudayev's people launching daring attacks on Russian units, inflicting heavy casualties on training units and noncombatants. Faced with heavy losses, Yeltsin ordered the military into hasty action in the dead of winter.

Throughout this period—from early summer 1993 to the fall of 1994—the Chechen military gained strength by enlisting the Chechen *Mafiya* into the nationalist cause. The Mafiya arranged for the flow of illegally purchased arms from the former Soviet Union and the Near East. Dudayev also used the Mafiya and its contacts with high-level Russian intelligence officials to threaten the oil and gas pipelines passing through the region—whose destruction would have devastated the Russian economy. Russia and the post-Soviet Commonwealth of Independent States (CIS) had been planning to modernize oil and gas resources in the Caucasus and across the Caspian Sea by building an elaborate network of new pipelines passing through Chechnya; Dudayev's threats now cast a long shadow over the plan.

By the fall of 1994, Dudayev and his Mafiya aides were confident that they could pressure the Kremlin through a combination of military and economic measures. Moscow was alarmed by the long-term ramifications of having the pipelines become hostage to Dudayev. But Yeltsin was plagued by indecision. Reluctant to confront the Chechens, but equally wary of forging an agreement with them, he opted for a "compromise"—namely, ordering Russian intelligence to arm and unleash

Chechen "opposition forces" on Dudayev's strongholds in the Grozny area. Moscow believed that the new Chechen "opposition" could retain support among the separatists and then turn popular opinion toward reaching an agreement with Moscow once Dudayev was defeated. In November 1994, however, a series of assaults by Moscow's Chechen forces failed both militarily and politically.

British journalist Sebastian Smith succinctly summed up Dudayev's role in this formative period in his 1998 book, *Allah's Mountains: Politics and War in the Russian Caucasus*:

> Dudayev, with his macho fantasies and total disregard for democracy, bore a particular guilt for what was about to happen. When there were chances for negotiations, he baited the Russian bear; when compromise might have kept peace, he threatened war. He began his rule with a threat, in November 1992, to blow up Russian nuclear power stations, and on the eve of war in December 1994 he threatened to kill prisoners, and his foreign minister repeated the threat about nuclear stations. A showdown, to Dudayev, was a self-fulfilling prophecy.

Goaded by his military to persevere, and promised a quick victory, Yeltsin lost confidence in negotiation as a way out of the escalating crisis. In late 1994 Moscow delivered an ultimatum, calling on all sides to lay down their arms, ostensibly because of mounting civilian casualties. When the ultimatum was ignored, as expected, the Kremlin had an excuse to involve Russian military and Interior Ministry forces. A hastily assembled and unprepared Russian force stormed the main concentration of Chechen separatist forces and began an advance toward Grozny. By December, the most forbidding time for military actions in the Caucasus Mountains, the Russian offensive peaked as more than forty thousand troops stormed Grozny. By the time the fighting subsided the following April, the Russians had bombed and shelled most of the urban centers of Chechnya, causing heavy destruction and civilian casualties. The fresh and untrained Russian troops became cannon fodder, suffering heavy casualties in street fighting at the heart of Grozny, with little to show for their trouble.

It was at this point in the fighting that Dzhokar Dudayev first invoked Islam as a premise for his ongoing war. Appealing for greater support from the Muslim world, he described the Chechen war against Russia as a jihad. This call to arms lured the first wave of Afghan and Arab mujahedin to join the ranks of the Chechen forces.

The Afghan and Arab "Afghan" mujahedin first appeared on the Caucasus scene in the neighboring nation of Azerbaijan. The government of Azerbaijan, based in Baku, was the first to recognize the importance of enlisting mujahedin irregulars to further Azerbaijan's strategic interests under the banner of jihad. In the early 1990s, Baku had attempted to exploit the mujahedin's Islamist zeal to escalate its war against the Christian Armenians without implicating Azerbaijan itself.

In the early spring of 1993, a group of covert Azerbaijani emissaries asked members of the Iranian government for help in establishing communications with Pakistani intelligence. They were also interested in working with Gulbaddin Hekmatyar, a key Islamist leader of the 1980s Afghan jihad against the Soviets—who was actually cooperating with Soviet intelligence for most of that period. The Iranians and Pakistanis arranged for a myriad of Afghan mujahedin forces, all loosely affiliated with Hekmatyar's Hizb-i-Islami, to deploy to Azerbaijan via Central Asia and Iran. In mid-August 1993, Azerbaijan's deputy interior minister, Roshan Jivadov, met secretly with Hekmatyar in Afghanistan and finalized the deal, which initially called for between one thousand and fifteen hundred mujahedin to be recruited under the banners of both Hizb-i-Islami (to operate in central Azerbaijan) and the Iranian-dominated Hizb-i-Wahdat (to operate near the Iranian border). The recruitment started soon thereafter: The mujahedin were promised a dollar a day for the duration of the war (a huge sum compared to the mujahedin's usual monthly payment, which ranged from ten dollars for ordinary soldiers to twenty for commanders) and a bonus of five thousand dollars after the victory over the Armenians.

The first mujahedin units, about fifteen hundred well-prepared elite troops, arrived in September 1993, toting Azerbaijani-Dari translation guides, artillery manuals, and a list of recognition signs and other measures to help them avoid friendly fire from their new Azerbaijani cohorts. By the time they arrived, the Azerbaijani armed forces were in a dire state. The Armenian armed forces had been advancing since spring; by late summer they had consolidated a safe corridor with Armenia and had begun capturing key Azerbaijani towns to the south and east of Nagorno-Karabakh, securing their connection to Armenia and all the way to the Iranian border. By the time the front stabilized, the Armenian armed forces controlled about a quarter of Azerbaijan's territory. Little wonder that Baku was determined to change the course of the war drastically.

On October 21, 1993, a mujahedin battalion three hundred strong went into action, taking part in a surprise attack in which the armed forces of Azerbaijan unilaterally violated a local cease-fire. The Afghan mujahedin were used as special forces, attacking Armenian positions in the Jebrail region of southwest Azerbaijan from the rear as Azerbaijani military units attacked from the front. At first the Afghans moved fast, capturing Goradiz, a town southeast of Nagorno-Karabakh's capital of Stepanakert, from the small Armenian garrison. Heavily armed with Soviet-originated infantry weapons, the mujahedin fought with tactical skill and discipline, sustaining casualties while removing their own dead and wounded from the battlefield. It took two days for the Armenians to stabilize the front, but soon they launched a counterattack, driving the Afghan and Azerbaijani units from the territories they had captured and going on to seize the Jebrail, Fizuli, and Zangelan regions bordering Iran. On October 25 the Armenians recaptured Goradiz, and by the end of the month they controlled a forty-kilometer stretch of land along the northern bank of the Aras River.

Despite such setbacks, the Azerbaijani high command was far more impressed by the mujahedin's battlefield performance than by the Armenians' ability to contain and reverse each new offensive. Baku resolved to use the Afghans as special forces in key sectors of the front, as well as behind enemy lines, and requested additional mujahedin, all of them terrorism-trained. By the summer of 1994, the Afghan force in Azerbaijan had reached a peak of around twenty-five hundred mujahedin. That spring Baku had begun sending small teams of mujahedin to the most rugged sectors of the front to seize tactically important positions, as well as mountain passes and peaks, from the Armenians. These mujahedin were determined fighters who held their positions despite heavy casualties. As their importance as an elite force increased, these mujahedin were organized into a distinct Afghan Brigade with a centralized command headquarters in Baku, answerable to the Azerbaijani general staff.

With the cease-fire agreement of May 12, 1994, Azerbaijan saw no further need for the Afghan Brigade as a distinct unit, and dissolved it within a few months. By then, however, Baku was more interested in using the Afghan mujahedin for deniable sabotage and clandestine operations against the Armenians. Though publicly it was committed to the cease-fire, Baku sponsored a series of deniable covert operations designed to destabilize Armenia, putting it at a disadvantage when negotiating with Baku. The period between the fall of 1993 and late 1995 saw a wave of sabotage throughout Armenia, including a sustained campaign to disrupt the lines of communication between Georgia and Armenia. Highly professional operatives bombed trains, rail lines, and fuel and gas pipelines, disrupting Armenia's supply lines from Georgia. When Georgian officials refused to join Azerbaijan's siege on Armenia, sabotage spread into Georgia. The Georgian government changed its mind after a series of such bombings, including several in Tbilisi, that killed a large number of civilians and inflicted heavy damage. Landlocked Armenia was forced to rely on Iran as its primary contact route with the rest of the world—a strategic relationship that put Armenia at odds with the United States.

By the winter of 1995–1996, Baku had exhausted its uses for the Afghan mujahedin. By now, many of them had made contact with local Chechen and Dagestani forces and joined a new jihad they were launching. A group of hardcore Islamist loyalists supported the ethnic group known as Lezghins in their fight against the Azerbaijani government, bonding over their resentment of Azerbaijan's transformation into a pro-Western secular state. A Sunni people based in northeastern Azerbaijan and southern Dagestan, the Lezghins wanted a united independent homeland. The Afghans adopted a nascent Islamist Lezghin movement called *Sadval* ("Union"), training them for terrorist operations against both Baku and the official Dagestani government in Makhachkala. A wave of terrorist strikes followed in Azerbaijan and especially Baku, coming to a halt only after Azerbaijani security arrested roughly twenty Sadval members in April and May 1996. Although the short-lived Lezghin terrorism campaign delivered few tangible results, it had a profound impact on the Azerbaijani government, which was coming to grips with the resolve and ruthlessness of the Islamist movement. Azerbaijan soon emerged as one of the first states committed to the struggle against Islamist-Jihadist terrorism.

Yet the target had shifted: By then the center of the Afghan mujahedin activity had moved to Chechnya, where it was already subverting and transforming the Chechen revolt from within.

The years between 1991 and 1995, the time of the First Chechen War, saw a decided shift in the nature of the conflict in Chechnya. The war had originally started as an anti-Soviet quest for liberalization and self-determination. In 1993–1994, it was taken over by organized crime—the Chechen Mafiya—in order to keep up with escalating costs. At the same time, and increasingly after 1995, the war in Chechnya began to be dominated by the efforts of the jihadists to Islamicize the conflict. And with the Islamists came a new reliance on terrorism—a development that took the battle all the way to the center of Moscow. The Chechens' use of terrorist tactics reflected the growing acceptance of the logic of the Islamist jihad, as advocated by Iran and Pakistan. A closer look at the evolution of external, state-sponsored support for the Chechen jihad reveals how fully the Islamists came to influence events in Chechnya in these years.

Through the mid-1990s, the Chechen revolt was gradually integrated into the state-sponsored international Islamist jihad. These years saw a steady flow of mujahedin from Afghanistan and Pakistan, Bosnia-Herzegovina, and the Middle East, and an increase in the training of Chechens in Afghanistan, Pakistan, Iran, and Turkey (northern

Cyprus). The leaders of the international jihad enlisted Islamist front companies and financial institutions in the West to launder and handle Chechen funds, and allocated proceeds from Golden Crescent drug sales to fund the Chechen war. The growth of Chechen drug smuggling and other organized crime activities in these years was immense. By the mid-1990s, the Islamists' network of smuggling routes, created to deliver mujahedin and weapons into Chechnya, was co-opted for the smuggling of drugs out of Chechnya, with proceeds going to finance the jihad and enrich its commanders.

One such channel, the Abkhaz route, had been operating since late 1993 under the control of Shamil Basayev and his brother Shirvani. Shamil Basayev, the leader of the Chechen volunteers in Abkhazia, established excellent relations with the Muslim rebels there. Using Mi-6 helicopters, the Basayev brothers shipped drugs acquired in Pakistan and Afghanistan from the Vedenskiy Rayon of Chechnya to the heliport in New Athens, using bases set up in Dzheyrakh Gorge in Kabardino-Balkariya as their intermediate landing points. Then, from New Athens, Abkhaz smugglers delivered the drugs by truck to Port Sukhumi on the Black Sea; Turkish ships carried them on to Famagusta, in northern Cyprus, delivering them into the hands of local drug dealers there. On the return routes, the ships, trucks, and helicopters carried arms and munitions acquired by Turkish intelligence for Basayev's forces.

The volume of illicit flights in and out of Chechnya also grew in these years, peaking at an annual rate of 100 to 150 unauthorized flights leaving Chechnya for cities in the former U.S.S.R. and elsewhere, including Turkey, Iran, Jordan, the United Arab Emirates, Saudi Arabia, and even Africa. Even this was only a fraction of the real traffic: Low-altitude flights went unrecorded, as did illegal flights from Chechnya into such Azerbaijani airports as Nasosnaya.

A marked increase in these aerial activities began in early 1995, the result of an arrangement between Dudayev and Pakistan. In early December 1994, Usman Imayev—a close aide of Dudayev who headed the National Bank of Ichkeria and served as Chechnya's minister of justice and prosecutor general—reached an agreement with representatives of Turkish intelligence to supply weapons to Dudayev's forces and to allow Pakistan to use its routes to do the same. Pakistan's Inter-Services Intelligence (ISI) agreed to provide the Chechens with stockpiles left over from the days of the Afghan jihad. The Turks purchased large quantities of Soviet-era weapons and ammunition from the arsenals of the former East Germany and transferred some of them to the Chechens.

To facilitate the shipment of these weapons, Turkey cleared Dudayev to run an airlift out of an airport in the region of Bitlis in Turkish Kurdistan. From there, several Chechen An-24 and An-26 transport aircraft started flying to Chitral, Pakistan, where mujahedin, weapons, and drugs had been loaded, and to the city of Nasosnaya in Azerbaijan. From Nasosnaya, the transport aircraft made night low-altitude dashes to Chechen airstrips in the Shatoy region and to the upper gorges of Belaya Shalazha. When fighting intensified in the spring of 1996, the Russian air force began flying interceptor patrols, and the base for the airlift was shifted to the Azerbaijani village of Zabrat-2, where Chechen forces established a forward base for a few Mi-8 helicopters. From there, the helicopters flew to Nasosnaya to load and refuel, then continued northwest to the Chechen bases in the Zakatala region.

Another key base for the export of drugs and import of weapons and ammunition was established in Shali, Chechnya, in 1995. Sensitive cargo such as nuclear materials, was transported via Shali, as were a number of Islamist leaders, and a special detachment of about one hundred Chechens, mujahedin, and mercenaries was dispatched there to stand watch over the operations. Shali was also used for the storage and shipment of weapons, arms accessories, and ammunition acquired from a wide variety of illegal sources, including ex-Soviet military depots and other sources in Mongolia, Germany, Lithuania, and elsewhere.

The relations between Chechnya and Iranian intelligence intensified in these years, and soon the Chechens were expanding their criminal activity accordingly. Excess cargo space in the planes operating between Bitlis and Chitral, and Bitlis and Nasosnaya, was targeted by Islamist commanders and emissaries in Chechnya as a potential profit center; they suggested that the Chechen Mafiyas use it to help distribute Tehran-made counterfeit U.S. hundred-dollar bills, and the

Pakistanis endorsed the idea warmly. By early 1995, the Iranians had at least $30 million in these counterfeit hundreds, which they offered on the black market at tempting rates. Chechen transport planes flew from Chitral to Zahedan, in southeast Iran, to pick up each new load, then returned to Chitral and on to Bitlis and Nasosnaya. Dudayev's aide Usman Imayev saw spreading these bills in Russia and throughout the CIS as a form of warfare, and by the end of the year the Chechen Mafiya had dispersed at least $26 million of these Iran-made bills. The success of this early venture emboldened the Chechens, and the volume of counterfeit loads collected in Zahedan increased. The growing involvement of Iranian intelligence in the training of Chechen cadres, as well as supplying further mujahedin and critical weapons, was added incentive for the Chechen Mafiya to expand its counterfeit distribution activities.

As the war progressed, the Chechens' support system grew increasingly dependent on Islamist sources and conduits for personnel, materiel, and especially financial resources. For example, throughout these years Libya clandestinely transferred tens of millions of dollars to the Chechens, shipping most of the money via Turkey, where Libyan intelligence used couriers and channels of the Refah Party to launder and deliver the cash. This was not a riskfree operation: At least one shipment, of $10 million, seems to have been embezzled by the Refah Party when a Libyan diplomat named Amar Hareba handed the money to a Refah courier and the cash promptly disappeared. Even after a group of Chechen officials, including a deputy prime minister, arrived in Turkey to discuss the missing funds with Refah officials, the latter denied any knowledge. It was assumed at the time that Refah used the Libyan money for its election campaign. (Moammar Qadhafi was so convinced of this that he publicly insulted the Turkish prime minister during his first visit to Libya.)

Chapter 4

FIRST STEPS TOWARD JIHAD

IN THE EARLY 1990S, WHILE DZHOKAR DUDAYEV WAS STILL FLYING A nationalist banner in Chechnya, the Islamists were taking a new look at the Chechen uprising and the future of militant Islam in the Caucasus. An indigenous Islamist elite was emerging in the region—an elite that offered guidance and inspiration for the nascent jihad. Among the most important results of this movement was the mid-1994 book *Our Struggle, or The Imam's Rebel Army* by Magomed (Muhammad) Tagayev, which served as a source of instigation to would-be Islamist mujahedin throughout the Caucasus. Significantly, Tagayev's thesis was written between early 1993 and the summer of 1994—that is, before the bloody climax of the war in Chechnya. Furthermore, Tagayev stressed in the original edition that he had consulted Dagestani, not Chechen, *Ulema* (scholars of Islamic law) while writing the book. Tagayev's avowed objective was to incite the entire Muslim population of Russia to join an uprising to be launched by the peoples of the Caucasus. Indeed, the book was written in Russian in order to make it accessible to all the local population, including the educated and the Russified.

Tagayev's book calls for the establishment of an Islamist *Imamat* throughout the Caucasus as a precursor for a strategic historic defeat of Russia. "The forthcoming struggle," he wrote, would belong not only to Dagestan and Chechnya but to "the entire Caucasus." Tagayev considered the Chechen state to be "the forward springboard for the liberation of the entire Caucasus." The current war in Chechnya would be the beginning of the regional drive. The Islamist state in the Caucasus, he warned, could be ensured only by the enslavement of all Russians and other non-Muslims living in the areas that were controlled by the Tatars at the height of their advance in the fifteenth century. Tagayev's declared aim went beyond the already aggressive ambitions of the Muslims of the Caucasus: His goal was to destroy and dismember Russia entirely. His work was marked by rhetorical extremes, colored by deep hatred and intense emotion. Tagayev found it impossible even to contemplate compromise of any sort in confrontation with the Russians. "Be scared, O Empire," he warned. "Nobody can withstand this kind of strike."

Tagayev used purportedly historical examples to argue that all of Russia's achievements—from the cultural to the scientific and technological—were actually the work of the Muslim peoples of the Caucasus, purloined by the Russians in their quest to establish their "Empire of rape and evil." To survive, he argued, the Muslims must "cleanse" themselves of everything Russian—seeking complete cultural and social, not just political, independence, a goal he felt could be achieved solely through the relentless and uncompromising use of "fire and sword."

In extending his call to jihad to the Muslims of the Caucasus, Tagayev exhorted the would-be mujahedin to take the war into the heart of the enemy—to strike out from Moscow to Yekaterinoslav to St. Petersburg, from the edges of the Arctic to Siberia's Pacific coast. "We will destroy everything—from Dagestan to Moscow, including the Kremlin," he declared. "We will write several bloody chapters in the new history, even if we have to destroy everything on Earth in the process." He predicted the formation of a mighty mujahedin army, with a specific objective: "Russians will [then] be given 24 hours to leave Rostov, Tsaritsyn [Volgograd], Astrakhan, Taganrog, the sea of Azov and the Black and Caspian Seas. Listen: the Yeltsins, Chernomyrdins, Luzhkovs, and Chubays, Communists and Democrats, we have risen up not just to free our land from you, but also to punish you. Winter and summer, fall and spring, morning and night, we will burn you, blow you up, slaughter and kill you, so that our retribution makes your blood curdle."

Tagayev goes beyond philosophy and rhetoric, offering strikingly specific recommendations on organizing and running the mujahedin's armies. Tagayev envisioned a total mobilization of society in order to establish an army of soldiers between the ages of fifteen and forty-five. Refusal to serve in the "Imam's Army" would be considered a "grave crime punishable by an Islamic military tribunal." The army would be optimized for terrorist operations and other forms of irregular warfare, both in the Caucasus and throughout Russia. "We will form special units that will be prepared and trained in special camps," Tagayev explained. Recognizing that today's Caucasian mujahedin might lack the necessary skills for his far-reaching jihad, he calls for a "foreign military legion" to carry out "terrorist acts in other states." With the requisite expert assistance, he stresses, "nothing will prevent us from blowing up a few nuclear power stations without even getting close."

Tagayev's thirst for revenge did not blind him to the challenges his Imam's Army would face. The current generation of mujahedin, he declared, would fight Russia's powerful armed forces to the death. The next generation would ultimately deliver victory, liberating the entire Caucasus. Only then would the Muslim state of the Caucasus be able to train a true Imam's Army, seizing power and forming an Islamist provisional government, finally establishing a Muslim-Caucasian state "liberated from the Russian-Muscovite empire."

However grandiose, Tagayev's treatise struck a chord: *Our Struggle, or The Imam's Rebel Army* became a major hit, nearly selling out its first print run within days after its publication in the Ukraine. By the fall of 1994, illegal copies were being distributed throughout the region. Though it's difficult to gauge how much genuine support there was for Tagayev's harsh and uncompromising message, it had one certain effect: No longer could any leader in the region afford to ignore the Islamist factor and the growing hunger for an anti-Russian jihad.

Among those who listened closely were the Chechen leadership in Grozny.

It was during that same fall of 1994 that Dzhokar Dudayev first raised the Islamic factor, describing the Chechen war against Russia as a jihad and appealing to the Muslim world for practical support during the heavy fighting in Grozny. Yet such support was slow to come: Iran and other key Muslim states were in the midst of strategic negotiations with Moscow, and support for the Chechens would have endangered the talks. By then, however, the Chechens' clandestine relationships with Pakistani intelligence (the ISI) were already burgeoning. While the ISI's first exposure to the Chechen issue occurred in the context of the Azerbaijan operation, practical considerations soon made cooperation with the Chechens a high priority for Islamabad.

Pakistan had been actively supporting the idea of a Chechen jihad as early as the spring of 1994, when an ISI-sponsored Taliban offensive had endangered the flow of heroin from Afghanistan—a critical source of funding for the Chechen revolt. Islamabad intervened to ensure the continued flow of drugs and to capitalize on the relationship between the Chechens and ISI-sponsored Afghans. The relationship was facilitated by Gulbaddin Hekmatyar's Hizb-i-Islami, which helped expand the ISI's direct relations with the Chechen leadership. The Chechens' key contact man in the transaction was a young lieutenant of Dudayev's named Shamil Basayev.

At the time, Basayev was an up-and-coming military commander with extensive combat experience, a taste for money and power, but no distinct ideological commitment. Born on January 14, 1965, in a village in Vedenskiy Rayon, he received an ordinary education and spent his military service in an air force fire brigade. In the mid-1980s he was accepted to the Moscow Institute of Land Engineers, and he and his younger brother Shirvani went to Moscow soon thereafter. Yet Shamil's heart was never in his studies, and soon he became involved with the Chechen community. Kicked out of school for academic failure in 1988, he soon cofounded a semicriminal trade-intermediary limited liability partnership. By this time Moscow's Chechen community was starting to throw political support behind the Russian reformers—a groundswell of activism that peaked during the coup attempt of August 1991, when Basayev was part of a group of Chechens who played an active role in the defense of Yeltsin and the Russian "White House." This was the first time that Basayev's activism and organizational skills attracted the attention of the GRU, the Russian military intelligence service.

By this time, however, Basayev was growing increasingly involved in the fledgling Chechen revolutionary struggle. He returned to Grozny in the early fall and soon started leading a series of vandalizing raids on Russian government and security-service buildings. Dudayev urged him to stop, lest his operations aggravate the impasse with the Kremlin over Chechnya's unilateral declaration of independence. Instead, on November 9, 1991, Basayev and two friends hijacked a Tu-154 from Mineralnyye Vody to Turkey to protest the declaration of a state of emergency in Chechnya.

By the end of the year, Basayev had returned to Chechnya, having made new contacts with Turkish intelligence. He found success as an arms trader, organizing the flow of weapons to various insurgencies and criminal entities throughout the Caucasus. Eager to capitalize on Basayev's growing popularity, Dudayev appointed him commander of a special-purpose regiment of the presidential guard, giving him the rank of colonel. In the winter of 1991–1992, Dudayev dispatched Basayev to command a group of Chechen volunteers to help the Azeri people in the liberation of Nagorno-Karabakh. It was on this mission that Basayev first encountered and interacted with Arab and Turkish Islamist mujahedin. He was deeply influenced by the experience—so much so that, by the time he returned to Chechnya in the early spring of 1992, he was referring to his unit as "my mujahedin."

The outbreak of fighting in Abkhazia, northwestern Georgia, in the summer of 1992 afforded Basayev a unique experience. A few days after the conflict began, he arrived in the town of Gudauta with a Chechen detachment and a desire to sell weapons. At the time, the Russians were supporting the Abkhaz, and Basayev once again came to the attention of the GRU: Though it didn't recruit him as an agent or asset, Russian military intelligence did train him and other would-be Chechen commanders in mountain warfare and special operations. Basayev and Dudayev both saw the 1992–1993 Abkhaz war as a training ground for fighters and commanders in the coming Chechen war with Russia. The Chechen battalion never exceeded five hundred fighters at any given time, but in total roughly ten thousand Chechens were rotated through Abkhazia to gain combat experience—commanded, in large part, by veterans of the Nagorno-Karabakh war (1988–1994), who had fought on the Azerbaijani side.

After August 1992, Basayev became commander of the Gagry front. In January 1993, he was appointed commander of the Confederation of Mountain Peoples (KNK) expeditionary corps in Abkhazia; soon thereafter, he became deputy defense minister of Abkhazia. In this capacity, Basayev was responsible for coordinating, uniting, directing, and supervising the flow of Muslim volunteers—a position that put him in touch with "Afghan" mujahedin who would later offer help in the Chechen jihad against Russia. The KNK volunteers became notorious for committing atrocities against Georgian civilians and prisoners in 1993; public beheadings were common—some victors even played football with their captives' severed heads. In late September, Basayev himself beheaded dozens of Georgian captives, most of them civilians, in the Sukhumi sports stadium. Eager to defuse the crisis with Tbilisi, Vladislav Ardzinba began neutralizing the KNK volunteers in late 1993, and Basayev was relegated to a largely administrative position. In February 1994, he returned to Grozny as the undisputed leader of the best-trained and most vicious force in Chechnya.

On his return, Basayev urged Dudayev to exploit the contacts he'd made with Afghans and Pakistanis to help in the Chechen war, which by now was escalating. Between April and June of 1994, Shamil Basayev led a high-level Chechen delegation on a visit to a network of ISI-sponsored terrorist training sites in Pakistan and Afghanistan. The Chechens visited the ISI's training facilities near Khowst in eastern Afghanistan, then being run under the banner of Gulbaddin Hekmatyar's Hizb-i-Islami. In Pakistan, the Chechens had a series of high-level meetings with Pakistan's leadership—particularly with its interior minister, General Naseerullah Babar; its defense minister, General Aftab Shahban Mirani; and the ISI's General Javid Ashraf, who was presented as heading up the ISI's sponsorship of Islamist causes. These three officials became the patrons of the Chechen jihad, arranging for the establishment of a comprehensive training and arming program for the Chechens in Pakistan and Afghanistan. The Chechens also met with a former ISI chief, General Hamid Gul, and an aide of his, one Colonel Imam, who would help the Chechens arrange for local connections and contacts for their drug- and weapon-smuggling operations.

The training of the Chechens in Afghanistan and Pakistan began immediately, with roughly one hundred Chechens added to the ISI training camps near Khowst, where two or three hundred Uzbeks and Tajiks were already being trained in guerrilla warfare, preparing to export the Islamist revolution to Central Asia. A select few Chechens were earmarked for command positions, receiving advanced sabotage and guerrilla warfare courses in the Markaz-i-Dawar center in Pakistan. In the fall of 1994, to expedite the flow of expertise to Chechnya, the ISI organized mixed detachments—including recently trained Chechens and veteran Pakistani operatives, most of them with long combat experience in the mujahedin in Afghanistan—to serve as the core of a significant, and heavily armed, new force buildup in Chechnya. Fighters from an ISI battalion of Afghan mujahedin stationed in northern Pakistan were also dispatched to Chechnya in late 1994 to bolster the Pakistani-Chechen detachments. These Pakistani-led detachments saw combat by the start of 1995. Significantly, the ISI retained combat and tactical control over these detachments: The units' Pakistani commanders maintained radio contact with their headquarters in Pakistan, much as the Islamist forces in Kashmir had kept in contact with their rear bases in Pakistan.

The end of 1994 saw the Chechen rebels gain new legitimacy within the Islamist world, as the international Islamist leadership—particularly the Armed Islamic Movement sponsored by Iran and Sudan—adopted the Chechen war as a jihad. Senior Islamist commanders and emissaries were deployed to Grozny, working from there to coordinate their activities with authorities in Tehran, Khartoum, and Islamabad. In their early reports back to the Islamist leadership, these commanders noted the growing Islamicization of everyday life in Chechnya, a campaign mandated by Dudayev himself. Between late 1994 and early 1995, several Islamist charities associated with the pursuit of militant jihad—charities then active from Kashmir and Afghanistan to Bosnia-Herzegovina—began establishing front offices in Chechnya. The flow of money and mujahedin, many of them veterans of previous jihads, began soon thereafter. "Mujahedin from around the world have been arriving at the Caucasus area to join with their brethren in the fight of the aggressor occupying forces of the Russian forces who came to subjugate Muslim Caucasia under their Christian Orthodox rule," the Chechnya-based jihadist commanders reported in late December 1994.

By the time fighting erupted late in 1994, then, the Islamist command-and-control system was already functioning in Chechnya. The Russian crackdown took the main mujahedin forces by surprise, but reinforcements of fighters and weapons were rushed to Chechnya, and by early January 1995 the Grozny-based commanders reported that "mujahedin are still arriving from all neighboring Muslim states to give the Russian enemy a taste of its own medicine." A week later, the commanders' reports offered an upbeat assessment of the Islamicization of the Chechen struggle and its impact on the foreign mujahedin. "Green Flags are flying throughout the city, marking high moral and strong Islamic spirit, and mujahedin are seen praying in groups on the charred streets of Muslim Grozny. Many held their *Qur'an* along with their personal weapons. The words of *Allah-hu-Akbar* in Arabic are seen written in some visible areas by some mujahedin arriving from other countries. Several mujahedin of the Martyrdom unit wore distinct black bands around their heads, [and] they vowed resistance to the end." In mid-January, at the height of the battle of Grozny, the commanders reported that "a number of Pakistani mujahedin have arrived in Grozny and are now fully mobilized along with their Chechen brothers."

In late January 1995, Islamist commanders reported the launch of terrorist operations against the Russians by a joint force of "Afghan" Arabs and Islamist Chechens. "Mujahedin have mounted an organized and disciplined Martyrdom operation which [has] instilled fear and terror among enemy ranks. Several mujahedin have dressed up in Russian soldiers' uniform[s], even [using] fake Russian ID, military ID cards, and are penetrating deep inside Russian lines, [to] ammunition supply and command centers. They are strapping their bodies with explosives and are blowing themselves up. Russian enemy special forces were reportedly firing at anything that moves, including animals, for fear of Mujahedin Martyrdom attacks."

At the same time, the Grozny-based jihadist commanders were worried by reports that Russian tanks were "filling the streets in neighboring Muslim Dagestan." The Russian army was said to have mobilized forty to fifty thousand troops at the border to block mujahedin and relief convoys from passing from Azerbaijan. Nevertheless, by mid-February the Islamists were able to expand the flow of mujahedin and equipment from Iran via Azerbaijan and Dagestan.

The Chechen jihadists received another injection of strength at this time with the arrival of an organized group of hardened Arab mujahedin from the Gulf states, including Saudis and Kuwaitis, and the Maghreb region of north Africa, including Algerian, Moroccan, Tunisian, and other troops. These fighters were commanded by one Ibn al-Khattab, often referred to as Emir Khattab or simply Khattab. Khattab, whose real name was Samir bin Salakh al-Suwailim, was a Bedouin from the Suwailim tribe of northwest Saudi Arabia and southern Jordan; over the years he has identified himself with both nations, depending on the circumstances. Born in 1970 to a fairly wealthy and well-educated family, Khattab received both Western and Muslim education, including learning English. In 1987 he was accepted to a college in the United States, but before continuing with his education, he decided to visit Afghanistan and briefly participate in the jihad.

Arriving in Pakistan in the fall of 1987, Khattab met some of the key leaders of the Arab "Afghans," including Sheikh Abdallah Azzam, Sheikh Tamim Adnani, and Osama bin Laden. Captivated by their call for jihad, he committed his life to the jihad. Khattab completed his training in the international camp in Jalalabad, under Hassan al-Sarehi, the commander of the 1987 Lion's Den operation in Jaji. Impressed with the zeal and skills of his young trainee, Sarehi invited Khattab to join his forces in Jaji. Between 1988 and 1993, Khattab participated in all the major operations in the Afghan jihad, including the capture of Jalalabad, Khowst, and Kabul. He also spent time expanding his knowledge of Islam and his military skills, while becoming conversant in both Pashto and Russian.

Khattab would later claim that he decided to join the Chechen jihad after seeing televised footage of Islamist mujahedin reciting *tak-*

birs (Koranic verses) before going into battle. But his status as a commander also played a role. By the early 1990s, Khattab had emerged as one of the most fierce and competent commanders, popular with both the Afghan and the Arab "Afghan" mujahedin. He also became one of bin Laden's key protégés. Khattab spent the years between early 1993 and early 1995 commanding a small Arab elite force in support of the Tajik Islamist mujahedin, particularly in the Fergana Valley. He returned to Afghanistan to train and lead one of the first elite forces to go to Chechnya.

When bin Laden and the Islamist-Jihadist leadership decided to escalate the jihad in the Caucasus, they summoned Khattab back from Tajikistan and dispatched him to Chechnya. Ali Hammad, a senior al Qaeda commander in Bosnia-Herzegovina in the mid-1990s (under the nom de guerre Abu-Ubeyda al-Bahraini), knew Khattab as a senior commander under bin Laden and considered him "one of the more important personalities in Al Qaeda." Ali Hammad confirmed that Khattab went to Chechnya on bin Laden's orders, and that he and bin Laden personally managed the subsequent flow of jihadist volunteers into the area.

Khattab arrived in Chechnya in the spring of 1995 with eight veteran Arab "Afghan" commanders, followed by a few dozen combat veterans. He soon became one of the most important commanders in Chechnya, quickly forming a close relationship with Shamil Basayev. One of Basayev's closest personal friends, Chechnya's onetime foreign minister Shamil Beno, reported that Basayev underwent a profound change in 1995 under Khattab's influence. Basayev "started moving from freedom for Chechnya to freedom for the whole Arab world," Beno said. "He changed from a Chechen patriot into an Islamic globalist."

The growing influx of Islamist-Jihadist volunteers and financial support was not lost on Moscow. In April 1995, convinced that the Chechen revolt would ultimately die down without foreign support, Yeltsin declared a unilateral cease-fire and the beginning of a new cycle of negotiations with Dudayev. Sporadic armed clashes between Chechen and Russian troops continued—increasingly a matter of independent Chechen elite formations attacking isolated Russian forces.

Nevertheless, during the summer, the situation in Chechnya stabilized, the level of violence and losses remained tolerable for Moscow, and Yeltsin saw no reason to make the painful concessions required to resolve the Chechen conflict.

When Dudayev began his tentative embrace of the Islamist-Jihadist cause in Chechnya, then, it was in part because he was feeling pressure to make progress in the conflict, knowing that the impasse could result in the collapse of his popular support. By the spring of 1995 he was embracing the cause, even though the attendant military buildup contradicted his declared commitment to a negotiated settlement.

The Islamist forces, meanwhile, were already preparing their escalation of the Chechen war into a regional jihad. The Islamist commanders in Chechnya reported that "Azeri and other Caucasus volunteers have been waiting in training camps in the mountains for their turn to carry out attacks on the Russian enemy and on the Georgian troops if they try to intervene." By June, these military measures developed into a strategic reorientation of the war. Throughout the summer, the Chechnya-based commanders reported that the situation in Chechnya was deteriorating: Though they were able to mount a few hit-and-run attacks, Chechen and Islamist forces were withdrawing to remote mountain sites and suffering acute logistical shortages.

Behind the scenes, however, the Islamists were preparing to launch a widespread rebellion against Russia. Following Tagayev's imprecations in *Our Struggle,* the Islamist leaders resolved to reverse the Russian presence in the area and the Russification that dated back to the days of Imam Shamil. "The Caucasus area does not belong to Russia," the jihadist commanders reminded the leadership in a mid-June report. "It belongs to its Muslim people, from the Black to the Caspian sea. The area was savagely suppressed and occupied by criminal Russia about 150 years ago[;] now is the time for every Muslim to share the reward of freeing the land of the free, Caucasia." And this was no mere rhetorical posture: The Chechen and Islamist-jihadist forces were preparing for a major escalation in the fighting. In mid-June, the Islamist commanders reported that "[Chechen] mujahedin have announced that they are preparing horses and mules, and that the coming months will require a lot of effort. Mujahedin have organized themselves in several Battalions, and have spread throughout the Muslim lands of the Caucasus."

Significantly, along with their military assessments, the on-site Islamist-Jihadist commanders also conducted a thorough review of the Islamicization of the population, and of the status of the Muslim forces operating in Chechnya. The commanders noted that the population was still secular in character and argued that while "[the] application of *Shari'a* [Islamic law] continues, there is a great need for *Da'wa* people [that is, Islamic evangelists] in the areas because of the many years of anti-Islam practices by both Russia and its former USSR."

The commanders were far more upbeat about the status of the Chechen fighting forces and their jihadist traits. They identified "three mujahedin groups which are all united under one banner:

> *'La Illaha Illa Allah, Muhammad Rasoolu Allah.'* The first group is the mujahedin themselves, led by their *Ameer;* they have strong knowledge of Islam. The second group is the Government troops, and they also practice Islam as much as they know, and are sure that victory will only come from *Allah*, but this group also has some minor *Bid'a* [deviations from Islamic practice] and that is why good Muslim *Da'wa* is needed. The third group is led by a brother called 'Shamil' [Basayev], named after the great *Imam* and *Mujahid* Shamil who defeated the Russians for over twenty-five years. This group is also abiding by Islam, and all three groups are working together and have the responsibilities divided among them. In short the banner of Islam and *Jihad* is now very clear: *Alhamdulillah* [Praise to God] in Chechnya."

The commanders endorsed the new offensive against the Russians, and recommended a series of terrorist strikes into the heart of Russia.

It was the first time that Shamil Basayev's name appeared in an Islamist-Jihadist report. He was already being singled out as a force to be reckoned with.

Chapter 5

THE ASCENT OF ISLAMIST-JIHADIST CHECHNYA

BY THE SPRING OF 1995, DESPITE HIS OWN CALLS FOR AN ANTI-RUSSIAN jihad, Dzhokar Dudayev was growing alarmed by the rise of Islamist fervor in Chechnya. Pakistan and Iran had begun sponsoring the Chechen revolt directly, and their active support was supplemented by the arrival of growing numbers of volunteers, including Arab "Afghans" and Islamist educators, along with weapons, supplies, and funds. The Chechen elite forces were now becoming increasingly Islamist, and the message of an all-Caucasian jihad, as distinct from a Chechen national liberation struggle, was spreading throughout the Caucasus and the rest of the Muslim world.

For Dudayev, however, the rising tide of Islamist sentiment posed a threat to his own control of the Chechen rebellion; to reestablish his position, he felt compelled to stage a dramatic strike. In June 1995, Shamil Basayev led a daring raid on the Russian town of Budennovsk. The Chechen force he commanded included ten mujahedin from Jordan, Saudi Arabia, and Syria, two of whom were killed in action. The Chechen raiders seized the local hospital, barricading themselves in the maternity ward along with some twenty-five hundred hostages. After a lengthy siege and a few failed assaults by the Russian security forces, Russian prime minister Viktor Chernomyrdin personally negotiated the safe withdrawal of the terrorists to Chechnya, as well as a framework for a new cycle of high-level negotiations between Dudayev's emissaries and Moscow. In July the adversaries reached an initial agreement that called for an end to hostilities, the withdrawal of Russian forces, and the self-disarmament of the Chechen forces. Both sides had a lot to gain from this agreement: For Moscow it was an opening to the end of the war, and for Dudayev it offered recognition and a guarantee of stability for his regime.

Intriguingly, a report from the Islamist-Jihadist commanders in Chechnya suggested that they were not directly involved in the Budennovsk operation, despite the fact that Arab mujahedin were part of Basayev's force and that the commanders supported it. The report, written in mid-June, recounted "the latest operation in the heart of what is called 'Russia,'" involving an elite mujahedin unit that penetrated Russian checkpoints using Russian uniforms and military trucks. The unit "rode out of the Caucasus into the city of Budiannovosk [sic] in the state of Strasvopol," a military base used for Russia's "criminal operation" in Chechnya. "We have not received confirming news from mujahedin," the commanders conceded, "but it appears that the elite unit was able to take over Russian heavy weapons. They then bombarded the military installation, damaging planes, equipment, and ammunition depots. As usual, the enemy Russian regime claimed that mujahedin killed local civilians. It is believed that mujahedin heroes are holding more than five hundred Russian enemy prisoners." A number of other units from this elite group had withdrawn to other positions, the report noted, to plan further attacks within Russia. "Mujahedin vowed months ago to reach Moscow, and force Russia to pay tribute for the Genocide it perpetrated on the Muslim Caucasians." Despite their enthusiasm, one wonders if the Islamist-Jihadist commanders were so apprehensive about Russia's response to such an audacious strike that they decided to keep their distance in case things went wrong for Basayev and his raiding party.

Yet Basayev's success in missions like this changed everything for the Islamist-Jihadist leadership and their representative commanders in Chechnya, encouraging them to become more directly involved in the conduct of military and terrorist operations, and likewise in policy and strategy formulation. The Chechnya-based jihadist commanders were now in constant touch and consultation with the Chechen leadership, including Dudayev and Aslan Maskhadov. They offered regular strategic and tactical analysis, made requests for assistance, and stressed the growing effectiveness and strategic impact of the terrorist strikes at the heart of Russia.

In another report, from July 1, the jihadist commanders noted that the newly negotiated cease-fire had already been violated, and they expressed no confidence in the ongoing negotiations with Moscow. "Talks continue; however, Jihad to liberate the Muslim land is escalating rapidly," they explained, and indeed the bulk of their report addressed military activities. "We have just received a report that Mujahideen forces have taken back the initiative and the element of surprise and are organized in several groups that operate throughout the region. The Muslim units are operating face to face with the enemy and behind its lines," the on-site commanders reported. "Muslim casualties were minor, *Alhamdulillah*."

The report also devoted a special section to the activities of Shamil Basayev. "*Ameer* Shamil sent some mujahedin inside Russia again to deliver a message to the enemy. Mujahedin who delivered the message were able to slip back unharmed to Muslim Chechnya. The message warns the Russian criminal crusade that any delay in negotiating the withdrawal of Russian occupation forces will result in instant strikes in the heart of Russia. Upon receipt of the message, Russian enemy was horrified and its forces inside Russia were put on full alert." According to the commanders, Basayev's message had a chilling effect on Russian security forces: "We received word from our brothers in Moscow that the criminal Russian police, with the help of Russian military, has been rounding up and arresting one thousand Muslims every week. More than one million Muslims live in Moscow; most of them are Tatar and Caucasian (Kafkaz). Our brothers said that the whole population is terrorized and talk all the time about the heroic operation carried out by *Ameer* Shamil. Public opinion now is totally against the Russian genocide in Chechnya." Again, the commanders recommended escalating the terrorism offensive.

The Chechen leaders in Grozny were also anxious to resume all-out fighting, but in a way that would make the Russians appear responsible for the collapse of the negotiations and the cease-fire. In August, in a conversation with the on-site commanders, "Mujahedin Commander Brother Aslan Maskhadov" stressed his profound mistrust of the "Kuffar and hypocrite" Russians, and foresaw the collapse of the agreement within two months. The jihadist commanders suggested launching terrorist strikes to provoke the Russians into violating the agreement. Highly trained Arab expert terrorists were sent to Grozny from Afghanistan/Pakistan and the Middle East to expedite the provocation. "Several training camps have been active in the mountains, and Chechen leadership vowed to fight until the last Chechen for complete liberation of Muslim land and independence from the criminal Russian occupation," the commanders reported in late September. The spate of car-bomb attacks against Russian officials in Grozny that started in early October was their handiwork. The commanders also updated Grozny on the exploits of "*Ameer* Shamil Basayev, the wolf of the Caucasus, [who] carried out another daring operation against Russian occupying forces," in direct defiance of the agreement with Moscow.

As expected, the Russian forces responded to these provocations with a massive use of force. But the mujahedin "were ready for this Russian move," the commanders reported, "and have been launching several raids on some enemy units. We [have] received information that Chechen Muslims are reorganizing their ranks for an expected major confrontation with the Russians." The flow of cash from Islamist sources was having an impact on the fighting, for "mujahedin were successful in buying heavy weapons for the first time."

As the Chechen and Islamist leaders had hoped, it was the massive Russian reaction—complete with shelling and bombing of civilians—that registered in the West as the violation of the cease-fire. High hopes for a resumption of the political process were soon dashed as guerrilla warfare continued in the mountains, and the plight of Chechen civilians grew steadily worse. To the on-site jihadist commanders, however, the only important factor was the Islamicization of the Chechen population. "The population is fully behind mujahedin and whenever they see them they shout '*Allahu Akbar!*' " they reported in late September. In mid-November, the commanders boasted about their contribution to popular political activities. "The Chechen Muslim population went out to the streets of Grozny and surrounding villages to protest these Russian plans and vowed not to participate in any elections until every Russian soldier is out of Chechnya. It looks like Moscow has not learned from its failure in Tajikistan." (The mention of Tajikistan was revealing: Khattab and his key aides were all veterans of the Tajik jihad.)

To Dudayev, perhaps the greatest challenge of this period was the Islamicization of his own elite forces—and the growing influence of the foreign mujahedin on the Chechen military. The conflict in Chechnya was being transformed by the infusion of foreign volunteers and the emergence of a coherent sponsorship and command structure that was effectively being run by several foreign intelligence services. Suddenly, Dudayev was under pressure from these foreign influences to further the struggle—now formally a jihad—by adopting a harsher and more uncompromising line.

Throughout the fall of 1995, the Chechen forces waged concerted guerrilla warfare against the Russians. As winter approached, Moscow resolved to escalate the war once again. The Islamist-Jihadist commanders recognized that a military end to the impasse was virtually inevitable. In early October, they reported on a new buildup in the number of Russian forces and skirmishes, which they saw as harbingers of a new Russian offensive. "We have just learned that Russian enemy forces have been reinforcing their positions inside Chechnya by sending fresh troops and equipment and digging winter trenches," the commanders reported. "We expect that the Russian military is preparing for a winter offensive against Muslim villages that are still free and protected by mujahedin. The Russian enemy has been using helicopters, laser-guided bombs, and airplanes to bombard mujahedin positions throughout the Muslim land; however, with the grace of Allah they have been unsuccessful. Mujahedin have been launching heroic attacks at Russian formations even inside the cities of Grozny and Argun, which are occupied by the enemy." The commanders anticipated that such indecisive clashes would continue, and even escalate, as long as the winter weather remained harsh. In later reports the commanders boasted that "powerful car bombs" in Grozny had served to derail and complicate Moscow's efforts to arrange elections for a pro-Moscow Chechen administration.

Still, the Kremlin was controlling the situation; the jihadists' reactions were attempts to wrestle the political and military advantage from Russian hands.

At this point, Dudayev decided to demonstrate that he was in charge of the Chechen revolt, still capable of outpacing his rivals and contenders. In late November 1995, using Dudayev's Mafiya contacts, Shamil Basayev organized the world's first known incident of nuclear terrorism. After arranging to have several packages of radioactive material (largely harmless Cesium-137) buried in the heart of Moscow, he notified Russian NTV of the location of one package, in Izmailovsky Park, so that it could be retrieved on live TV, to the horror of the public and the Kremlin alike. The jihadist commanders in Chechnya were elated by the dramatic results of this terrorist strike and its impact not just on Moscow, but also on Basayev's standing in the Chechen power structure. Their report on the incident, though clearly embellished, reflected the story they wanted their Islamist-Jihadist leaders to hear. "Shamil Basayev, the Chechen hero who instilled terror in the Russian army, has done it again, but this time inside Moscow and with nuclear material. Shamil was able to purchase highly radioactive nuclear material from Russian military sources; then he smuggled it to Moscow, where he buried several of these packages around the city. Shamil immediately called for a Russian withdrawal from Chechnya and for Russia to pay compensation for war victims. Shamil said that Chechen Muslims will use all available weapons to defend themselves against Russian genocide and that included nuclear material." The commanders' report also noted, "It was reported early this week that Shamil Basayev was in Moscow preparing for another operation deep into Russian territory." It was a telling claim, but a false one: Basayev was actually with the commanders themselves in the Grozny area, planning additional terrorist strikes with his ally Khattab.

By December 1995, the Chechen forces had suffered a series of setbacks. Russian forces evicted Chechen rebels from Gudermes—Chechnya's second-largest city—by destroying it in a hail of artillery and rocket fire. The partial success of the Chechen elections suggested a certain level of public support for the Russian authorities. The military operations of the Chechen insurgency were limited largely to ambushes along roads used by the Russian forces and officials, the mining of those roads, and a few car bombs they detonated in Grozny. The Islamist-Jihadist leadership was compelled to review the situation in Chechnya—and in the process they made a particular study of the prospects of further Islamicization of the Chechen culture.

The Islamist leaders dispatched a few senior Imams to assess the religious landscape in Chechnya. In late December, the Imams reported the conclusion that "Chechens are Muslims with strong faith." Their report also made clear that the Islamicization process in Chechnya was following the tenets of neo-Salafite (Wahhabi) "Arabization"—a path favored by Saudi Islamists, even though it was ill suited for the Chechen culture.

"During our last visit to Chechnya," the Imams reported, "we noticed that Friday prayers [*Juma'a*] and *Khutba* [sermons] were being conducted in the Arabic language, even though 99 percent of the audience did not speak or understand Arabic. The Chechens are very strongly attached to Islam, to the point that they refuse to deviate from the way the Prophet, peace be upon him, conducted his prayers. Although the Chechens of course are not obliged to hold the *Khutba* in Arabic, they insist on doing it." Though their own history and culture were more than eight thousand years old, and their language "distinct and unique," the Chechens had "changed their writing to Arabic and changed their names to Muslim ones, especially those after Prophets and the Companions of the Messenger of Allah, peace and blessings be upon him."

This question of Arabization would prove crucial. Although Muslim, the peoples of the Caucasus had always followed the traditions of their tribes and nationalities, relying on codes of behavior and Muridism (Sufi Islam), rather than on "classic," Arab-dominated Islam. The fierce rebellions that marked the region's history were motivated largely by tribal and Muridist sentiments, most notably in Imam Shamil's revolt. There was no local tradition of Islamism or strict adherence to Islamic codes in the Caucasus. In the absence of the international Islamist movement, the population would probably have reverted to its traditional social and legal system after the collapse of the Soviet and Russian legal system. Now, instead, it was targeted as a new breeding ground for jihadists.

Meanwhile, Dudayev was still grappling with the ongoing challenge from Moscow. In early January 1996 he sent his son-in-law, Salman Raduyev, to head the Lone Wolf terrorist force in raiding the Russian air base in Kizlyar in nearby Dagestan. The operation was planned by a Pakistani expert and veteran "Afghan" known by the nom de guerre Fakh. Reflecting the changing face of the Chechen revolt, this Lone Wolf unit also included several non-Chechen Islamist terrorists in senior positions. Having attacked the air base, the unit then continued, seizing a nearby hospital and some two thousand hostages. This time, after a brief siege, the unit withdrew with numerous hostages to the village of Pervomayskoye. For several days they were besieged and bombarded by Russian forces, yet key elements of the Lone Wolf unit managed to break through the Russian lines with several hostages and reach Chechnya.

The jihadist commanders' report on the incident stressed its importance in mobilizing Dagestanis and other Muslims of the Caucasus to join a regionwide jihad. "The mujahedin unit that carried out the latest Kizlyar operation, which is called 'Lone Wolf,' were well aware of the Russian attempt to divide Muslims in the mountains. For that reason, many of the fighters in the 'Lone Wolf' unit were not Chechens, but Dagestani Muslims." Although the Russian forces had subjected the local population to intense fire attacks, the Lone Wolf raiders benefited from the support of the local population; the report claimed that "hundreds of Dagestanis volunteered to join" the Chechens, and "the original mujahedin unit actually increased in number after leaving Kizlyar and releasing two thousand hostages, because many Dagestanis and Chechens in Dagestan joined them on their trip back to Chechnya." The commanders took this as further evidence that the entire Muslim population of the Caucasus, despite their ethnic and national diversity, could be united under the banner of neo-Salafite Arabism.

At the same time as the Lone Wolf raid, an Islamist terrorist force that included a few Chechens, but was led by a Turk named Muhammed Tokcan, seized a Russian ferry in Trabzon, Turkey, with 120 passengers and forty-five crew members. The terrorists identified themselves as part of Shamil Basayev's force. After a few days of negotiations, which served to highlight the Chechen cause in Turkish and Western media, the terrorists surrendered to the Turkish authorities in Istanbul.

As 1996 began, as the conflict with Russia dragged grimly on, the nationalist movement faced a transformation. By now, the mujahedin had completed the training of Chechen elite units in tactics based on the lessons of Afghanistan. These units included both veteran mujahedin, mostly veterans of Afghanistan, and Chechen combat veterans. And the influx of new blood continued: In preparation for the anticipated escalation, some six hundred mujahedin—mainly Afghans and Chechens trained by Iranian intelligence and the HizbAllah in Sudan—deployed to Chechnya in late 1995/early 1996.

To control the flow of these and other foreign mujahedin into Chechnya and the North Caucasus as a whole, Ibn al-Khattab had organized a system of safe houses in Makhachkala, Dagestan, where ISI-trained security experts could manage the reception, vetting, and transfer of volunteers into Chechnya. "All the interested mujahedin who want to go to Chechnya and join Ibn al-Khattab's unit must first travel to Dagestan," a report from al Qaeda's Ali Hammad noted, "where they wait for the approval of Ibn al-Khattab to continue to Chechnya, if they are appraised as precious warriors. [Otherwise they] could be ordered to return." Khattab's Dagestani allies also controlled the smuggling routes into Chechnya—ensuring that only al Qaeda–approved mujahedin would reach the front.

Another influx of jihadists came with the end of the war in Bosnia-Herzegovina. After the Dayton Accords were signed on November 20, 1995, several dozen Arab mujahedin, including highly experienced commanders, were redeployed to Chechnya starting in early 1996. Khattab rejected many "ordinary soldiers" from the Bosnian conflict, insisting on securing high-level commanders and experts; as a result, Hammad noted, "it is highly probable that many commanders or highly positioned personalities in mujahedin forces during the B-H war have escaped to Chechnya, for they are wanted for various crimes."

By early 1996, the arrival of these reinforcements, along with funds, weapons, and other equipment, emboldened the Chechen leadership to order a marked escalation in the war. In the spring, Basayev and Khattab triggered a series of devastating raids, ambushes, and sabotage operations conducted by Pakistani-trained mujahedin using experience they had gained in Afghanistan. In a report, the Islamist-Jihadist commanders stressed the series of strategic sabotage operations throughout the region:

> Mujahedin have been working steadily at eradicating all Russia's vital links in the Caucasus. Last Saturday [February 17, 1996], mujahedin blew up the Baku-Stavropol natural gas pipeline near the village of Sholkovskaya. Another explosion destroyed Grozny's pumping stations, along with the main telephone exchange used primarily by the enemy and its collaborator mercenaries. On 22 February, mujahedin blew up the gas pipeline from Chechnya to Dagestan and on 20 February in the city of Novogroznenskii, the Grozny oil refinery was attacked, setting ablaze a storage tank containing three thousand metric tons of fuel, which was the main source for enemy vehicles.

In early March, the on-site commanders reported that "the brave Mujahedin Wolves of Muslim Chechnya" had launched a carefully planned offensive on Grozny, which had been under Russian control for a year. "Mujahedin approached the city from four fronts, including clandestine units who were ready inside the city and booby-trapped several important enemy installations." After weather conditions improved in mid-March, the Russians launched several retaliatory sweeps into the mountains of Chechnya. "Mujahedin were ready for the latest offensive and allowed columns of enemy tanks, artillery, and vehicles to advance deep into the mountains before closing in on them and destroying them," the on-site commanders reported. In April, Basayev led one of these units in the lethal ambush of a Russian convoy in the mountains near Vedeno; this was followed by another ambush in the mountains near Yaryshmardy, this time led by Khattab.

Then, in late April 1996, Dzhokar Dudayev was killed when the Russian air force bombed his mobile headquarters. The nationalist war he had helped to initiate had already begun its transformation into a combination of popular guerrilla struggle in the mountains of Chechnya (and, increasingly, other parts of the Caucasus) and daring terrorist strikes by international Islamist-Jihadist forces in Russia and elsewhere. Moscow was willing to cope with a low-level insurgency in the Caucasus, but international terrorism was another story. The Islamist international leadership and its sponsor states were now increasingly dictating the duration and intensity of the Chechen war, at the expense of the Chechen nationalist leadership. Younger commanders like Basayev and Raduyev, who were associated with the Islamist-Jihadist forces and terrorist operations, were rising to prominence.

After Dudayev's death, his deputy, Aslan Maskhadov, a former colonel in the Russian armed forces and a professional officer, emerged as the de facto Chechen leader. In summer 1996, Maskhadov resolved to break the deadlock in Chechnya with a dramatic outburst of fighting. On August 6 he ordered the infiltration of more than fifteen hundred fighters, mostly Islamist Chechens and foreign mujahedin, into the heart of Grozny, to launch a wave of terror fighting that flagrantly challenged Russia's hold over the city and, indeed, the nation. "Mujahedin have been in complete control of Grozny since 8 August, after advancing from three fronts, and suddenly attacking Russian enemy troops in their own camps, buildings and trenches," the commanders reported in late August. "In the cities of Argun and Gudermes to the east, Russian soldiers were so shocked they dropped their weapons and gave up immediately to the Muslim forces, who loaded tens of them on trucks and herded them to safe prisoner areas deep inside Chechnya. Both cities fell without a fight." The commanders also noted that Basayev, who was directing several fronts, had announced he would pay one hundred dollars to any mujahedin who would volunteer to count and tag Russian corpses left on the streets of Grozny.

The violence of mid-1996 led one Russian leader to reach out and negotiate a cease-fire with Chechnya's new prime minister—one that would mark a fateful turning point in the Chechen conflict. To General Aleksandr Lebed, Russia's newly appointed Security Council secretary, the conflict in Chechnya was swiftly becoming a true war, one that could not be won at the price Moscow was willing to pay. Resolving to stop the futile carnage and end the war swiftly, he embarked on an intense new round of negotiations with Aslan Maskhadov, and within a couple of weeks a cease-fire agreement had been reached.

The agreement, signed in Khasavyurt, Dagestan, on August 31, 1996, outlined a contingency plan that promised the Chechens a provisional independence. Between early October and late November, Maskhadov—now serving as prime minister—signed a series of agreements with Russia that ensured a unique autonomous status for Chechnya, while postponing resolution of the sticky question of full independence until 2001. In turn, on November 23, 1996, Boris Yeltsin ordered all Russian troops withdrawn from Chechnya.

For Moscow, the Khasavyurt agreement may have looked like an end to the painful war in Chechnya. For some of the Chechen leaders, however—and their new Islamist-Jihadist allies—it would fan the flames of a new jihad, one bent on establishing not just an Islamic state in Chechnya but a new regional order in the Caucasus. In the long run, this new war for the soul of Chechnya—still unfolding today—would be more crucial than the violent war that had just ended.

Chapter 6

INTERLUDE

AS 1996 CAME TO A CLOSE, THE JIHADISTS WERE PREPARING TO REVIVE the Islamist violence in Chechnya. Throughout the summer, even as the cease-fire negotiations were ongoing, they had relied on a series of terrorist acts—including a series of bus bombings in the center of Moscow—to exact further concessions from the Russians. From Ankara, Chechen commander Solta Ersanov claimed responsibility for the Moscow bombings, calling them "a warning to the Russian authorities" designed to force an end to Russian attacks. In truth, the Chechen leaders never intended to reach a meaningful agreement with Moscow: Goaded by the jihadist commanders, the leaders in Grozny continued to hint at the possibility that fighting would resume. Chechen leader Zelimkhan Yandarbiyev declared that he "cannot affirm that the war has really ended." Basayev referred to the war as merely "suspended." Good faith was thin on the ground. Since "Russia has proved that it does not keep its promises," Basayev stressed, "the possibility of the war being resumed remains on the cards."

The international Islamist leadership (including senior leaders in Iran, Pakistan, and Sudan) saw the cease-fire as no more than a convenient interlude to replenish personnel and supplies before resuming the fight. The agreement postponed the question of Chechnya's independence until 2001—a condition the Chechen commanders found unacceptable. In mid-December, as the cease-fire was being signed, Raduyev warned that he had "put my troops on red alert" pending new developments. "If they [the Russians] want to fight with me, let them come and we see who wins." When Yeltsin fired General Lebed in early October, the Islamists took the move to signify that the resumption of war in Chechnya was virtually inevitable—a posture they soon announced to the Islamist international leadership in the West.

One figure who would emerge as crucial to the next phase of the Chechen conflict was the newly appointed Islamist-Jihadist senior commander in Chechnya—a Saudi known as Abu-Sayyad, who had previously held a senior command position in the mujahedin forces in Bosnia-Herzegovina. Abu-Sayyad's command position within the jihadist network—especially his relationship with the Sunni mujahedin, who were now open followers of Osama bin Laden and Ayman al-Zawahiri—had been confirmed in the spring of 1996 with the establishment of the Tehran-sponsored HizbAllah International, of which bin Laden was a most senior leader. Now he was given the title "Amir of the Ansar [Forces] in Chechnya"—that is, commander of Iran's 4th Ansar Legion in Chechnya. The 4th Ansar Legion is supervised by the Islamic Revolutionary Guard Corps (IRGC)'s al-Quds Forces, Iran's network for training and commanding Sunni Islamist terrorists, and is specifically tasked with exporting the Islamic revolution to the independent republics of the former Soviet Union.

In mid-September, less than a month into the cease-fire, Abu-Sayyad issued a communiqué in his capacity as Amir of the Ansar in Chechnya. The message charged that the Russians were not abiding by the cease-fire and suggested two possible interpretations, both designed to provoke renewed hostilities. The more likely possibility, he suggested, was that the Russians were switching out their military units for newer, more efficient soldiers and regiments, and that "the fighting will return more severe than it was before." The second interpretation was that the Russians were planning to consolidate their forces over the winter, "carry[ing] out a series of assassinations of the mujahedin leadership in the combat zones trying to bewilder the mujahedin." Then, once the weather improved, the Russians would launch an offensive "stronger than the previous time," including "violent attacks on mujahedin positions and cities where mujahedin were to be found."

The Islamist leaders' decision to resume the jihad was heavily influenced by assessments from the Iranian and Pakistani intelligence services, based on reports from their on-site commanders in Chechnya. The Islamist leaders soon arranged for an increased flow of logistical and financial support from the West. In late August—even as the cease-fire was being negotiated—plans for escalation in Chechnya were being discussed at a summit of senior commanders of HizbAllah International, held in Mogadishu, Somalia. The summit was attended by either Osama bin Laden or a trusted emissary, as well as the Iranian commander of the al-Quds Forces based in Sudan, Ethiopia, Somalia, and Yemen, and other senior al-Quds officers from both Tehran and Khartoum. The summit decided to deploy between five and seven hundred "Afghan" mujahedin (including Arabs and Pakistanis as well as Afghans) to Chechnya during the fall. These mujahedin would be drawn mainly from ISI-run camps in Afghanistan and other camps in Sudan and in Lebanon, which would supply recently trained Chechen HizbAllah and HizbAllah veterans from Persian Gulf states and Bosnia. The ISI was also directly responsible for the transfer of "special weapons" to Chechnya. General Ashraf of the ISI, who was heading the ISI branch in charge of support for Islamist causes including Chechnya, was put in charge of the Pakistani part of the operation. Additional funds were moved to Chechnya from Saudi Arabia and Persian Gulf states via Western Europe. At a second Mogadishu summit in late September, the senior HizbAllah International commanders pushed further to accelerate the schedule.

In early October, an inspection team including several senior Islamist commanders was dispatched to assess the prospects for their expanding jihad firsthand. The team included bin Laden's most senior military commander Ayman al-Zawahiri, Zawahiri's deputy Salah Shahatah, and a few bodyguards and expert terrorists. Leaving Afghanistan by land, they traveled through Turkmenistan, Kazakhstan, Russia (specifically the Kalmyk Republic, Chechnya, and Dagestan), and Azerbaijan. At each stop the terrorist commanders met with local Islamist leaders, studied the strength of their forces, and assessed their ability to contribute to the planned escalation.

In late December, roughly a month after Yeltsin ordered his

troop withdrawal, Zawahiri and his team completed a lengthy survey of Chechnya and Dagestan and left for Azerbaijan, where they were slated to link up with a Baku-based jihadist command cell led by Ahmad Salamah Mabruk. But Russian security authorities on the Dagestani-Azeri border noticed that the passports of most team members, including Zawahiri and Shahatah, were false, and the group was promptly arrested and transferred to Makhachkala. The Arabs insisted they were simply illegal aliens looking for work in the oil fields. The Dagestani authorities sent the pictures to Interpol but got no identification. Shortly after the new year, bin Laden was notified of Zawahiri's arrest and immediately sent a key operative to Makhachkala. There, bin Laden's aide arranged for both legal aid and "appropriate payments" to the local authorities. Zawahiri and his team were soon released on bail, and by June 1997 they were back in Afghanistan.

Despite their ordeal in jail, Zawahiri and his team were full of praise for the Islamist networks they inspected. They were encouraged by the growing capabilities of the Islamist forces in Chechnya and the rest of the Russian Caucasus. With further expert training and mujahedin reinforcements, they felt, the Chechens would be able to lead a regional jihad against Russia and its ally Armenia. Zawahiri was also optimistic about the impact his discussions with top leaders in Chechnya would have on that country's Islamic character and commitment to the escalation of the anti-Russian jihad.

By now, foreign-trained mujahedin were pouring into Chechnya. After the Mogadishu summits and Abu-Sayyad's report, at least two hundred were deployed to Chechnya from camps in Pakistan and Afghanistan. The Pakistani minister of the interior, retired major general Naseerullah Khan Babar, personally arranged for the fighters to pass safely through both the Taliban and Ahmad Shah Massud lines, as part of the Pakistani mediation effort in Afghanistan. More than one hundred Arab "Afghans" arrived from bases in Sudan and Yemen, reaching Chechnya via Iran or Afghanistan. The first class of native Chechen *HizbAllah* completed a six-month advance training in the HizbAllah camps run by the Pasdaran (Revolutionary Guards) in the Biqaa in mid-November before returning to Chechnya early in 1997;

roughly four hundred Chechen graduates would be dispatched from the Biqaa, Lebanon, by the spring of 1997. A year later, the flow of mujahedin along these routes was still ongoing.

Native Chechen jihadis were being trained in ISI-sponsored camps in Afghanistan near Warsaj (in the province of Takhar), Jabal ol-Saraj (in Parwan), Khowst (in Paktia), and smaller sites. Two hundred and fifty Chechens were undergoing clandestine training in a camp near Peshawar, Pakistan, by ISI operatives and expert terrorists from Egypt and Sudan. In the Lahore area, roughly one hundred were receiving ISI training in sophisticated terrorism and urban warfare. A terrorism training base run by VEVAK (the Iranian Ministry of Intelligence and Security) in Ziarat Jah, Herat, Afghanistan, was transferred to Gorgan, Mazandaran, Iran, in the fall of 1995 because of the fighting. In Iran, Chechens were now attending Islamist terrorism schools controlled by al-Quds Forces. Several hundred mujahedin, mainly Afghans and Chechens, were being trained by Iranian intelligence and the HizbAllah in Sudan. Moreover, in the spring of 1996, in anticipation of a marked escalation, roughly four hundred Chechens were sent to HizbAllah training camps in the Biqaa to undergo six-month advance courses run by Iranian Pasdaran instructors. These training programs were still operating in early 1998.

With so many mujahedin training and fighting in Chechnya, the Chechen jihad was swiftly becoming a kind of all-purpose boot camp for the entire Islamist movement—particularly as it prepared a new jihad against the U.S.-led West. The ISI ordered the Taliban to hand terrorist training camps in eastern Afghanistan over to the Harakat-ul-Ansar, which it tightly controlled. The Harakat-ul-Ansar had successfully mounted terrorist operations in Kashmir, Burma, Tajikistan, Bosnia, Chechnya, Canada, and the United States. The ISI was tightening its control over jihadi training, upgrading its training staff to include several mujahedin described by a Pakistani official as being "battle-inoculated, [having] fought in Afghanistan, Kashmir, Tajikistan and Chechnya." This staff included roughly three hundred mujahedin, mostly Pakistanis and Kashmiris, as well as Afghans and Arab "Afghans." The camp they ran offered both a forty-day basic training program in small and heavy arms and basic guerrilla warfare tactics and a lengthy advanced-training program that ranged from several months to a couple of years. In August 1998, the United States would launch cruise missiles against these training camps.

The intelligence services of Iran, Pakistan, and Turkey continued to train and dispatch foreign volunteers to Chechnya. In Afghanistan, Sheikh Muhammad Ali Akhund organized a Taliban force for deployment to Chechnya. The Islamist commanders and instructors from Afghanistan, Pakistan, and numerous Arab states who nurtured this new generation of Chechen and other Caucasian mujahedin were veteran "Afghans" and "Balkans," as well as Middle Eastern terrorists, and they also served as the core of the Chechens' elite terrorist and special operations units. The Islamist-Jihadist leadership had a new objective: They would make Chechnya a center for Islamist regional and international terrorism as potent as Afghanistan or Lebanon.

The Islamist leaders' effort to further the process of Islamicization within Chechnya—with its emphasis on the Arabization of society at the expense of the local Sufi-based traditions and heritage—continued on its ruthless and increasingly violent course. Many of the Saudi-funded neo-Salafite/Wahhabi textbooks in Russian that were distributed in the Caucasus (as well as Central Asia) openly advocated imposing "the right Islam" by force. One Saudi-originated Russian language manual, *Teaching to Pray,* by Dr. Abdallah bin-Ahmad al-Zayd, explicitly called for the slaughter of Sufi Muslims as an Islamic requirement. "He who of his own will performs an act of worship (prayer, fast, supplication, vow, sacrifice or plea for salvation) addressed to anyone besides Allah, to a monument or benefactor, falls into polytheism . . . one of the greatest sins," the manual warned. "He who does this may be killed, [and] his property may be taken away." This call to eliminate those who prayed differently could only have been directed against the Sufis—those residents of the Caucasus who insisted on preserving their own traditional ways rather than succumb to the religious and political dictates of the Islamist-Jihadist movement.

In the long run, this Islamicization and Arabization campaign in Chechnya and the northern Caucasus was fiendishly successful. During the cease-fire that began in late 1996, the Islamist-Jihadist influence came to dominate everyday Chechen life. The most conservative variant of the *Sharia* (Islamic law) was promulgated by Chechen jihadists, who imposed the Sharia by force, or through Islamist revolutionary courts. The Islamic scholars known as *Ulema*, most of them educated in Islamist schools abroad, oversaw the establishment of Councils of Ulema as the only judicial bodies in Chechnya. The Chechen Ulema were aided by religious experts from the Arab world, Iran, and Pakistan. Their enforcement of the Sharia was draconian. In the rebel-held parts of Grozny, the Sharia became the law of the city. Campaigns to enforce a strict "modest" dress code for women and a ban on alcohol were launched, and people detained for public consumption of alcohol were flogged with a cane. Thieves could have a hand cut off. These were Arabist social norms, a by-product of the Islamist-Jihadist absorption of the Chechen revolt. And they represented a profound and dangerous shift in the culture of the Caucasus.

It did not take long for the Russians to discover how serious the Chechens were.

On November 16, 1996, a major explosion in a housing complex in Kaspiysk, Dagestan, destroyed an entire residential block housing senior officers of the Russian border guards and their families. Many of these officers were from units patrolling areas on the administrative border with Chechnya. There were at least sixty fatalities, including ten children. The explosion involved at least two bombs, containing up to fifty-five pounds of high explosives, that had been hidden in the basement of the building.

The bombs were most likely planted by Chechen terrorists, operating from the Vedeno area under the command of Shamil Basayev. The unit of border guards targeted in the blast had recently managed to block a number of routes used to transport arms, ammunition, and medicine into Chechnya, disrupting Basayev's efforts to build strategic stockpiles in the Chechen highland districts; they had also cut off a major smuggling route for Central Asian drugs used to finance the

Chechen war effort. From the time of this operation until shortly before the explosion in Kaspiysk, several Basayev associates visited Kaspiysk repeatedly, allegedly in order to purchase food and medicine for Chechnya, and doubtless to lay the groundwork for the bombing.

After the bombing, as the extent of the civilian casualties became clear, several Chechen leaders denied involvement in the explosion. Shamil Basayev, in contrast, merely wondered out loud about the usefulness of such an attack, neither condemning it nor flatly denying involvement. The secretary of the Russian Security Council, Ivan Rybkin, called the explosion "an allergic reaction to the peace process in Chechnya."

In mid-December 1996, the Chechen Islamists launched a new series of terrorist operations, designed both to derail the ongoing negotiations with Moscow and to compel foreign visitors to leave Chechnya. The first blow came when roughly sixty Chechen mujahedin under the command of Salman Raduyev assaulted a Russian post near Penza, taking twenty-one border guards as hostages. The Chechen leaders seized the moment to demand political concessions from Moscow in exchange for the hostages' release, only to have a confident Raduyev flatly refuse to consider releasing anybody. (Raduyev had coordinated his strategy in advance with other Chechen leaders.) Indeed, only after receiving unspecified guarantees from both Chechen and Russian negotiators did Raduyev finally release the hostages.

While this drama was unfolding, a few masked gunmen penetrated the International Committee of the Red Cross (ICRC) facilities in Nvoye Atagi, Chechnya, and assassinated six Western relief workers, five of them women. It was a highly professional job—the victims were killed while asleep and the assailants used handguns with silencers. The undeclared but clear objective of this assault was to force Western relief agencies out of Chechnya so that the Islamist regime could be imposed more efficiently. The plan worked: Within a day, the ICRC and all Western humanitarian organizations announced they were pulling their personnel out of Chechnya and freezing all humanitarian work there. Chechen leaders attributed the assassination of the Red Cross personnel to "forces interested in frustrating the peace settlement and the coming elections in Chechnya, as well as isolating the republic from international humanitarian organizations which are rendering medical and food aid to its population." Russian officials attributed the assassination to "radical separatist fighters" determined to "undermine peace" (citing Salman Raduyev in particular). It was another step in the Islamists' consolidation of control over the population in Chechnya.

The wave of terrorism against foreigners continued with the assassination of six Russian civilians in central Grozny. The victims, members of two families who elected to stay and live under Chechen rule, were also assassinated by professionals, using a silenced handgun and an assault rifle. The jihadist leaders had long considered Chechnya's local Russian population a hindrance to their plans for an Islamist regime there, not to mention a potential future excuse for Russia to intervene within Chechen borders, ostensibly to protect native Russians in distress. With strikes like these, they were acting to eliminate the problem.

The leaders of the Chechen jihad were increasingly focused on their wider objectives, looking beyond the goal of an independent Islamist Chechnya. Their new goal was a regional liberation campaign. As one leader, Zelimkhan Yandarbiyev, explained, the Chechens were determined to establish Chechnya as "a state based on Islamic values and laws," to serve as a base for expanding the Islamist revolutionary movement throughout the Caucasus. "The Chechen people are today playing a central role in the process of determining the basic [course] of development in the Caucasus region, particularly its northern parts," Yandarbiyev declared. "They are defining the nature of Russian-Caucasian relations."

In late November 1996, Aslan Maskhadov refined the point, suggesting specifically that Chechnya could become a launching pad for a Caucasus-wide campaign against Russia. While pledging that Chechnya was committed to establishing "a normal policy and balanced relations between Russia and the peoples of this region" that would ensure their independence as Muslim states, he warned that if Moscow chose not to accept this vision, it could expect a resumption of armed struggle. "If anyone thinks that they can subjugate [our] peoples by the power of the gun," Maskhadov warned, "then this collapse [of Russia] is bound to start."

Toward that end, Maskhadov said, Chechnya was willing to provide shelter for "citizens from Arab and Islamic states who fought in Chechnya" and offer them a base for future operations. Maskhadov hailed the contribution of "mujahedin from many Islamic states [who] fought by our side," he declared, because "they took up arms and defended our people," and "welcome[d]" them "among their brothers in Islam, if they wish to stay." Behind the innocent language, it was an invitation to establish a new Afghanistan, Bosnia, or Sudan in the Caucasus.

The Chechen leaders were also expecting the next stages of the war to be waged mainly through terrorism. Movladi Udugov, the spokesman for the Chechen government, suggested that Moscow deserved blame for instigating the latest round of terrorism and violence as a cynical way to avoid granting Chechnya independence. Udugov charged that Russia's actions were turning Chechnya into "a criminal cesspool, which should be isolated from the rest of the world"—a characterization that might better have been leveled at the jihadists.

Chapter 7

ESTABLISHING A TERRORIST STATE

STARTING IN MID-DECEMBER 1996, CHECHNYA OPERATED AS A QUINTessential terrorist state. On one hand, Prime Minister Aslan Maskhadov was pursuing an overt policy concentrating on two priorities: getting the Arab world to offer overt support for an independent Muslim state in Chechnya and furthering negotiations with Russia in the hope of ultimately attaining formal independence. On the other hand, the Chechen Islamist-Jihadist leaders intensified their campaign of terrorism and subversion, both at home and within Russia, to influence their Russian "interlocutors" into concessions beyond what the Chechens' legitimate position could have achieved on its own. Significantly, this dual posture essentially followed recommendations made by Ayman al-Zawahiri during his December 1996 visit to Chechnya.

The Chechen leaders had scheduled elections for January 1997—a swift timetable that was hastened by the regular occurrence of terrorist strikes. With preparations for the elections heating up, the entire political leadership of Chechnya was growing anxious to remove Western observers from the country, apprehensive that journalists would file damaging reports on election rigging and intimidation (and, more generally, that Western humanitiarian workers and activists would have a "moderating influence" on potential voters). The spate of terrorist attacks against Westerners that began with the killing of the six Red Cross workers, had a chilling effect on firsthand Western journalism in the country. ICRC officials have charged that the murders were committed by Khattab, who, in turn, was serving under Salman Raduyev. Yet pointing a finger at Khattab and at Chechen commanders like Raduyev (who was often described as a rogue operative) also gave the Chechen jihadist leaders a cover of deniability as they continued to carry out terrorism as needed by the Grozny leadership. Whatever the case, the killings had their desired effect: By the time the election campaign peaked, very few Westerners (including Russians) remained in Chechnya.

By this time, in the aftermath of the war, much of the real power on the Chechen side lay with the foreign mujahedin commanders, who were serving as increasingly influential advisers to key Chechen military and intelligence commanders-turned-politicians. One of these was Ibn al-Khattab. With a résumé that included jihadi experience in Afghanistan and in several Persian Gulf countries, a number of terrorist strikes he claimed to have executed against Israeli and French citizens, and the destruction of a Russian armored convoy near the villages of Serzhen-Yurt and Yaryshmardy in the spring of 1996, Khattab had also by now established his bona fides in Chechnya. He was a revered commander, considered harsh but caring and fair.

In the winter of 1996–1997, having consulted with Zawahiri and his team, Khattab converted to peacetime operations. He established a terrorist-commando training school near the village of Serzhen-Yurt, Vedeno Rayon, where he and several of his senior veteran "Afghan" and "Bosnian" mujahedin served as instructors. The same unit also served as the core of an operational unit for terrorist and other clandestine operations. The school trained and indoctrinated some of the most promising young Chechen war veterans, preparing them to populate Chechnya's future intelligence, special, and terrorist forces. Khattab's school also featured a special department training Algerians and French *Maghribis* for terrorist operations in France. Like the entire terrorist-mujahedin infrastructure, Khattab's camp remained under the supervision and control of the international Armed Islamic Movement, al-Qaeda, and the key terrorism-sponsoring states.

As 1997 began, a number of key Chechen leaders—particularly Maskhadov and Zelimkhan Yandarbiyev, under whose command Khattab fought—started building a deniability gap with Khattab; in this period, Shamil Basayev would emerge as Khattab's primary patron. The reason for this shift was that Khattab was embarking on a major project: organizing, training, and preparing a fierce new terrorist force, including both native Chechens and foreign mujahedin, to conduct spectacular terrorist strikes within Russia and throughout the West. Whether completed or simply threatened, the Chechen leaders knew that such attacks could help pressure Moscow to compromise in their coming negotiations—even though the new Chechen government would never be identified as the instigators of the terrorism.

In the late January 1997 elections, Aslan Maskhadov emerged as president of the Chechen Republic of Ichkeria. His inauguration, in mid-February, brought Chechnya as close as possible to unilaterally declaring independence without burning all bridges to Moscow. Grozny was convinced that the Muslim world would now rally to support Chechnya, both financially and politically. To underscore the Islamic element in his leadership, Maskhadov immediately went on hajj to Saudi Arabia. During his stay in the Middle East, he made a concentrated effort to demonstrate his Muslim identity and win Arab support. To his surprise, the gambit didn't work. As several area leaders explained to Maskhadov in no uncertain terms, the Arab states were still too concerned about alienating Russia, and too much in need of their weapons and military assistance, to support the Chechens in public. By the spring of 1997, several Arab and Muslim governments quietly notified Grozny that, with the Chechen people's "liberation struggle" completed and no threat of war in sight, they could no longer side with the Chechens in their ongoing struggle with Russia.

Even in the face of such rejection from the Islamic states, however, Maskhadov remained committed to his ultimate objective: complete independence from Russia. "I fully intend building an independent state," he declared in late February. His stubborn posture put Grozny at a disadvantage as negotiations with Moscow grew more complicated and challenging. By late March, Maskhadov conceded that the negotiations with Moscow "have reached a dead end." If the "political and economic blockade" imposed by Russia could not be breached, he warned, Chechnya would collapse. In an unmistakable signal of the possible consequences of the stalemate, Maskhadov nominated the terrorist commander Shamil Basayev as his first deputy.

Indeed, Grozny was brandishing the "terrorist alternative" to concessions by the "moderate" Maskhadov in the form of statements by the increasingly belligerent Salman Raduyev. As early as mid-January, Raduyev began building pressure by promising to launch a campaign of spectacular terrorism against Russia on April 21, the anniversary of Dudayev's death. "We are declaring April 21 a day of national revenge," Raduyev declared. "At least three Russian towns will go up in smoke. . . . Revenge is inevitable." Raduyev's appetite for revenge was only increased in early April, when he was badly wounded by a car bomb just outside Grozny. His aides attributed the assassination attempt to Russian intelligence, and vowed to intensify the April 21 terrorism campaign in retaliation for Raduyev's injuries.

Behind the scenes, the terrorism-sponsoring states' public neutrality didn't keep them from continuing to support the jihad. In mid-April, the highest authorities of the key terrorism-sponsoring states authorized the use of their assets in Chechnya—senior mujahedin and specialized equipment—in the strikes planned for Russia. With Raduyev out of commission, a close aide named Adam Deniyev was put in command of the planned terrorism campaign. A momentary setback emerged when Chechen security chief Abu Movsayev decided to win some political capital for Grozny by warning Moscow about the coming attacks, triggering a security alert that temporarily stayed the jihadits' hands. Still, between late April and early May the Chechens blew up two bombs in Armavir and Pyatigorsk in southern Russia, causing a few fatalities and injuries, as well as widespread damage. Chechen fighters also clashed with Russian security forces in the border areas. These operations were a far cry from the wave of terrorism Raduyev had promised for April 21, but they proved that Chechen terrorism was still alive and capable of initiating at least a regional eruption of violence.

Though Chechen officials denied any connection to the escalating terrorism, they warned Moscow that it would be increasingly difficult for Grozny to restrain the building frustration and despair in Chechnya, or to contain the ensuing violence, without progress in the negotiations.

Reinforcing the warning, Raduyev reemerged in early May and repeated his promises to escalate the violence. "The explosions in Armavir and Pyatigorsk were carried out on my personal order," he declared. "This is a new phase in the Russian-Chechen war." Grozny still condemned terrorism, even alleging that Russian intelligence may have been involved in the strikes—and yet Shamil Basayev continued his back-channel support of Raduyev.

At the negotiation table, Moscow and Grozny were trying to reach an arrangement to fix the major oil pipeline crossing Chechnya, which was damaged in the war, so that oil transfers from the Caspian Sea to Russia could resume. In early June, as the oil negotiations were progressing, Maskhadov made another play for political support from the oil-producing states of the Middle East, dissolving Chechnya's secular courts and replacing them with Islamic tribunals, based on the *Sharia*, as the core of the country's legal system.

But the Arab world was increasingly dependent on Russian weapons supplies, and once again Maskhadov's bid failed, leaving the Chechens to fall back on new terrorist strategies. This time, official Grozny went to exceptional steps to increase its deniability: Maskhadov ordered the disbanding of all private armies, Salman Raduyev would be permitted to retain only a personal guard, and the rest of his force would be compelled to join either the national guard or the largely ceremonial presidential guard.

In late June, Ibn al-Khattab was driving a jeep near Benoy, some seventy kilometers south of Grozny, when a remote-controlled land mine blew up seconds after the car had already passed it. Khattab was not hurt in the blast, and given the expert activation of the high-quality bomb, some wondered whether the near miss was intentional, a signal to Khattab and a convenient ruse to enhance deniability. A marked escalation in the violence followed, soon spreading into nearby Dagestan. In a series of raids, Chechen jihadists planted mines and ambushed government vehicles. In early July, seven Russian policemen were killed and thirteen injured near Khasavyurt in Dagestan, near the Chechen border, when the truck they were traveling in was blown up.

In mid-July, Russia made a major concession to Grozny, resolving the negotiations over the oil pipeline by agreeing to pay for the project. That same day, masked gunmen opened fire on a Russian government car in the center of Grozny. The Chechen leaders denied any link between the pipeline deal, signed only a few hours earlier, and the shooting. But the skirmishes and bombings would continue as the Chechen-Russian negotiations dragged on.

Weeks later, Salman Raduyev became the next Chechen to survive a near miss, when a van filled with explosives was detonated by remote control as Raduyev's car was about to pass by. Three others were killed in the blast. "They missed killing me by mere seconds," he said, blaming Moscow and its local allies. Seizing the moment, Raduyev also threatened to attack the oil pipeline, preventing its refurbishment or use, unless Moscow recognized Chechnya's independence. "Russia needs the pipeline, but we'll explode it," Raduyev said.

Around this time Shamil Basayev resigned as vice-premier of Chechnya, complaining that the Maskhadov government had "failed to implement all my plans," and that the authorities were driving Khattab out of Chechnya (which some found a spurious claim).

By the end of the summer, two trends were dominating events in Chechnya. First, Moscow was increasingly apprehensive that the United States might intervene in the Caucasus, using conservative Arab regimes as go-betweens in a quest for oil while challenging Russia's vital interests. Second, the leaders in Grozny were demonstrating an unprecedented level of self-confidence about Chechnya's economic future. In early September, Maskhadov declared that the nation would build a new capital city—Dzhokar—rather than attempting to rebuild the devastated Grozny; it was an ostentatious boast, yet Grozny gave no explanation about the source of its funds, and Moscow concluded that the money was coming from the Muslim world under U.S. influence.

Together, these trends cast a long shadow over the next round of negotiations, which were slated to begin in Moscow in late September.

Maskhadov was optimistic that he could convince Yeltsin to sign an acceptable treaty recognizing Chechnya's independence. At the same time, the Chechen leader signaled to Moscow that new terrorist strikes within Russia were still an option. In mid-September, he awarded medals to Shamil Basayev and more than one hundred other fighters who participated in a 1995 hostage incident in the Russian town of Budennovsk, hailing the raid as the turning point in the Chechens' war of independence. He publicly suggested that such raids might resume if Moscow failed to deliver what Maskhadov considered legitimate claims (and Moscow considered to be unacceptable secessionist demands).

As this defiant award ceremony suggested, the Chechen leaders' faith in the negotiations with Russia was fading. A new round of talks failed to produce any progress on resolving Chechnya's political status, with Moscow still reluctant to recognize Chechnya's secession and independence. After late September's round of negotiations, Chechnya's first deputy premier, Movladi Udugov, declared that Moscow's position could not be reconciled with Grozny's. The Chechen leaders responded by expelling Russia's representatives in Grozny on short notice, ostensibly because of a disagreement over a number of disputed flight paths through Russia (paths used frequently for the smuggling of drugs, weapons, and other contraband).

Chechnya continued planting flags for its own independence—announcing the introduction of Chechen passports—and for its status as a Muslim state, declaring that all women working for the government or studying in institutes of higher learning must wear traditional Muslim garb or risk losing their positions.

The Chechen leaders' open defiance peaked with a series of public executions in Grozny, which were broadcast on Russian TV despite protests from Moscow. Grozny used the executions to showcase the emergence of the Sharia law in Chechnya in place of Russian law. "I spit on Russia," Chechnya's vice president, Vakha Arsanov, declared when Russia denounced the display. "Russia means nothing to us; we are an independent state." A week later, as a crowd shouted *"Allah Akbar!"* around them, a Chechen firing squad publicly executed another couple of convicts in defiance of Moscow's threats. "It does not matter how much Russia shows its indignation," said Magomed Magomadov, Chechnya's deputy prosecutor general. "We are living in an independent state, we have our own *Sharia* courts, and we shall punish criminals according to *Sharia* law." In mid-October, regardless of the already shaky negotiations with Moscow, Maskhadov reiterated that Grozny would agree to nothing less than full independence. He added that the Chechen leaders had decided that it would treat Moscow no differently from any other foreign power. "Chechnya is ready to discuss with Moscow the need to establish full-scale diplomatic relations of a friendly nature," said Maskhadov's spokesman, Kazbek Khadzhiyev.

Then, in a new show of audacity and brinkmanship, Maskhadov demanded greater presidential powers and reshuffled the Chechen government when the Chechen parliament failed to grant his request. Soon thereafter, as he left on a trip for Turkey, Maskhadov appointed Shamil Basayev acting premier until his return. (Maskhadov had refused to accept Basayev's resignation as vice premier.) Maskhadov also appointed a special committee, led by Deputy Premier Ruslan Gelayev, to expedite the purging of the Chechen leadership. Reflecting Chechnya's real power structure, the committee comprised representatives of all of Chechnya's military departments and religious leaders.

In early December, Maskhadov announced he was transferring some of his authority as premier to Shamil Basayev. Ziyavdi Aybuyev, the Chechen secretary of state, explained that Basayev would "conduct sessions of the Cabinet of Ministers, examine correspondence received by the government and make decisions in its regard, administer the entire economic complex of the Chechen Republic, and issue decrees and directives pertaining to the activity of ministries, divisions, and departments that fall under the Cabinet of Ministers." By the turn of the year, meanwhile, the results of the restructuring committee's work were revealed: On January 1, 1998, Maskhadov dismissed his cabinet and tasked Basayev with forming a new government, replacing the cabinet's forty-five ministers with twenty-two. Maskhadov had expected to remain both president and prime minister of Chechnya, but Basayev let it be known that he wanted the prime minister chair, and Maskhadov knew that Moscow would greet the avowed terrorist's rise in power as a belligerent action. And that was exactly the signal Maskhadov wanted to send.

In short, Grozny was spoiling for a crisis with Russia, highlighting its Islamist aspirations with each new step it took. "Russia has not met its commitments in respect of Chechnya so far, and nor will it do so in the future," vice premier Movladi Udugov said in early January 1998. He announced the formation of a Sharia Security Committee to address the new challenges facing Chechnya. "The *Sharia* Security Committee will become a major power-wielding body within the republic and will be guided in its activities only by *Sharia* laws and norms," Udugov explained.

Chapter 8

ESCALATING AND WAITING

A WEEK BEFORE MASKHADOV DISMISSED HIS CABINET, HIS ADMINISTRATION had demonstrated its approach to deadlocks in the negotiations with Moscow. On the night of December 22, 1997, a large Chechen-Dagestani terrorist force, led by senior Arab mujahedin, attacked the tank battalion of the 136th Motor Rifle Brigade of the 58th Army based in Buinaksk, Dagestan, roughly one hundred kilometers from the Chechen border. At the same time, a number of predominantly Chechen detachments launched diversionary attacks on the Chechen-Dagestani border, in such places as Khasavyurt and Pervomayskoye, where five Chechen fighters were seized by the Dagestanis in a nighttime attack on a police post. The plotters had also apparently targeted the Kizlyar bridge over the River Terek, a piece of infrastructure key to rebuilding the oil pipeline in the Caucasus: Russian sappers would later find fifteen kilograms of explosives, three howitzer shells, and three hundred-meter spools of detonator cable near the bridge.

The main attack, on a Russian base at Buinaksk, was personally commanded by Ibn al-Khattab. The on-site Islamist commanders filed a dispatch on the attack, ascribing it to "the foreign mujahedin in Chechnya" under Khattab's leadership, but the main report in Dagestan was sent by Khattab himself on December 25, 1997. Khattab, too, labeled himself "Ameer of the Foreign Mujahedin in Chechnya," but he left no doubt that he was in command of the Chechen forces as well. This report was especially significant—not just because Khattab's participation in the raid confirmed the importance of the strike to Grozny, but also because Khattab's detailed account of the strike reflected the growing influence of bin Laden/Zawahiri devotees as key decision makers in strategic and political planning.

Khattab began his report by outlining the strategic logic behind the audacious strike. One of the main objectives of the raid, he stressed, was to break the political impasse in Grozny's favor—a development that was proving beyond the Chechen negotiators' abilities. "During the last eighteen months since the treaty declaring peace between Chechnya and Russia was signed, the Russian government has been constantly harassing the Chechen Muslims," Khattab explained. "Until now, the Russian government has not, and nor does it have any intention to recognize Chechnya as an independent country." He added that the Russian army had technically violated the agreement on occasion by operating on Chechen territory.

However, Khattab intimated with great candor, the real imperative for the raid was not frustration over the negotiations themselves, but a fear that Grozny was losing support among the Chechen population. Khattab accused the Russian government of conducting a relentless campaign "to create animosity amongst the Chechen people against the Chechen government." As he conceded, their efforts were proving successful. "Some of the Chechen Muslims have even begun to say that they were better off during the war than they are now," Khattab reported to the leadership. Moreover, Moscow's policies had even led to declining support for Islamist causes throughout the Caucasus. "The Russian government has been running a similar colonialist harassment campaign upon the other two Muslim nations in the Caucasus as well, namely Dagestan and Inguishetia," Khattab warned. "They even had the courage to carry out two assassination attempts of the mujahedin leadership inside Chechnya itself." To Khattab, it seemed imperative to reverse the trend before it was too late.

Khattab's report suggests that he and his key allies—most notably Basayev—were the key decision makers at the strategic-political level. "In response to this naked aggression by the Russian government, and after several warnings to stop, the group of foreign mujahedin in Chechnya decided to put an end to this harassment themselves," he writes. Yet his subsequent description of the military raid makes clear that the force was largely composed of mujahedin from the Caucasus—that is, forces normally not under his direct command. Khattab may have claimed responsibility for the decisions in order to preserve Maskhadov's deniability even within Islamist-Jihadist leadership circles. Moreover, though Khattab hints that the raid was the result of a last-minute decision, advance intelligence preparations for the raid indicate that the strike was in the works earlier in the fall, as Maskhadov and the Chechen leaders were losing faith in the negotiations with Moscow.

Khattab's description of the strike itself matches those of other Islamist-Jihadist reports, as well as later investigations by Russian security and intelligence services. The strike force comprised three mujahedin platoons, a total of 115 Chechen, Ingush, Dagestani, Central Asian, and Arab fighters—all of them Islamists-Jihadists, and many of them veteran "Afghan Arabs." The force had been trained in the special terrorist and combat training camps run by Khattab and Basayev. In preparation for the strike, the mujahedin had been drilled by Khattab and his close aides—mostly Jordanians, Egyptians, and Gulf Arabs—in three training centers on Chechen territory. All the Chechen-Dagestani combat detachments were under the command of Arabs. Russian security services later identified five Arab mujahedin (Egyptians and Gulf Arabs), as well as three mujahedin from Tajikistan and three from Afghanistan, in the raiding force. During the fighting, Russian intelligence intercepted radio commands given in three different languages: Arabic, Chechen, and Russian.

The attack on Buinaksk was a highly professional operation, the subject of extensive intelligence and reconnaissance preparations. For several months before the strike, local Russian troops noticed suspicious people milling around the area where armored equipment was parked, some taking notes and pictures. Though local sentries took down car registration numbers, neither the police nor the interior ministry managed to crack the plot in advance—a severe failure for Russian intelligence.

In the few days before the strike, around a dozen mujahedin commanders arrived in the Buinaksk area, arriving one or two per day over different infiltration roads. After making some last-minute observations, they prepared for the arrival of the main strike force.

That force arrived at the last minute, at around midnight of December 21–22, just hours before the strike was scheduled to begin, according to Khattab's report. Despite official repeated denials from Grozny, Khattab stated that the main strike forces "crossed the Chechen border and entered Dagestan (in Russian territory) in a number of buses and trucks." The vehicles converged on the forward organization and concealment point in the heavily Wahhabi village of Karamakhi, 30 kilometers from Buinaksk and 150 from the Chechen border. Mukhammed-Shafi Dzhangishiev, the leader of the rebel website Kavkaz-Tsentr, played a major role in the support system.

The terrorists arrived in several KamAZ trucks and a few buses, taking the military by surprise. Khattab stressed that "the mujahedin were able to reach the border gates of the base without being noticed." Equipped with assault rifles, machine guns, and rocket launchers (RPGs), they broke into a few small groups, then surrounded and sealed off the camp. Then the mujahedin opened up heavy fire, and at around two o'clock in the morning they launched a full-scale raid into the Russian base. "A large number of Russian army soldiers, many of whom had just awakened from deep sleep, were also killed in this initial attack," Khattab reported.

Only twenty-five Russian soldiers and officers were on duty at the vehicle parking area when the attack was launched, and they were helpless to defend the base. Within five minutes the mujahedin were in control of the parking and storage area, where Khattab reported finding roughly three hundred army vehicles, including tanks and armored personnel carriers (BMPs). The mujahedin claimed that they tried to drive some of the tanks away, but found that their batteries had been removed, so they started blowing up and burning the vehicles instead. According to Khattab's report, the mujahedin destroyed "each and every one of the three hundred vehicles, including more than fifty brand new Russian T-72 battle tanks. They also destroyed the weapons stores of the base in addition to burning over 260,000 kilograms of fuel." This was an exaggeration: In truth, only four tanks were destroyed—two by the mujahedin's RPGs and two from secondary explosions—along with two BMPs and two other vehicles. In addition, five tanks, six armored personnel carriers, and four trucks were damaged, as were four cisterns of lubricating oil.

More significant was the damage inflicted on the town's civilian population. Within a few hours after their attack, the raiding force went on to sabotage and disable all the local power substations, completely cutting off electricity to the town of Buinaksk. When they withdrew, they left the town's two power substations burning and the town itself shrouded in darkness.

The losses on both sides were relatively low. According to Russian data, two citizens and a paramilitary guard were killed and thirteen wounded (including six servicemen, one militiaman, one paramilitary guard, and five civilians) during the attack. There are no reliable data about the losses incurred by the mujahedin, who carried most of their wounded and dead with them. In his report, Khattab acknowledged that two Egyptian mujahedin (commander Abu Bakr Aqeedah and a fighter named Abu-Ammar) were killed. Three mujahedin were injured. Russian intelligence sources reported that Khattab himself was seriously wounded during the attack, while Islamist reports insisted that he was only slightly wounded. Russian sources also noted that the Chechens buried the bodies of roughly twenty people involved in the attack in the settlement of Zandag.

The mujahedin left in the predawn hours, driving toward the border in a convoy of vehicles they had brought with them. By now the Russian security forces throughout the area were on alert, and just before 5 A.M., near the village of Dylym, the mujahedin convoy divided into two groups and dashed for the border. One group of roughly fifty fighters, apparently including some of the commanders and several wounded mujahedin, seized a Dagestani bus. Using the passengers as human shields, they drove unmolested through the security forces' checkpoints all the way to the Chechen border, where they released their hostages and vanished into Chechnya. The second group, traveling in a large KamAZ truck in the Kazbekovskiy Rayon, was intercepted by MVD (Russian interior ministry) units. Gunfire was exchanged, and the radiator of the terrorists' truck was hit, but the mujahedin managed to escape and reach the Dagestani village of Almak, on the border with Chechnya. On the outskirts of the village, the mujahedin burned their truck and radioed their headquarters for transport. Meanwhile, they took four hostages—two policemen and two local residents—seized a bus and drove toward Chechnya with the hostages in tow.

In his report, Khattab hotly denied that the mujahedin seized any hostages. Instead, he wrote, the mujahedin "asked the local villagers for assistance." He stressed that "once the Dagestani Muslim villagers realized that these mujahedin were returning from attacking a Russian army base, they shouted *Allahu-Akbar* in joy, provided the fighters with food and drink, and lent them one bus and one truck, to assist them in their return to Chechnya." By all accounts, a few kilometers from the border, the mujahedin abandoned the bus, released the hostages, and started walking toward Chechnya. Around midday, security forces caught up with the mujahedin as they were walking in a ravine. "Just before the Chechen border, the mujahedin hid in undergrowth and set an ambush for the pursuing Russians," Khattab wrote in his report. A firefight erupted, lasting three or four hours before the mujahedin escaped across the Chechen border. Two mujahedin and one policeman were killed in the standoff. Khattab claimed the mujahedin fought for some ten hours "until they finally broke the siege, killing many enemy soldiers in the process."

By now, other Chechen forces were becoming involved in supporting the raiders. Even as Russian security forces were closing in on the trapped mujahedin, a force of twenty Chechen fighters opened fire on a guard post near the railway bridge near Kizlyar, killing one of the guards. This Chechen force deliberately crossed the border from Chechnya to relieve their surrounded comrades, advancing straight into the main body of the security forces, which turned around and moved on the advancing Chechens. Five of the Chechens were cornered, isolated from the main force, and ultimately surrendered to the Dagestani authorities. Among them was Magomed Khambiyev, a brigadier general in Raduyev's Army of General Dudayev. Back in the canyon, the rest of the force exploited the confusion and managed to flee in the darkness and thick fog of the canyon, crossing back into Chechnya on foot.

Official Grozny disassociated itself from the attack in Buinaksk, even insisting that no terrorist had crossed the Chechen border in either direction. Raduyev was more circumspect, ascribing the raid on Buinaksk to the "*Jihad* Army of Dagestan." As he explained, "Mujahedin troops united with the Dagestani *Jihad* movement are likely to have staged the attack since they are opposed to the presence of [Russian] occupational forces [on their territory]." He did not try to conceal his support for the terrorist strike. "In the event the reports on destroying the armored equipment are confirmed, the [Dagestani] servicemen who hit the targets will be awarded military decorations of the Dudayev Army," Raduyev added. Raduyev stressed that none of the fighters of "General Dudayev's Army" under his command was in Dagestan, and that he had not given the order to attack Buinaksk. But he also reiterated that the attacking mujahedin of the Dagestani jihad organization were "friendly" to his forces, and that neither army "intend[s] to tolerate the presence of occupation troops in Dagestan." Raduyev even acknowledged that some Dagestani mujahedin had been trained by his men, and that the entities had a mutual assistance pact in place.

To the Islamist-Jihadists, the raid on Buinaksk demonstrated the close relationship and operational cooperation between the Islamist elements in Chechnya and Dagestan. These relationships were clearly based not just on ideological convictions, but also on the elaborate clan and family ties between the two groups. The same could even be said of the foreign mujahedin: Indeed, one of Khattab's wives lived in the village of Karamakhi in Dagestan. In Khattab's assessment, news of the raid would have a mobilizing effect on the entire Caucasus. "Hundreds of mujahedin from the surrounding Russian regions are arriving at the base of the mujahedin. We are opening a number of training camps in order to train these mujahedin and, *Insha-Allah*, we hope to teach the Russians another lesson in the approaching summer of 1998, should they not stop their harassment of the Muslims." He then asked for additional logistical support in order to facilitate the buildup and escalation. "Winter has now arrived and a request is being put out to the Muslims all over the world to help us financially in order to fulfill our objectives," he wrote. Khattab even conceded another unexpected result of the raid: "Surprisingly, since this attack, the Russian government has been acting very nicely to the Chechen people," Khattab observed. Yet he repeatedly stressed that there would be no withdrawing from the forthcoming military escalation.

The terrorist strikes of late December 1997 were the opening shots in the Chechens' bid to draw the region's many separatist and Islamist movements into an "all-Caucasian" jihad against Russia. And there were other skirmishes: Around the same time, Saratov police arrested three armed Chechens for planning a terrorist attack in the city. When arrested, they were carrying a Kalashnikov assault rifle, a pistol, a bomb equipped with remote-control devices, and two detonators. Arrested while secretly surveying a potential bombing site around a central food market, they resisted arrest and one of them was wounded in a brief fight.

The Chechen jihadist leaders realized that instigating a region-wide rebellion would prevent Russia from mounting a single, concentrated military response to crush the Chechens. The Chechens offered help to virtually any Muslim separatist group in the entire North Caucasus; in the next stages of the jihad they were counting on the support of the Akkin Chechens living in areas of Dagestan bordering on Chechnya, and the events of late December 1997 proved them right. As the conflict continued, the Chechen commanders fully intended to drag Dagestan into the armed conflict, turning it into an "eastern springboard" for Chechen expansion. Grozny had begun openly touting the idea of creating a single Muslim state in the North Caucasus, with Chechnya in a leadership position and a Chechen-controlled Dagestan serving as an eastern front in the ongoing crisis with Russia.

Seizing Dagestan would also give the Chechen leaders unfettered access to the Caspian Sea, something Aslan Maskhadov considered "vitally necessary" for Chechnya. Dagestan controlled two-thirds of Russia's access to the Caspian coastline; if it should slip from Russian control, Russia's southern border would be pushed four hundred kilometers north, marginalizing Russia's influence in a strategically and economically crucial region. Losing Dagestan would also push Russia's borders with the increasingly radicalized and volatile Muslim world right up to Astrakhan and Kalmykia, in the Volga region—namely, to the heart of the country. And these regions all had significant Muslim populations, sure to prove susceptible to Islamist-Jihadist and separatist incitement. The Caspian Sea could turn into a Turkish-Iranian-Muslim body of water, reducing Russia's influence on the entire Caucasus area and in Central Asia as well. The Chechen leaders also expressed interest in reaching the Black Sea via Georgia and Abkhazia, though in this case they were content to rely on nonviolent methods, capitalizing on the Chechens' strong relations with the Abkhaz rebels.

Naturally, Moscow viewed this potential realignment as unacceptable, and the rhetoric between Russian and Chechen leaders escalated. "Someone apparently intends to push Russia into a new Caucasian war," Russian officials explained. Citing the attack on Buinaksk as "a precursor of this war," they pointed to the instability in Dagestan alone as "the biggest threat to the integrity of the Russian State since World War II."

The Islamists and their state sponsors made no effort to conceal their position on the matter. At the time of the Buinaksk raid—but without ever mentioning the incident—the official Tehran media urged Moscow to grant independence to the entire "independence-seeking North Caucasus region." One editorial claimed that "the Caucasus region's historical struggle against Russian hegemony, which has been going on for four hundred years, cannot linger on forever." Tehran argued that Moscow should immediately "realize that there is no other way but to end the dispute and grant Chechnya an outright independence."

Before a formal political decision on the region's future was reached, several states embarked on new cooperative economic initiatives in the area. The most important of these was a U.S.- and U.K.-endorsed effort to create a so-called Caucasian common market, designed to concentrate on energy development without Russia's involvement. Conceived in the fall of 1997, this effort was supported by all the states of the

Transcaucasus region except Armenia, by major Western oil corporations, and by organizations lobbying their interests, in both the United States and Britain. Moscow was alarmed by this Caucasian-American chamber of commerce, not least because it was led by Khozh-Akhmed Nukhayev—a leader of a Chechen criminal group in Moscow in the early 1990s, and later first deputy prime minister in Zelimkhan Yandarbiyev's government in Grozny.

Moscow was also annoyed to learn of an October 1997 protocol of intent regarding the establishment of a Transcaucasian energy company, which a group of prominent businessmen and politicians from Britain, Pakistan, and Hong Kong had signed with Aslan Maskhadov as if he were the president of a sovereign state. Under the agreement, Chechnya would allow the consortium to rent part of the Baku-Grozny-Novorosiisk oil pipeline. At the same time, regional powers were making moves of their own, designed to preserve their security and their claim on oil in the area. Pakistan had several ISI-controlled Afghan security detachments stationed in Azerbaijan since 1993, and the ISI deployed several hundreds of Hizb-i-Islami forces to Azerbaijan to help fight the Armenians and guard the oil pipelines. Turkey had designs on parts of an air base at Incirlik that was being abandoned by the United States. The new Turkish forces would use both stationary and mobile forces to secure the Baku-Tblisi-Ceyhan (BTC) oil pipeline.

To Moscow, all these activities seemed to be premised on a radically diminished Russian presence in the Caucasus. And Moscow had good reason to be apprehensive. On January 6, 1998, Pyotr Marchenko, a plenipotentiary representative of the Russian president in Adygeya, Dagestan, Kabardino-Balkariya, Karachayevo-Cherkessia, and the Stavropol territory, noted that the Russian security services had evidence that "the Northern Caucasus is a region of special and enhanced interest for foreign secret services" and the terrorist organizations they sponsored. The Russian security services "had detained and were investigating cases of a number of citizens from the West and the Middle East, who took part in reconnaissance and sabotage operations," he noted. "These operations are aimed at destabilizing the situation and, in particular in Dagestan, at kindling internecine strife." The operatives' activities weren't limited to gathering intelligence, either. "Overseas secret services all but openly organize, train and equip militants at semi-clandestine centers, which is not always actively resisted in the localities," Marchenko stressed. According to Marchenko, the December 1997 Buinaksk strike "had been provoked by precisely such militants."

On January 8, 1998, Russia's interior minister, Anatoliy Kulikov, blamed Grozny for the Buinaksk raid and warned that Russia had "the right to deliver preventive strikes against bases of bandits, wherever they are located, including the territory of Chechnya." Even though senior Russian officials distanced themselves from Kulikov's warning, Chechen leaders—led by Shamil Basayev—seized on the comment. Assuming his position as acting prime minister, Basayev was quick to capitalize on Kulikov's warning to justify a reevaluation of Grozny's priorities vis-a-vis Russia. Former Chechen spokesman Movladi Udugov, newly promoted to first deputy prime minister, observed that "it is impossible to trust Russian politicians, even if important documents are signed at the highest level." He warned that Grozny would consider any unilateral use of force by Russia as the resumption of "a full scale war." Basayev himself threatened to strike blows "at the places all over the Russian Federation where the military men are concentrated."

Moscow made urgent overtures to calm the mounting crisis, offering economic incentives, but the effort proved futile. Nevertheless, Russian deputy prime minister Ramazan Abdulatipov led a delegation to Grozny for talks with Maskhadov and Basayev. On January 10, Abdulatipov announced a "serious breakthrough" in Russian-Chechen relations, reporting that Russian and Chechen officials had agreed to pool their funds in order to revive Chechnya's economy.

But the rapprochement was short-lived. On January 9, thousands of Chechen troops began advancing toward the Russian border. Udugov explained by accusing Moscow of moving large forces toward the same border and noting that Maskhadov was augmenting patrols to protect the border. Neither Russia's denials nor its economic overtures had made any impact on Grozny: The Chechens' military buildup continued.

On January 11, General Anatoliy Kvashnin, chief of general staff of the Russian armed forces, made a one-day inspection trip to Dagestan, accompanied by his chief of operations and the chiefs of all general staff departments. The visit did little to ease tensions, especially when Kvashnin and his delegation made Buinaksk and the 136th Motorized Brigade part of their itinerary. "There is no need at present to reinforce the federal contingent in Dagestan," Kvashnin told RIA Novosti, the Russian press agency. Kvashnin promised to introduce a new network coordinating federal and local security forces throughout Dagestan, particularly in the areas bordering on Chechnya. Grozny ignored this assurance as well.

By early January 1998, then, the Chechen leadership was using Moscow's reaction to the Buinaksk terrorist strike as a pretext to heighten anxiety in the region. Basayev's threats to launch terrorist strikes within Russia were aimed to blackmail Moscow into granting further concessions, but they weren't empty threats: The elaborate terrorist infrastructure nurtured by Basayev, Khattab, and their Islamist allies was ready to strike if Moscow held firm.

Chapter 9

THE SLIDE TO CRISIS

AS 1998 BEGAN, A NUMBER OF GRASSROOTS DEVELOPMENTS SUGGESTED that the slide toward crisis in the Caucasus was irreversible. Large segments of the population were being lured by the populist Islamic movement, and by calls for an all-Caucasian revolt against Russia. The Islamicization process was expedited by a set of changes to the constitution in January 1998, formally declaring Chechnya a Muslim state and discriminating against non-Muslims. Selim Beshayev, the first deputy chairman of the Chechen parliament, noted that under the new constitution all citizens had the "right to profess any religion in accordance with the norms of Islam."

The establishment of the law of Islam in Chechnya had profound political and military ramifications. With the Chechen public proving surprisingly receptive to Islamism and militancy, a series of extremist Islamist theologians were appointed to top state posts. A new edition of Tagayev's *Our Struggle, or The Imam's Rebel Army* sold out overnight, its distributors unable to meet the growing demand.

It didn't take long for populist Islamicization to spread into Russian-controlled Dagestan, where its effects were decidedly political and military, not just religious. The terrorist organizations began issuing leaflets explaining their attacks on Russian security forces, a strategy first observed with the distribution of a leaflet signed by the Central Front for the Liberation of the Caucasus and Dagestan. "We wanted to capture a division of Russian aggressors," the leaflet read. "We captured one within five minutes of launching the operation. . . . We wanted to do this to show all the peoples of the Caucasus that no force can withstand the sword of *Allah*. We struck deep inside, at the very heart of the Russian aggressors and returned safely, without incurring any losses ourselves. . . . This strike marks the beginning of a jihad on the path of Allah in Dagestan . . . we will never leave the Russian atheists alone, wherever they may be (even in Moscow), until they free Dagestan and get out of the Caucasus. The Caucasus must be a free great power [*derzhava*] of all Muslims."

By now, militant Islamism was the fastest-growing and most influential political force in Dagestan. The public increasingly supported the launching of "campaigns of conquest" (ghazavat) against Russian authorities and particularly troops—attacks intended, as another leaflet stressed, to "create a free Caucasian power to which Russia will pay taxes for the damage it has done since its empire came into being." Russian efforts to contain such rebellion by arresting key activists like Dzhangishiev of Kavkaz-Tsentr and now the editor of the journal *Khalifat* (arrested for illegal arms possession), failed to dampen the burgeoning Islamist movement. "It is better to die *Shahid* [martyr] than rot in jail in Makhachkala," became a rallying cry in many religious gatherings.

Meanwhile, the Russian investigation of the Buinaksk strike continued, exacerbating the situation and pushing the public into the arms of the Islamists. In an interview with the Russian newspaper *Nezavisimaya Gazeta*, Khalif Atayev, the Karamakhi Rural Council leader, warned that Russian authorities had "begun to commit real excesses against Muslims" and that "our only choice in this situation is to take up arms." He stressed that he and most other local leaders were "in total solidarity" with the Central Front for the Liberation of the Caucasus and Dagestan. "Dagestan must break away from Russia and become an independent Islamic state," Atayev declared. The public was primed for the mujahedin to launch a strike on Russia from Chechnya. Chechen border posts were now flying the black flag of jihad; local troops explained to passersby that "this means the war with Russia is not yet finished." On February 5, Ibn al-Khattab personally inspected Chechen positions along the Dagestani border. "Khattab is one of us now, and we are in very close contact with him," explained a Dagestani militant.

Another crucial development was the return of maverick terrorist Salman Raduyev to the Islamist fold. Raduyev had been outspoken in his criticism of Wahhabism—that is, heavily Arabic Islamism—on the Caucasian Islamic movement. On February 13, Raduyev's "Dudayev Army" claimed responsibility for the assassination attempt on the Georgian president Eduard Shevardnadze, who had failed to support the Chechen cause. Raduyev further threatened a wave of "Tbilisi-style" strikes in Moscow against "Russian politicians and army officers" involved in the killing of Dudayev. Then, a week later, Raduyev announced his support for the Chechen Islamist leadership, including Khattab. "We will work as a united team to strengthen our state and construct an Islamic republic," Raduyev declared on February 23 to a standing ovation from ten thousand veterans at the Grozny sports stadium. "We are preparing for war. As long as the main issue of the war—Chechnya's independence—is not settled, what we have is just a temporary agreement on a cease-fire," Raduyev declared. The coming war, Raduyev promised, would see the Chechens launch bombing campaigns against Russian cities.

The grassroots jihadist campaign was also having an impact. By April, there were distinct "pockets" of militant Islam in Dagestan, well-organized and agitated, helping to spread militant Islam throughout economically deprived rural areas such as the Chechen communities in Khasavyurtovskiy District. In the spring, the call to arms reached such a fever pitch that Dagestani authorities warned that the Islamicization "could destabilize the republic and, simultaneously, all of the North Caucasus." They attributed the growing popular support primarily to "the Chechen factor." The Moscow-Grozny conflict was clearly spreading beyond Chechnya's borders, and the Dagestanis had now officially lined up alongside Chechnya. (It helped that a number of increasingly popular Islamist leaders—including Ibn al-Khattab, who has family settled in Karamakhi, and the spiritual leader Mullah Bagauddin—had ties to Dagestan).

Political activity in Chechnya was an increasing source of incitement to the Dagestani population. On April 26, the immensely popular propagandist Movladi Udugov chaired the Congress of the Islamic

Nation, billed as a "socio-political movement." In a written statement, Udugov maintained that the Congress had only one issue on its agenda: "Ichkeria [Chechnya]-Dagestan: Strategy of Cooperation." Neighborly relations between Chechnya and neighbor states such as Dagestan and Ingushetiya, of course, were a noble goal. But the speeches at the Congress betrayed the gathering's real objective: to call for "the re-creation of Shamil's Great *Imamat* within its historical borders," as one fiery speaker put it. The speakers noted that Shamil's nineteenth-century Imamat had stretched from Georgia to the Caspian Sea, then considered a springboard for uniting the entire northern Caucasus into a single, anti-Russian Islamic state. With the disintegration of the Russian Empire in 1918, leaders in the region had briefly established a Mountainous Republic of the Northern Caucasus on Islamic territory from the Black Sea to the Caspian Sea; that entity was soon absorbed into the U.S.S.R., but with the Soviets' disintegration, the Islamic Nation called for a renewal of the republic. As early as 1991, a new Confederation of the Mountain Peoples of the Caucasus (since renamed the Confederation of Caucasian Peoples) had announced a new attempt to revive Shamil's Imamat. The Confederation had made little contribution to the first war in Chechnya; now its president, Yusup Soslambekov, tried to explain away his group's low profile, claiming that "Russian special services worked on our people."

Even as the Congress was deliberating, Chechen-based Islamist terrorists, many of them mujahedin commanded by Khattab, began launching cross-border attacks against Russian security forces, aimed at provoking a wider conflagration. Dagestani Security Council secretary Magomet Tolboyev told Russia's *Kommersant-Daily* newspaper that a series of shootings were most likely organized by "international terrorists, religious fanatics who believe that the more 'infidels' they kill, the better." One of Khattab's goals was to boost morale among his Islamist force—composed of both foreign "Afghans" and a rapidly swelling force of locally recruited fighters. Khattab's mujahedin had been trained and organized in three mini-*Imarat* school-camps in the Vedenskiy and Nozhay-Yurtovskiy districts of Chechnya, run in accordance with a rigid interpretation of the Sharia. A team of veteran "Afghans" hand-picked by Khattab had trained the recruits in ambush, sabotage, mine laying, and subversive tactics including rumormongering and bribery. A special effort was made to recruit local fighters with no known Islamist history, training them as spies and assigning them to infiltrate state institutions and law enforcement agencies within Dagestan, Ingushetiya, and other neighboring republics. As a result, the mujahedin strike forces operating in that spring of 1998 had access to real-time intelligence from these spies in the Russian security system.

In one incident, unidentified gunmen opened small-arms fire on a police checkpoint near Chechnya after police stopped a vehicle inside Dagestan and asked for documents. The five passengers responded by shooting at the officers. One officer and three of the gunmen were killed, and two police and two of the gunmen were injured. The car was registered to a resident of Chechnya; police later discovered that two of the gunmen were carrying ID cards for members of Chechnya's Sharia courts.

At first these skirmishes failed to elicit Russian reaction, however, and by April the Islamists were escalating their efforts. On April 17, just as top Russian police officials were slated to meet Chechen leaders in Moscow to discuss anticrime efforts in the region, a Chechen detachment of roughly a dozen fighters ambushed a Russian army convoy inside Ingushetiya near the Chechen border, subjecting it to an intense barrage of assault rifle and rocket-propelled grenade fire. Major General Viktor Prokopenko of the general staff, two colonels, and three privates were killed in the ambush, a professional operation based on real-time intelligence specifying which cars to target. The attackers disappeared back into Chechnya, and official Grozny denied any involvement in the incident. Given the timing, the political significance of the ambush was impossible to ignore.

This latest series of incidents culminated in early May when Valentin Vlasov, President Yeltsin's envoy to Chechnya, was kidnapped while traveling in Ingushetiya and taken across the border into Chechnya. It was another professionally planned and executed operation: Vlasov was traveling in a single unmarked car to ensure anonymity, suggesting that the kidnappers had intelligence about his travel plans.

Wearing masks and camouflage clothes, the five kidnappers used three cars for the attack: Two cars surrounded Vlasov's vehicle, and the team pushed him into a third, a black BMW sedan, before speeding off into Chechnya. Russian and Chechen officials alike considered the kidnapping a political act; one Chechen official, Islam Khalimov, charged that it was "designed to aggravate the tensions in the region and in the relations between Russia and Chechnya when a new Russian government is taking over." But official Grozny was quick to blame Moscow, charging that the kidnapping was intended to give Yeltsin an excuse to attack Chechnya, diverting attention from his political problems. Lecho Khultygov, the chief of the Chechen National Security Service, blamed Russia's Federal Security Service for the kidnapping. "According to our information, Vlasov is being kept at their secret base outside Chechnya," Khultygov said.

Grozny was apprehensive about potential Russian reaction to this series of incidents, and after the Vlasov kidnapping the cross-border strikes ceased virtually overnight. Within a few days, however, it became clear that Moscow had no plans to invade Chechnya to find Vlasov, and the cross-border skirmishes resumed. Leaving nothing to chance, the Islamists also made a series of internal purges. Suspecting that Aslan Maskhadov was maintaining separate lines of communication with Moscow, they assassinated the two key officials involved—First Deputy Security Minister Shamsudi Uvaisayav, who was investigating Vlasov's kidnapping, and former Chechen foreign minister Ruslan Chimayev, Maskhadov's contact man—during a meeting at Uvaisayav's home. (The officials' bodyguards were also killed.) The Chechen government officially labeled the assassination the work of a deranged individual, but in mid-May the pattern was confirmed when a powerful roadside bomb destroyed a car in the center of Grozny, narrowly missed Assistant Prosecutor Magomed Magomadov, the head of Chechnya's antikidnap unit and the official responsible for finding Vlasov. Magomadov was traveling in a white Volga, but this time the terrorists targeted the wrong car: An identical vehicle in front of his was hit instead, killing three or four other officials.

By now, four Arab commanders had emerged as the primary voices of religious and terrorist instigation for the Chechen jihadist movement. Together, Ibn al-Khattab and Abdul-Malik in Chechnya, and Salakh-Uddin and Mohammad-Sharif in Dagestan, commanded the loyalty of more than two thousand local devotees. Khattab himself wrote and delivered a number of key sermons to the trainees, including an April 1998 message titled "Your Task Is to Sow Terror among Those Who Have Betrayed Allah":

> Right now Moscow is trying to convince everyone that it has bestowed peace upon us. I do not believe this, and nor do Shamil [Basayev] and Aslan [Maskhadov]. All Moscow's promises of funding are no more than words to gull fools. Aslan is a fine fellow. He leads the Russians by the nose, and will probably manage to obtain money for our National Bank. And if he does not, there is no big problem. Certain European countries, and also Pakistan, Iran, and Afghanistan, will give us it. Moreover, we have received arms and equipment from them. Indeed, even among senior Russian functionaries there are many who are ready to sell us all this. We have only one demand: full political independence. Russia, as the loser of the war, is obliged to pay an indemnity—to the last kopek and without any conditions.
>
> For one month you have been learning the art of sabotage, bribery, and rumor-spreading. Your task is to sow deadly terror among those who have betrayed *Allah*. They must feel the cold hand of death at every hour. A special task is assigned to those who will settle in Russia and neighboring friendly republics. You must infiltrate the structures of power and the administrative and financial organs. Set up bases, recruit people. If Ichkeria does not receive independence by the spring [of 1998], we will strike a blow against all the major industrial cities. Pay particular heed to the Cossacks—they are our long-standing and deadliest enemies. You must fling mud at all patriotically minded Russians. Accuse them of fascism. But those who wish to rise up under the holy banner of the Prophet must be bound with bonds of blood.

As summer neared, the Islamist attacks grew more audacious. On May 20, a detachment of Chechen gunmen engaged police within Dagestan, killing two policemen and wounding six bystanders in the capital city of Makhachkala. Authorities rushed in reinforcements and sealed off the city center, chasing the gunmen into a building and surrounding it. But the strike had been a setup: The building was the home of Nadirshakh Khachilayev, the outspoken leader of the Muslim Union, a leading Islamist group (and member of the Russian Duma) and his radical brother Magomed. By morning, more than two thousand armed and unruly supporters of the Khachilayev brothers had surrounded the building, defending their leaders from the police onslaught. After mounting a loud cry for the ouster of the Dagestani government and new elections, the rioters moved to Makhachkala's central square, where a core group of two to three hundred militants wearing distinctive green armbands seized the State Council building and took the prime minister hostage. Shouting "*Allahu Akbar!*" and firing in the air, they tore down Dagestan's flag and flew the green banner of Islam from the building's rooftop. With two thousand demonstrators outside keeping police at bay, the rioters looted and destroyed the building. The spectacle, covered on television, riveted the entire region: By the time the authorities negotiated an end to the rioting, the Islamists had proven their violent power.

By mid-June, the third anniversary of Basayev's raid in Budennovsk, the Islamist campaign had escalated into an almost nonstop succession of kidnappings (including those of numerous Russian soldiers), ambushes, roadside bombs, and mines. Shamil Basayev, Chechnya's acting prime minister, encouraged the strikes, claiming that "massive bombing and destruction of civilians" by Russian armed forces had led to these strikes. "We wanted to stop the process of destruction of the whole [Chechen] people," Basayev declared, promising that more drastic terrorist strikes would be necessary to prevent further threats to the Muslim peoples of the Caucasus.

By that summer of 1998, the transformation of Chechnya into a terrorist-sponsoring rogue nation was virtually complete. On June 24, Osama bin Laden convened an international conference in Qandahar, Afghanistan, to address the latest doctrinal and theological developments and issue a declaratory policy statement. For the first time, Chechen representatives joined the roughly one hundred Bosnian, Somalian, and other Islamist leaders and activists at the conference, along with representatives from most Arab states and organizations and Afghan and Pakistani notables. One attendee, a senior member of al-Muhajiroun, a major Islamist organization in the West, then led by Sheikh Omar Bakri Muhammad, from London, held lengthy discussions on mobilization with the Chechen representatives. Bin Laden and Zawahiri chaired the conference and delivered the key addresses. The deliberations on the future direction of the jihad continued for a day and a night; after a few private consultations, bin Laden and Zawahiri concluded the formal conference by announcing a new "plan of action for all the members of the World Islamic Front," with the Chechen jihad an integral component of the grand strategy.

In Chechnya itself, a last-minute effort by the Chechen government to reverse the Islamist trend backfired in late July. Maskhadov was planning a trip to Ankara and Washington, D.C., in early August to secure political and economic aid, but he knew that the increasingly violent Islamist-Jihadist influence in his new republic was unsettling to his hosts. Lured by the promise of participation in Western-dominated regional energy development projects, Maskhadov mounted a campaign to suppress the Islamists, especially the Arab mujahedin, by publicly accusing them of responsibility for the prevailing lawlessness in the region. The "Wahhabi" Muslims were "coming to Chechnya from Arab countries," he charged, "calling for war and trying to justify abductions."

The next day, when Chechen security services attempted to seize Islamist officers in Gudermes, riots broke out. About a thousand Islamist and government troops engaged in brief but fierce street fighting, leaving six people dead and fifteen injured. Maskhadov immediately accused "Wahhabite militants" of orchestrating the clashes. A few days later, he ordered the disbandment of the Islamic Regiment and the Sharia Battalion, the two main Islamist units involved in the Gudermes fighting. He also sacked the units' commanders, Arbi Barayev and Abdul Malik Mezhidov, and stripped them of their ranks. Characterizing the riots as "a mutiny," Maskhadov ordered more than five thousand military reservists to help in "the crusade against crime," and declared a state of emergency.

But Maskhadov was hardly oblivious to the Islamists' strength—or their prevalence in the Grozny power structure—and he immediately rushed to placate them before the Gudermes incident evolved into a real challenge to his regime. Maskhadov appointed Shamil Basayev as deputy chief of the republic's armed forces, and put him in charge of relations with the militant Islamists-Jihadists—the "Wahhabis." In retrospect, Basayev's return to a senior government position was a turning point. At Maskhadov's specific request, he brought his own battalion of two hundred well-trained veteran fighters from Dagestan to serve as the regime's Praetorian Guard—a move that underscored official Grozny's commitment to Basayev's regional "causes."

Still, Basayev's elevation came too late to placate the enraged Islamists. The next day, Maskhadov narrowly survived a powerful car bomb, which severely damaged his four-wheel-drive vehicle and destroyed other vehicles, including two cars in the presidential motorcade. Maskhadov received a minor wound to the leg, and his driver and senior bodyguard were killed. In a signal of Grozny's new political line, Maskhadov immediately suggested that Moscow was responsible, insisting that the incident was the work of "foreign special forces . . . acting from afar with the hand of local provocateurs." Though he made no further attempt to identify these "special forces" specifically, he quickly noted that Russia had tried to kill him in the past. In later discussions with his confidants, Maskhadov made his accusations against Russia explicit, explaining that Moscow employed Chechen "traitors" for such purposes. Movladi Udugov also blamed Russia, calling the attack part of a Russian conspiracy to destabilize Chechnya.

The power structure in Grozny had settled by now, with the advantage going to the Islamists. Maskhadov found himself increasingly isolated at the top, more a figurehead than a real power broker. After his narrow escape in the car bombing, Maskhadov realized that his very survival depended on retaining favor with the Islamists. He resolved to support their militant cause, even at the expense of his agreement with Russia—sacrificing Chechnya's chances for economic recovery as well as a political settlement that promised broad autonomy and even independence by 2001. From the summer of 1998 on, Maskhadov would stand aside as Basayev and his allies prepared for the all-Caucasian Islamist jihad.

Yet soon Maskhadov would leave for the United States—traveling with a Russian passport—to attend the second International Islamic Unity Conference and meet with members of the U.S. Congress and American businessmen. While Maskhadov's visit could not be canceled, the Islamist-Jihadists resolved to provide him with a constant reminder of their power in the region. Even as Maskhadov was arguing that peace and stability had returned to the Caucasus, the Islamists began a new round of terrorist strikes.

First, Chechen fighters opened heavy fire on a Russian military patrol on Dagestani territory. When the Chechens kept firing, Russian armored vehicles and helicopter gunships attacked their positions until they were silenced. Movladi Udugov immediately accused Russia of an unprovoked attack, but claimed that two Russian armored vehicles had been destroyed by the Chechen forces. "All responsibility lies with the Russian leadership," Udugov warned, calling the incident "an infringement of our peace treaty." Russian officials correctly identified the skirmish as "an attempt by Maskhadov's opponents to use his absence . . . to ruin the promising results of recent talks between the Chechen leader and Russian Prime Minister Sergei Kiriyenko."

Then, a couple of days later, an explosion rocked the center of Makhachkala. The bomb, planted in a police van parked not far from the Dagestani capital's central market, was detonated by remote control, tearing through the roof of the van and setting it on fire. Immediately after the explosion, an unidentified person hurled a grenade at the van and the nearby crowd. Maskhadov was in Washington when news of the bombing reached him; there was little he could do to dispel the impression these incidents left on his hosts, and he returned to Grozny in mid-August empty-handed.

The Islamist-Jihadists were also expanding their "hearts and minds" operations in Dagestan, offering widespread assistance to the Dagestani civilian population in areas near the Chechen border. From their bases in Chechnya, the Islamists helped local farmers struggling to reclaim their failed *kolkhozes* (agricultural communes) in the border areas of Dagestan by providing seeds, trucks, and markets. They also helped build new mosques, clinics, and schools. Funded by Saudi and Gulf state charities, these activities had a visible impact on the culture of the region: Increasingly, local citizens signaled their acceptance of the strict Islamic lifestyle by adopting beards and traditional Arab dress. The Islamists' humanitarian campaign had a widespread impact on the Dagestani population, replacing its fear of the movement with popular support—a shift in allegiances that would prove crucial in the Islamist surge of the following summer.

The success of these humanitarian activities was reinforced by a series of ruthless terrorist strikes against perceived enemies of Islamism, particularly local clerics who resisted the abandonment of traditional Islam in favor of Arab-influenced Wahhabist Islamism. The turning point was the assassination of the immensely popular mufti of Dagestan, Said Mukhammad-Khadzhi Abubakarov. Abubakarov had been engaged in "an implacable struggle" with the Islamists over the character of local Islam; the jihadists saw his elimination as crucial to the spread of Islamism in the region. In late August 1998, a crowd of followers watched as his Volga exploded near Makhachkala's central mosque immediately after Friday morning prayers. The blast completely disfigured the bodies of Abubakarov and his brother and driver. Russian security forces arrested Dzherula Gadzhimagomadov, a resident of the village of Karamakhi and a leader of the local Wahhabi movement, and accused him of organizing the hit.

The Russian security services responded by cracking down on the Wahhabite leadership, increasing tensions in the region. On both sides of the border, Islamist forces began preparing for combat operations against the Russians. In Chechnya, Chechen fighters and foreign mujahedin were sent to forward positions near the Chechen-Dagestani border. Operating under the command of Basayev and Khattab, these forces established bases of operations from where they could seize key villages in a few hours. A larger force of about twenty-five hundred mujahedin, with a core of volunteers from Afghanistan, Pakistan, Turkey, Iran, Jordan, Sudan, and Egypt, was put on permanent combat alert, and one unit was deployed just a three-hour march away from the village of Karamakhi. The Islamists also supplied these troops with money and weapons to be provided to the Dagestani population once the fighting started.

At first the Islamists reacted to the Russian crackdown with terrorist attacks on Russia's regional allies and interests. This campaign peaked in early September, when the mayor of Makhachkala, Said Amirov, was targeted with a car bomb. Seventeen persons died and approximately sixty were injured in the explosion, but Amirov himself was unharmed. Then, after Dagestani Islamist leader Magomed Khachilayev was arrested for organizing mass disturbances, Salman Raduyev issued an ultimatum to the Russian and Dagestani authorities: Unless Khachilayev was immediately released, Raduyev's forces would "carry out a series of terrorist acts." When the Russian authorities ignored the warning, Raduyev repeated the warning at a news conference in Grozny: "If Magomed Khachilayev is not released from custody by midnight on 13 September, we reserve the right to carry out retaliatory strikes against the pro-Moscow Dagestani government, and Dagestani State Council Chairman Magomedali Magomedov will bear responsibility for this."

In the early fall, the Islamist-Jihadists demonstrated their new power in an open and humiliating challenge to the Chechen government. On September 23, Chechen front commanders forwarded a letter to the parliament of the Chechen Republic, demanding the impeachment of President Aslan Maskhadov. The letter, signed by jihadist commanders Shamil Basayev, Khunkar Israpilov, and Salman Raduyev, accused Maskhadov of violating the constitution of the Chechen Republic by failing to provide sufficient support to jihadist causes throughout the Caucasus and failing to implement the Sharia law strictly enough in Chechnya. The deputies of parliament immediately convened to discuss the letter—but when he was summoned to the session, Maskhadov told the deputies that he agreed with the Islamists' criticism and promised to follow their recommendations. The weeks that followed were charged with the possibility that Maskhadov would be overthrown and replaced with a more militant regime, but Maskhadov's loyalists repeatedly defused the crisis by promising additional concessions to the Islamist-Jihadist commanders.

By the time the crisis ended, the militant Islamist-Jihadists had confirmed their position as the dominant power in Chechnya. Maskhadov managed to survive as president, but he was increasingly required to demonstrate his fealty to the jihadist cause in both rhetoric and action. Around this time, Maskhadov began adopting the anti-Semitic rhetoric common to Islamist groups in the Arab world. Since the mid-1990s, the foreign mujahedin in and out of Chechnya had given rise to numerous extremist Islamist-Jihadist organizations with anti-Semitism as a core part of their ideology. These groups insisted on banning the activities and even mere existence of Jewish organizations, calling them part of "a global Zionist conspiracy directed against the Islamic religion." Local elites with no history of hatred or fear of the Jews were increasingly influenced by this strain of Islamist ideology. In an early October 1998 speech at a Congress of the Chechen People in Grozny, Aslan Maskhadov suggested that subversive activities in his nation were part of a "Jewish plot against Chechnya as the only Islamic state in the world." No one in the audience challenged the assertion. Anti-Semitism soon become a key element of Maskhadov's rhetoric. By mid-March 1999, he was arguing that "International Zionist centers throughout Saudi Arabia [sic] are coordinating the activities of anti-government groups in Chechnya," to facilitate U.S. control over the region's oil.

In late August 1998, the Islamist-Jihadist leaders in Grozny had a new opportunity to demonstrate their political allegiances when the United States launched a series of cruise missile attacks on terrorist bases in Sudan and Afghanistan. The morning after the attack, vice premier Movladi Udugov announced "the establishment of diplomatic relations between the Chechen Republic of Ichkeria and the Islamic Emirate of Afghanistan, and the opening of official missions in Grozny and Kabul." A couple of days later, vice president Vakha Arsanov called for retaliatory strikes against the United States, labeling the U.S. attacks "the start of an undeclared World War Three," and citing President Bill Clinton as the "number one terrorist, who needs to be severely punished." "The Islamic world should oppose the U.S. dictatorship," Arsanov said. He then claimed that Grozny was in control of "bases, though not very large," all over the world, and that the Chechen leadership had already issued orders to these bases to "launch appropriate strikes against the United States." Arsanov then made an intriguing comment concerning the relationship between Grozny and Arab Islamist-Jihadist terrorism, as well as the extent of Grozny's influence over their operations. "I am not saying that Chechen citizens will carry out these orders. The Arabs are the ones that are going to carry them out, but I do not want attacks to be launched against Baku or Moscow or any city with Chechen people in it. This matter concerns Afghanistan and Sudan, but Muslims are brothers, and we cannot be silent over what happened," Arsanov concluded. Udugov's and Arsanov's statements were further attempts to bolster the impression that Chechnya was an integral player in the worldwide Islamist-Jihadist movement.

And this was more than a posture. By mid-1998, Chechnya- and Dagestan-based Islamist-Jihadist networks were up and running in Baku, Azerbaijan, one of bin Laden's most important clandestine communications centers. It was relatively easy to travel from Baku to Chechnya and back, and the Islamists exploited this convenience to arrange clandestine meetings between key Western operatives and senior commanders from Afghanistan and Pakistan. The first leader of the Baku cell, an Egyptian called Ibrahim Eydaruz, would be arrested in connection with the August 1998 bombings of the U.S. embassies in Nairobi and Dar es Salaam. One of the key operatives in Nairobi, Khalid Salim (a Yemeni whose real name is Mohammad Rashid Daoud al-Owhali) had received several phone calls from Baku at his hotel prior to the attack, and faxed alerts from al Qaeda about the impending strike were sent to London from Baku. The Baku Islamist network was also planning to bomb the U.S. embassy there, until the CIA notified Azerbaijani security forces, who foiled the attack and rolled back the Baku network in September. Ahmad Salamah Mabruk, a prominent commander of Zawahiri's Islamic jihad, and operatives named

Ihab Sakir and Esam al-Din Hafiz were captured in connection with the plot and extradited to Egypt. The Azerbaijani security forces also recovered important intelligence concerning jihadist activity in several countries, including internal communiqués of the "Islamic Army for the Liberation of the Holy Places" and correspondence between the Baku network and the Islamist-Jihadist leadership in Chechnya that clearly demonstrated the Chechnya-based leaders' central role in sponsoring international terrorism.

But the Chechen leaders' greatest service to Osama bin Laden and the jihadist movement was helping al Qaeda acquire a number of nuclear "suitcase bombs." Between 1996 and 1998, bin Laden had spent well over $3 million trying to purchase an ex-Soviet nuclear suitcase bomb, but his efforts were futile. The Chechen Mafiya, on the other hand, found better sources for such weapons. Back in 1994, a Moscow-based Palestinian Islamist named Shaaban Khafiz Shaaban claimed to have purchased two suitcase bombs with the help of the Chechen Mafiya. The Chechens claimed to have acquired a few such nuclear weapons, and with the economic crisis in the former Soviet Union nearly out of control, it was not inconceivable that Chechnya could have used its substantial resources to arrange such a sale. Indeed, by 1997, General Aleksandr Ivanovich Lebed, Russia's former security tsar, acknowledged that several nuclear suitcase bombs had disappeared from Russia's arsenal.

In early October 1998, a senior Arab intelligence official asserted that "Osama bin Laden has acquired tactical nuclear weapons from the Islamic republics of Central Asia established after the collapse of the Soviet Union." This assessment was shared by Russian intelligence and several Arab intelligence services on the basis of diverse and multiple-source data. By the fall of 1998, there was little doubt in the intelligence community that bin Laden had succeeded in his quest for nuclear suitcase bombs. And the key to his success was that the negotiations, purchase, and delivery of the weapons were conducted by the members of the Chechen Mafiya committed to the Islamist movement, who had unparalleled entrée into the Soviet-era defense system. In return, bin

Laden's emissaries paid the Chechens $30 million in cash, along with two tons of Afghan heroin, worth about $70 million in Afghanistan and at least ten times that on the streets of Western Europe and the United States.

Estimates of how many nuclear suitcase bombs the Chechens secured for bin Laden vary from "a few" (according to Russian intelligence sources) to "more than twenty" (according to conservative Arab intelligence services). Most of the weapons were purchased in four former Soviet states: Ukraine, Kazakhstan, Turkmenistan, and Russia. A Western-educated Arab nuclear scientist, who had worked for Saddam Hussein's nuclear program before converting to fundamentalist Islam, supervised the acquisition for bin Laden and by the fall of 1998 was running the program for him. He was assisted by five Muslim Turkoman nuclear experts and a team of engineers and technicians, all of them Central Asian Muslims. For security reasons, they were preparing the weapons in two clusters of facilities: one in deep tunnels in the Khowst area, and the other in deep caves near Qandahar.

The few details that have emerged make sense when seen against the background of the Chechen Islamist movement. In fall 1998, leaders of the Chechen Mafiya approached Ukrainian and Balt mercenaries in Shali, Chechnya, on bin Laden's behalf, looking for veterans of the Soviet SPETSNAZ (special-purpose forces) trained in using the suitcase bombs. Back in 1995, the Chechen commanders under Basayev established a special detachment of roughly one hundred Chechen natives, Arab mujahedin, and mercenaries—mainly Ukrainian and Balt veterans of SPETSNAZ, OMON (special police units), and similar units—to secure the shipment of highly sensitive cargo (such as nuclear materials) and passengers between Afghanistan/Pakistan and Europe via Shali, a city frequently used by both al Qaeda and Chechen jihadists for smuggling purposes. The Arab mujahedin in this force were all veteran "Afghans," and thus loyal to bin Laden. It was only natural that bin Laden would try to recruit SPETSNAZ veterans at Shali. By fall 1998 Russian intelligence sources believed that bin Laden had succeeded in recruiting a few nuclear-qualified former SPETSNAZ troopers.

The cooperation of such experts would enable the Islamist-Jihadists to trigger a suitcase bomb successfully. Colonel Boris Alekseyev, the chief of the Russian Federation Ministry of Defense Ecological Center, has noted that once authorized by a coded radio transmission from Moscow, a single SPETSNAZ trooper could prepare a suitcase bomb for explosion within half an hour. In late 1998, bin Laden's nuclear experts were able to hot-wire one of their suitcase bombs so that the coded transmission was no longer necessary. Now it would require only a single would-be martyr to activate the nuclear bomb.

In early November 2001, the Pakistani intelligence service (ISI) captured an al Qaeda messenger who confirmed that a few suitcase bombs "acquired by al Qaeda from Central Asian rogue groups" were in bin Laden's arsenal. After "intense treatment" by ISI interrogators, he described in detail the two compartments comprising the bomb and the charging mechanism placed atop them. He noted that a new charging mechanism that "can be activated through a timer or even through a cell phone command" had been added to the Soviet-era mechanism. He described the Soviet-era markings and the code indicating that one of the suitcase bombs was produced in October 1988: General Lebed confirmed the accuracy of the messenger's description. Senior ISI officials who had studied the interrogation material concluded that "at least two briefcase nuclear weapons may have reached U.S. shores" by fall 2001.

Both Osama bin Laden and Ayman al-Zawahiri also confirmed to the well-connected Pakistani journalist Hamid Mir that al Qaeda had nuclear weapons. In an interview on November 7, 2001, bin Laden told Mir that "we have chemical and nuclear weapons as a deterrent and if America used them against us we reserve the right to use them." On March 21, 2004, Mir elaborated on al Qaeda's claims to a nuclear arsenal, citing a late 2001 conversation he had with Ayman al-Zawahiri. "Mr. Mir, if you have $30 million U.S., go to the black market in Central Asia, contact any disgruntled Soviet scientist, and a lot of dozens of smart briefcase bombs are available," al-Zawahiri told Mir. "They contacted us; we sent our people to Moscow, to Tashkent, to other central Asian states; and they negotiated and we purchased some suitcase bombs." Zawahiri's reference to a $30 million price tag confirmed the sum reported by Arab intelligence sources in the fall of 1998. And his contention that "our people" were dispatched to conclude the deal fit the original description of the Chechen emissaries.

And so by the fall of 1998 the Chechen Islamist-Jihadist leadership had facilitated a further important step in the empowerment of al Qaeda—by playing a key role in giving them access to nuclear weapons.

Chapter 10

THE SLIDE TO WAR

IN RETROSPECT, THE FALL OF 1998 WAS A TURNING POINT IN THE Islamists' ascent in the Caucasus. Beginning with the bombing of a military train in Dagestan on October 9, the Islamist-Jihadists mounted a series of major challenges to Aslan Maskhadov's power, intensifying the crisis with Russia so effectively as to paralyze the diplomatic process favored by Maskhadov.

The political landscape in the Caucasus remained chaotic. On October 17, the Coordinating Council of Chechens in Dagestan and the areas of Terek and Sulak, an organization of ethnic Chechens living in Dagestan, urged widespread support for Maskhadov and what it called "his great Jihad task of establishing a free Chechen country." Significantly, the council stressed that its own constituency should be an integral part of that Chechen state. "We link our future with the people of Chechnya, and we will gain freedom and independence together with the rest of our Chechen brothers," the council decreed, introducing one of the key themes that Shamil Basayev and Ibn al-Khattab would use in justifying their surge into Dagestan the following year.

From within and without, the political pressure on Maskhadov was mounting. Basayev, Salman Raduyev, and Khunkar Israpilov continued to curtail Maskhadov's powers by threatening his removal by force and increasingly relying on the Sharia court, rather than the state's institutions, as the ultimate legal and arbitrating instrument. Maskhadov issued a series of brave critiques of terrorism and crime, including a denunciation of the Grey Wolves of Islam, but his calls for the various nongovernment armed forces to disarm and disband were brazenly ignored. Maskhadov's war of words with Shamil Basayev proved both futile and demeaning. In an extraordinary challenge to Maskhadov's Grozny administration, the Sharia court decreed the laws of the state illegal. Maskhadov had lost any leverage he may have had against the Islamists.

The struggle for power in Grozny peaked in October 1998, when a series of bombings and assassination attempts shook the capital city. The first was another well-planned car bombing that killed General Shadid Bargishev, a high-level official of the Ministry of Sharia State Security who had been nominated by Maskhadov to lead the struggle against crime, kidnapping, and private armies. A couple of days later, Akhmad-Hajji Kadyrov, the mufti of Chechnya and a major ally of Maskhadov in the quest for Islamic legitimacy, was almost killed by a similar car bombing. Kadyrov's car was thrown twenty meters by the explosion, but he was only slightly wounded. Another powerful bomb was defused near the Chechen Sharia State Security Ministry in Grozny. "The bomb was similar to those used in assassination attempts on Chechnya's renowned officials," observed Nasrudin Bazhiyev, the deputy Sharia security minister.

Forensic evidence—including the structure and materials of the bombs—pointed to either Arab mujahedin or mujahedin-trained Chechens as the perpetrators. Ivan Rybkin, the former head of the Russian-Chechen negotiating commission, explicitly declared that "certain political, financial, and religious forces from Middle East countries" were behind the latest violence. The wave of bombings "was the job of outsiders from Jordan and Saudi Arabia, who came to Chechnya back in the time of fighting [the First Chechen War]." Their objective, he warned, was to prevent the revival of the diplomatic process, and "to unleash a civil war in Chechnya." But other officials pointed fingers elsewhere: Maskhadov's foreign minister, Movladi Udugov, blamed "foreign secret services"—his euphemism for Russian intelligence—for the terrorist acts in Grozny, claiming that the services wanted "to destabilize the situation and unleash a civil war in Chechnya."

The violence in Chechnya wasn't limited to political assassinations. On October 3, a Russian national and four employees of Granger Telecom (three British citizens and one New Zealander) were kidnapped by Arbi Barayev—known as "the Wahhabi" because of his commitment to the Islamist-Jihadist cause—and held under inhumane conditions. The hostages were subjected to near starvation, repeated beatings, and torture, as well as violent "interrogations" designed to elicit confessions for spying for the United Kingdom, United States, and Israel. A veteran of ransom kidnappings, Barayev sent emissaries to negotiate a $10 million ransom with Granger Telecom. But the real motive of this kidnapping was never financial; it was part of an arrangement Barayev had made with the Islamist-Jihadist leadership to terrorize foreign workers—with both commercial and humanitarian nongovernmental organizations—into abandoning Chechnya so that they could not challenge the Islamicization and jihadist movements. The proof of this came in early December, when the four English-speaking hostages were brutally beheaded the moment the new level of cooperation between the Islamist-Jihadist leadership and the Chechen leaders was reached.

Besides, the potential ransom from Granger Telecom paled in comparison with the money Barayev stood to collect from other sources. Barayev acknowledged to several fellow Chechens that he had close ties to the Arab mujahedin and the Taliban, and bin Laden's organization paid Barayev and his partners—mainly, the five Akhmadov brothers—$30 million for the kidnapping and beheading. In November 2001, Abdurakhman Adukhov, the Russian hostage who had been held with the four but ultimately ransomed, told the BBC that Barayev himself had revealed why the four were beheaded. "Now we'll get $30 million, not $10 million. We are helping the Taliban. Our brothers from the East wanted it to be done. They will pay us," Barayev reportedly said.

When the severed heads of the four hostages were found in a sack on December 8, the news shocked the Westerners still working in Chechnya and the northern Caucasus, and the vast majority of them abandoned the area to the mercy of the Islamist-Jihadists.

No longer able to ignore reality, Maskhadov recognized that the only way to retain his presidency, and most likely his life, was to continue his tacit support for the Chechen Islamists, the mujahedin in Chechnya, and the Islamist supreme leadership. At the end of the month, Maskhadov continued his administration's alignment with state sponsors of terrorism when he announced that Grozny would recognize the Taliban's regime in Afghanistan. This burgeoning new relationship between Islamist leaders in Chechnya, Afghanistan, and Pakistan would drive the events of the next several months.

In the late fall of 1998, Chechen and Afghan officials launched a new set of comprehensive discussions about cooperation on a host of clandestine projects—including the expansion of the network they had built to smuggle drugs, weapons, and strategic materials throughout the Islamic world. Another key issue was the securing of safe refuge for key leaders in case of a future crisis with Russia. The Chechens expressed their desire to send some of the leading Arab mujahedin—in particular, Ibn al-Khattab and his men, who were involved in the kidnapping for ransom of Westerners—to hiding places in Afghanistan. In return, the Taliban broached the idea of bin Laden receiving temporary asylum in Chechnya if the international pressure on Afghanistan to extradite him became unbearable.

In early December, Abdul-Wahid Ibrahim, the head of the Afghanistan and Central Asia office in the Chechen foreign ministry, arrived in Qandahar for a few days of negotiations with the Taliban leadership that led to a comprehensive agreement on cooperation. "The Chechens are indebted to the mujahedin of bin Laden and the other Arab 'Afghans' for fighting along with [them] against the Russians," Ibrahim acknowledged in Qandahar. But Grozny was still dependent on its support from the West, and it was reluctant to be publicly identified with bin Laden. Should the need arise, Ibrahim suggested, perhaps bin Laden could be granted asylum as a "guest" of Salman Raduyev, whose private army controlled certain areas of Chechnya that were beyond the reach of the Chechen government. (These areas had already served as a safe haven for numerous Arab "Afghans," including bin Laden's men.) This arrangement would allow Maskhadov's government to deny that it was supporting bin Laden. A few days later, Chechen deputy prime minister Yusup Soslambekov declared that Chechnya "refuses to become a swamp for terrorists," and denied that bin Laden had been offered asylum there. However, Soslambekov did confirm part of its deal with the Taliban—that if Ibn al-Khattab was forced to escape Chechnya, he might relocate to Afghanistan.

These negotiations in late 1998 would have unexpectedly far-reaching consequences for the global Islamist surge. By this time, the Islamist leadership was contemplating a major new strategic surge, targeting the vital interests of the United States and the West as a whole. This mounting Islamist threat was a direct function of the growing importance of the Caucasus and Central Asia to the key terrorism-sponsoring states. The ultimate objective, furthered by Pakistan and Iran and actively supported by the Taliban, was to evict the United States from the region, whose untapped energy resources were an increasingly appealing substitute for Persian Gulf oil. The leaders of Iran and Pakistan were convinced that applying pressure in the form of Islamist subversion and terrorism would convince the governments of the Caucasus and Central Asia to favor partnerships with Western European and East Asian companies, to the detriment of America's strategic interests. Ultimately, their goal was to evict the United States and Russia from the region, to destroy Armenia altogether, and to draw Turkey into the Islamist fold.

As if to confirm the Islamists' analysis, in mid-December Chechnya directed a new series of terrorist threats toward Russia. A number of armed detachments operating under the umbrella command of the Supreme Council of Islamic Jamaats, led by Arbi Barayev and Ramzan Akhmadov, announced their intent to cross into Russian territory and carry out terrorist bombings. (The term *jamaats* is Arabic for "group" or "society," and is used in the Caucasus to refer to a distinct Islamist community usually involved in jihadist activities.) The jamaats' warning was tied directly to the evolving power politics in Grozny: If Maskhadov's forces moved against the jamaats' armed detachments, the detachments would not engage in fighting fellow Muslims but instead cross the Russian border and launch terrorist strikes within Russia itself. The Islamists explained that they were unwilling to spill the blood of their Muslim brothers in order to please those who opposed Chechnya's independence. It was another signal that the Islamists saw further terror strikes in Russia as part of their evolving strategy to advance their objectives throughout the Caucasus.

In mid-February 1999, Chechen president Aslan Maskhadov once again insisted—this time to Sheikh Muhammad Hisham Qabbani, the chairman of the Supreme Islamic Council in the United States—that Chechnya would "not offer any form of refuge or settlement to Osama bin Laden, whatever this decision costs the Chechen government, even if that means war." Whatever its public attitude toward bin Laden, however, the Chechen leadership was already harboring a new buildup of bin Laden-sponsored terrorist forces within its borders, including Arab, Afghan, and Pakistani mujahedin. Moreover, bin Laden's mujahedin were activating Islamist units that had served as the key Chechen commanders' elite strike forces during the war against Russia. Among these units were the Soldiers of the Orthodox Caliphs (which had served under now-president Maskhadov), the Abd-al-Qadir Forces (which had fought under Shamil Basayev), and the Islamic Liberation Party forces (which had been commanded by Salman Raduyev). These "Afghan"-dominated forces had played a crucial role in Chechnya's war against Russia; no government in Grozny would have dared question their right to operate on Chechen soil. And there were other forces preparing for future combat: In early February a headquarters for the liberation of Dagestan was established in Khattab's main training camp, and soon, at least five hundred young Dagestanis were training under its banner.

These Chechnya-based networks operated throughout the region—ambushing Russian patrols in Dagestan near the Chechen border, for example. These attacks were coordinated with a propaganda campaign recently unleashed against the Russian forces in Dagestan. In mid-January 1999, Shamil Basayev's Congress of the Peoples of Chechnya and Dagestan and Ibn al-Khattab's United Muslim Communities of Dagestan and Chechnya distributed a quantity of leaflets in the area of Khasavyurt, a major center of Russian forces. In the leaflets, Basayev and Khattab made threats against the troops, demanding that they leave Dagestan; they also vilified "the crafty methods which the special services of Russia and Dagestan use against people they do not like" and threatened to take "commensurate steps" if the Russians refused to release supporters of theirs who had recently been arrested.

* * *

The early spring of 1999 saw a series of spectacular new terrorist strikes in the Caucasus and Central Asia. The most significant of these, in strategic terms, was a series of six car bombings in Tashkent, Uzbekistan, on February 16, 1999, killing 15 and wounding 128. The plot was an elaborate effort to assassinate Uzbekistan's president, Islam Karimov—a key opponent of Islamist radicalism and militancy in the region. Investigators determined that the types of high explosives and fuses used in the Tashkent bombings were identical to those used by Chechens, and particularly by mujahedin trained in Khattab's schools. Subsequent arrests proved that the operation was run by a network whose expert terrorists were Chechnya-trained Islamist Uzbeks. Uzbek security services noted that "all the detainees had received special training in sabotage in Chechnya, Tajikistan, and Afghanistan," and specifically that the bomb makers had been trained in Chechnya. Karimov himself briefed foreign ambassadors that "most of the detainees underwent training in sabotage in Chechnya and Afghanistan, and some were able to 'distinguish themselves' in the fighting in Chechnya, Afghanistan, and neighboring Tajikistan." A few days later, acting on information provided by the detainees in Tashkent, Russian security services operating on the administrative border between Chechnya and Dagestan captured three Uzbek trainees of a Basayev-Khattab school.

Another audacious terrorist operation was the March 6, 1999, kidnapping of Major General Gennady Shpigun, the Russian interior ministry's top envoy to Grozny. Shpigun was seized by masked men who boarded his plane when it was preparing to take off from the Grozny airport. Members of the Chechen secret services were involved in the plot, but the Chechens had no intention of maintaining secrecy about the operation forever: After the incident, Salman Raduyev ordered that all participants be awarded high state honors and decorations. At the ceremony, Raduyev told the audience that "your task is to seize hostages and kill. *Allah* will forgive all." Still spoiling for a Russian military response, Raduyev then announced his intent to strike at Russia's cities and industrial facilities "if Russia starts bombing our bases and cities" in retaliation for the ongoing Chechen terrorist strikes.

The following weeks saw a marked escalation in border clashes around Chechnya. In early March, Russian forces had to use helicopter gunships to silence Chechen fire. But terrorist operations also continued in and around Chechnya—including a series of bombings in Vladikavkaz, North Ossetia, the jihadists' latest bid to instigate a major confrontation. The first site to be bombed was the local marketplace, where a single blast led to more than sixty fatalities and several hundred injuries. (Two other bombs were discovered by security and defused.) The bomb that was detonated was made of seven kilograms of caseless plastic explosive, activated by a conventional but sophisticated firing mechanism based on an ordinary mechanical alarm clock. This fuse pointed to foreign involvement because "Khattab's people"—the Chechen-trained bomb makers—usually work with comparable organic fougasse (improvised land mines) or bombs. Indeed, in mid-March Russian interior ministry (MVD) intelligence learned that six groups of mujahedin identified as "specialists in sabotage" had graduated from a special course instructing them in techniques not commonly used in the Caucasus at a training base just fifteen kilometers from the Vedeno station in Chechnya. Four of these groups had already been dispatched to link up with Islamist networks in Volgograd, Saratov, Samara, Voronezh, Krasnoyarsk, Moscow, and St. Petersburg.

Two days later, Aslan Maskhadov was the target of an assassination attempt—which used another sophisticated bomb—in Grozny. Maskhadov had shown increasing discomfort with the Islamist-Jihadists' use of Chechnya as a base for international terrorist strikes, but the near miss apparently convinced him to reduce his complaints about the cross-border operations. In return—ostensibly in response to Moscow's threats over Major General Shpigun's fate and the bombing in Vladikavkaz—Islamist-Jihadist forces throughout the Caucasus expressed support for Grozny, promising to launch a jihad against not just Russian forces but all Western presence in the area. The rhetoric was backed by a flow of reinforcements from Afghanistan, Pakistan, and the Arab world, traveling through Azerbaijan and Georgia to Chechnya.

To official Moscow, a military confrontation with the Chechens seemed inevitable. On March 7, 1999, Sergey Stepashin, then head of the MVD and later prime minister, delivered a stern statement, threatening the Chechen terrorists with "strong-arm action" if the attacks on Russian checkpoints, patrols, and officials continued. The Duma approved Stepashin's statement, and soon the Russian military and the MVD began preparing for battle. As Stepashin later told *Nezavisimaya Gazeta*, "The plan of active operations in the republic had been in development since March [1999]. We had planned to approach the Terek in August-September. . . . I worked actively to fortify the borders with Chechnya, preparing for an active onslaught."

It was a timely decision. By early April 1999, as the weather became more conducive to both logistical buildup and fighting, the situation in the northern Caucasus was deteriorating further. In one incident, four policemen were killed on the border between Stavropol Krai and Chechnya; soon after, three Ingush policemen were killed at the Chechnya border. Moscow urged Grozny to crack down on such cross-border attacks, but to no avail. In retrospect, this final ambush was the turning point in the Kremlin's evolving debate about how to deal with Chechnya. The events of early April convinced the Russian government that Aslan Maskhadov's power was a fiction—that he was "afraid of," and thus incapable of confronting, Shamil Basayev and the militant Islamists. The MVD's Sergey Stepashin told a high-level meeting that "the official authorities of Chechnya are losing the possibility of keeping the situation under control, despite the efforts being made by Aslan Maskhadov." To the Kremlin, it seemed increasingly clear that it was only a matter of time before widespread violence returned to the Caucasus.

In the early spring of 1999, Chechen-trained fighters were dispatched to yet another front in the ongoing jihad outside former Soviet territory. Anticipating an escalation in the crisis in Yugoslavia, bin Laden's Islamist-Jihadist leadership dispatched numerous expert terrorists to Albania to help prepare and lead key UCK (National Liberation Army) units for special operations. The task of preparing the UCK for battle against heavy forces equipped with modern Soviet-style weaponry was entrusted to a few teams of Chechen "commandos"—all veterans of the war against the Russian forces and later terrorist raids. On April 10, Movladi Udugov formally announced that the Congress of Peoples of Ichkeria and Dagestan was sending "a volunteer peacekeeping battalion" to Kosovo. "We are speaking about several hundred people. Only military chiefs and commanders know the exact number," Udugov explained. The mission of the Chechen force would be to defend Kosovo's Albanian population against "genocide at the hands of the Yugoslav authorities." Udugov hinted that the Chechen decision was based on consultations with the Islamist-Jihadist leadership, not the UCK itself. He acknowledged that the Chechen leaders had yet to inform the UCK of its decision. "We have not had an opportunity for direct contact with the Albanian authorities and the commanders of the Kosovo Liberation Army," Udugov explained, but "official documents and appeals are to be sent tomorrow."

By the time the Chechens consolidated their forces in Albania, the Kosovo crisis was running hot, and the Western intelligence services—mainly the American CIA and the German BND (Federal Intelligence Service)—were busy organizing their own Albanian forces to support the UCK. The Islamist high command held sway over much of the UCK, but now it faced the challenge of maintaining that control while outperforming Western-sponsored Albanian "commando" teams. The Chechens would play a central role in this new campaign. The senior jihadist commanders in Afghanistan and Pakistan reasoned that ethnic Chechen fighters could be introduced more successfully into the UCK than Arabs, who were likely to be distrusted by U.S. personnel as Osama bin Laden's troops. (This judgment was confirmed by Muhammad al-Zawahiri, the senior jihadist commander in Albania/Kosovo and the brother of Ayman al-Zawahiri.) The Islamist cannily resolved to send teams of Chechen mujahedin to augment the UCK even as the U.S.-led NATO intervention intensified.

In the summer of 1999, once Russia deployed troops in Kosovo as part of the UN's KFOR peacekeeping force, the Chechen presence became more overt. Chechen patrols began flaunting their presence in the vicinity of the Russian forces. In late July and early August, armed Chechen fighters in camouflage uniforms and green bandanas drove Toyota jeeps bearing the green banner of the Chechen Republic—and Shamil Basayev's banner, with its picture of the *Bozkhurt* (the steppe wolf)—along the main street of the Kosovar village of Bresje. In at least one case, the Chechens engaged directly with Russian forces: On August 1, a column of Russian troops heading from Leskovac, Serbia, to the Russian base at Slatina was approached by a group of people identifying themselves as Chechens and asking the Russians to "send their greetings" or "just say hello" to their commanders.

The very presence of such a Chechen terrorist network in Kosovo—operating in cooperation with, and drawing support from, bin Laden's Islamist-Jihadist terrorist infrastructure in the region—constituted a major threat to the Russian and other Western contingents in KFOR, as well as to humanitarian organizations and other entities operating in the area. Such proximity to Chechen jihadists made them vulnerable to terrorist attacks in retaliation for Russian actions in the Caucasus, and even to preemptive strikes meant to deter such action. The UN peacekeepers' inability (or unwillingness) to confront the jihadist terrorists in their midst has kept the threat viable to this day.

Chapter 11

THE SECOND CHECHEN WAR

AS THE SUMMER OF 1999 APPROACHED, IBN AL-KHATTAB AND SHAMIL Basayev started dispatching terrorists and operatives into the neighboring republics and Russia itself, assigned to establish new sleeper cells and networks and bolster existing units. More Caucasian terrorists were graduated from advanced courses at Islamist-Jihadist schools in Chechnya, and several experienced Arab commanders arrived in Chechnya from the Middle East and Afghanistan. Fresh supplies and funds were pouring into Chechnya from the Middle East and Central Asia, mainly via Azerbaijan and Georgia. These preparations were grounded in the Islamists' theological-ideological framework: The Chechen jihadist leadership, led by Shamil Basayev, articulated its regional and global aspirations clearly, and the so-called Chechen nationalist leadership, led by Aslan Maskhadov, failed to challenge them, let alone stand in their way.

The stage for the Second Chechen War had been set.

One authoritative voice that expressed the reasons for the coming war was a new book by the author of *Our Struggle, or the Imam's Rebel Army*. One hundred pages long, Magomed Tagayev's new work—*Holy War, or How to Become Immortal*—surfaced in the spring, supplementing his influential earlier book with more explicit and practical instructions for aspiring terrorists. By spring 1999 both books enjoyed wide distribution in Chechnya and Dagestan.

Holy War, or How to Become Immortal suggested that imminent violence in Dagestan would open the door to a fateful regional confrontation with Russia. "Our tasks are specific and as clear as day. They are the creation of our own state, the Islamic Republic of Dagestan, as part of a Caucasian Confederation," Tagayev wrote. He was sure that the anti-Russian revolt would be supported by most of the region's population. "All the Muslims populating these lands, and not only the Muslims, believe that they were enslaved," he contended.

Under the current circumstances, Tagayev argued, war with Russia was an urgent imperative: "As long as the [Russian] empire is ruled by upstarts and half-wits, war is inevitable. So we must prepare for it. As well as constructing new military bases, we must work on creating research centers to develop and manufacture arms enabling us to wage an effective war and eliminate the problem of supplying captured weapons." He called for the kidnapping of experts who could be forced into producing weapons for the peoples of the Caucasus, and offered a clear and precise prescription for the coming war, right down to the timetable:

> We must finish forming the Imam's Rebel Army not later than midsummer 1999. It absolutely must not prize any valuable assets in cities and settlements if Russian armed formations seek protection in them. Every street, every district, and every house must be fought for. All forces, starting with Russian armed hordes, bandits, looters, rapists, mindless types who have deliberately betrayed the homeland, senile post-Soviet *nomenklatura* types, and cowards who would hide behind their sisters' or mothers' skirts, must picture the approaching victorious liberation struggle. No one in this world holds authority for us in this sacred struggle for honor, dignity, and a place in the sun other than almighty Allah and the holy writings of the Koran.

Tagayev predicted that the coming war unleashed by the Muslims of the Caucasus would spread throughout Russia, stressing that the key to victory lay in terrorizing the Russian population until the Kremlin was forced to abandon the Caucasus in deference to its own peoples' fears:

> We must create sabotage groups and organize high-precision sabotage operations in locations where there are high concentrations of gangs of colonialists and their rear services, blow up sewer systems, heating systems, and duct and cable communication systems, and set fire to plants, factories, and woods near industrial facilities and processing industry facilities. We must blow up bridges, airfields, railroad stations, elevators, and city transport fleets. The families of serving officers, generals, and special services personnel, and others too, must be kept under constant surveillance. To that end we must organize individual groups to simultaneously supply the enemy's Internal Affairs Ministry, Federal Security Service, and army with misinformation in the form of bomb and other explosives warnings. One in twenty warnings would take effect, leaving the people in a state of constant anxiety. This kind of action will be our common practice until the Russians see sense.

Tagayev saw the coming war as a historic confrontation that could not be resolved in a few years. "The question of a grand campaign against Moscow will still be relevant as far as the Caucasus is concerned in a hundred years' time. But first we must destroy their foundation," Tagayev concluded.

The region's Islamist-Jihadist leaders shared his worldview. In an early April interview with *Moskovskiye Novosti,* Shamil Basayev signaled that his differences with Aslan Maskhadov were irreconcilable, and articulated a vision that was as much anti-American as anti-Russian. Basayev explained that his "basic disagreements with Maskhadov had to do with the fact that the country needs radical reforms and changes. After all, everyone understands that we cannot go on living like this. But Maskhadov wants to return to the old way of life, the Soviet way, and therefore he is inclined toward Russia. The budget, salaries, ministerial posts—everything depends upon Russia." At this point, Maskhadov was still seeking a negotiated agreement with Russia that would lead to the rebuilding of war-torn Chechnya, even if it meant that the nation would initially remain legally connected to Russia. In contrast, Basayev insisted on a dramatic break with Russia, and the creation of a regional

Islamist Imamat as a necessary first step in establishing an independent Chechnya.

Echoing the Islamist-Jihadist ideological line, Basayev considered the United States as great an enemy as Russia. He argued that "so much evil has accumulated in the world that it has to break out somewhere. And the best thing would be for America and Russia to get together and leave the rest of the peoples in peace." Later, he commented that "it would be more advantageous to create a Russian Republic and eliminate the empire. Then it would be possible [for the peoples of the Caucasus] to unite under new conditions." The *Moskovskiye Novosti* interviewer noted that there was "a widespread idea in Chechen society that the world would be a better place if three capitals were destroyed—Moscow, Tel Aviv, and Washington." In response, Basayev said, "Why destroy them if you can capture them and establish Allah's laws there? Islam is the most tolerant religion." He then singled out Russia for its anti-Islam posture. "It is only in Russia that the cross rises above the vanquished crescent. But the crescent will never be vanquished," Basayev assured his interviewer.

On April 17, the Islamist-Jihadists offered a new public display of their power when the Second Congress of the Peoples of Ichkeria and Dagestan convened in Grozny, in open defiance of Maskhadov and his government. The very size of the gathering demonstrated the Islamists' reach, with more than seven hundred participants—including 195 "official representatives" from Chechnya, 297 delegates from twenty-five Dagestani village-based jamaats in the rayons of Buynakskiy, Gunibskiy, Tarumovskiy, and Khassavyurttovskiy, and more than two hundred invited guests from throughout the Caucasus and the Muslim world. Shamil Basayev chaired the congress, and delivered its keynote report. Abdulla Parkuyev, the deputy chairman, interacted with the media. Other prominent participants included "Amir al-Khattab," Zelimkhan Yandarbiyev, Bagautdin Magomedov (the Dagestani Wahhabi leader), Magomed Adallo (a prominent Avar poet and intellectual), and Magomed Tagayev, who emerged as the most prominent Islamist-Jihadist spiritual leader in the Caucasus.

The Second Congress of the Peoples of Ichkeria and Dagestan

proved to be more than just a public show of force. It was during this gathering that the Islamist-Jihadist strategy for the next phase of the jihad against Russia was formulated, declared, and legitimized. In his opening speech, Basayev announced the formation of a military-political council and a security council to manage the forthcoming struggle, as well as an "Islamic Legion" and a "Peacekeepers Brigade" comprising of a few thousand well-trained mujahedin to lead the imminent fighting. Basayev stressed that "these troops [were] needed to implement the resolutions of the Congress, the major goal of which [was] the creation of an independent Islamic state." Moreover, he explained, the mujahedin forces were in urgent need of expansion, and several military training camps were already operating on Chechen territory toward this end. Basayev also noted that the Congress had sponsored comprehensive assistance programs for "brothers in need" throughout the Caucasus—ranging from social and economic welfare programs (based on the Islamicization model) to new "peacekeeping Caucasian forces" to be used in regional military intervention.

Abdulla Parkuyev, the deputy chairman of the Congress, told the newspaper *Nezavisimaya Gazeta* that the body was committed to the liberation and Islamicization of Dagestan as a high priority, even if it meant military confrontation with Russia. "All religious jamaats of Dagestan are in favor of introducing Muslim legislation in the republic, but the constitutions of Russia and Dagestan stand in the way of this. Therefore, as the Amir of the Congress, Shamil Basayev, said at the Congress, the center of political gravity has now shifted to Dagestan." Parkuyev disavowed any specific plans to use force: "Our aspiration to unity and desire to devote our lives to the will of Allah certainly does not mean war is inevitable," he said, claiming that the Congress was "against the use of violence." But he acknowledged that a crisis was forthcoming. "The Congress [did] indeed set as its goal the demilitarization and decolonization of Dagestan and the establishment in the republic of the Shariah as the only form of administration and way of life for Muslims," he noted. "Ultimately this means the unification of Chechnya and Dagestan into a single Islamic state as it was quite recently, in the last century under the Immamat of Shamil. The Chechens always used to say that they were from Dagestan, our common homeland, not from Chechnya."

Other speakers stressed the importance of reversing the legacy of the Russian occupation that dated back to the days of Shamil. In one passionate speech, the writer Magomed Adallo equated the current Russian rule of the Caucasus with the "occupation of the Immamat." Noting that "the state Shamil created" during the Caucasian War had survived for as long as Shamil managed to "unit[e] the highlands of Dagestan and Chechnya," Adallo concluded that the current "Russian occupation" could only be broken once Chechnya and Dagestan were reunited under a jihadist banner.

Several speakers, including Basayev, noted how the Islamicization of the Caucasus had transformed society. Zelimkhan Yandarbiyev devoted his speech to the need to "quell the animosity between Tariqatists and Wahhabis" throughout the Caucasus, to better unite its people for the forthcoming struggle against their common foe, the Russians. Yandarbiyev urged "all Muslims" to "abandon their [own separate] ambitions" and to "stop wasting their strength looking for enemies among brothers in faith." Parkuyev emphasized the political realities that still confronted the Caucasus, acknowledging that "proclaiming an Islamic state does not solve all the problems. For more than four hundred years the Chechens and Dagestanis have been in the 'secular field' of Russia. Islam was forgotten, its norms were violated, and therefore the Shariah is now perceived in various ways. Certain hotheads are trying to enter on the path of revenge, including blood revenge, when they are punished according to the Shariah. But there is no turning back, just as there are no alternatives to Islam." Every one of the speakers insisted that the process of Wahhabi Islamicization throughout the Caucasus was irreversible.

Speaking to *Nezavisimaya Gazeta*, Parkuyev attributed the conflict between Basayev and Maskhadov to their profound disagreement over the role of Islam in the life of Chechnya:

> At the end of last year and the beginning of this year the Congress and Amir Shamil Basayev personally had differences of opinion

> with President Maskhadov, which could have led to serious conflicts. Maskhadov was trying to depart from the principles of Islam, relying on support from Moscow. All the people who thought as he did left him at the time, and when he finally came to believe that there was no hope for support from Russia, he gradually had to introduce Islamic order in Chechnya. If the president introduces the complete Shariah in the republic, he will have no disagreements with Shamil Basayev. Otherwise I do not think that he will hold on to his post for long.

Maskhadov could not ignore the Congress's flagrant challenge to his leadership. By this point, however, he was so besieged and threatened that his only reaction was to send his "representatives" to brief the Russian Television Network (RTN) on his administration's response to the Congress. Maskhadov's spokespeople called "the meeting of the Congress of Chechnya [sic] and Dagestan a routine event in the life of Ichkeria." According to RTN, they acknowledged that "the leaders of the Congress, Basayev and Udugov, [carried] no less political weight than Aslan Maskhadov." Maskhadov's representatives noted that "the leadership of Dagestan has already called the Congress illegitimate and its meeting yet another act of interference in the republic's affairs," and that Russian analysts considered Basayev's and Udugov's speeches attempts to "reduce the influence of President Maskhadov." But the president himself refused to comment.

The Congress only confirmed the public sense that violence was imminent. Participants in the gathering made clear to friends and foes alike that the current rate of low-level skirmishes and political impasse could not last long. A hectic new military mobilization now began on all sides—even among those, like Maskhadov's own forces, who had no desire to go to war. Maskhadov realized that if he stayed out of the forthcoming war he would effectively become irrelevant. The only way for him to remain a "leader" in Grozny was to assume leadership in a war he had tried for months to avoid. And so Maskhadov committed wholeheartedly to preparing for the war the Islamists were about to instigate—and, in so doing, aligned himself once and for all with a radical force he had long struggled to keep at bay. For him, there would be no turning back.

One sign of Maskhadov's personal involvement in preparations for the coming war was his adamant refusal to nominate a minister of defense. In mid-April, he abruptly ended negotiations with two leading candidates: Ruslan Gelayev, the former commander of one of the fronts during the First Chechen War (1994–1996), and Magomed Khambiyev, the well-known commander of the National Guard, two individuals with both political and armed loyalists of their own. At the risk of alienating both men, Maskhadov insisted on serving as his nation's own ultimate defense authority, taking command of the regular units and subunits of the Chechen Army, the National Guard, the Border-Customs Service, and a myriad of regional regiments and battalions.

A close examination of the actual military buildup under Maskhadov reflects his commitment to a regional war involving other parts of the Caucasus. Although the entire population of Chechnya by spring 1999 was likely no more than four hundred thousand people, Maskhadov assembled an arsenal comprising between three and five hundred thousand units of fairly modern small arms: Kalashnikov assault rifles, several types of machine guns, and various models of grenade launchers and flamethrowers. These stockpiles were amassed, at great effort and cost, to arm the new Caucasian Muslim armies in the soon-to-be-liberated parts of the Caucasus. In addition, Grozny's forces included an impressive array of artillery pieces—including Grad multiple-barrel rocket launchers, guns and howitzers of various calibers, and a myriad of heavy mortars. The armored formations included several dozen T-62 and T-72 tanks with various modifications (although only a few of them were truly combat ready) and about a hundred Soviet-produced armored personnel carriers (APCs), of varying degrees of operability. Most of the heavy weapons were deployed in defensive positions around Grozny and other centers of power.

However, the best-equipped and most combat-effective forces in Chechnya were the "private armies"—the armed groups who answered to the region's leading commanders and religious authorities. These detachments, usually recruited on the basis of the ancestral clans of their leaders and deployed at their traditional places of residence, were cohesive and fiercely loyal to their commanders. In spring 1999, the most effective private armies were Basayev's forces, located mainly in Vedenskiy Rayon; Raduyev's army, based near Gudermes; and the forces of Arsanov, Udugov, and Gelayev, which were based in Urus-Martanovskiy Rayon. All of these detachments had good personnel staffing, their key personnel had combat experience, and they had ample modern weaponry, combat equipment, means of communication, and transportation resources. Another private army was the reappearance of Chechnya's Sharia Security Regiment as an opposition force, soon after Maskhadov decided to disband them. Now led by Chechen and Arab Wahhabi militant clergy, this force remained in the Grozny area. Taken together, these private armies could easily challenge Maskhadov's forces for control over Chechnya, while shielding the Islamist-Jihadist terrorist detachments with impunity.

Of singular importance were the Islamist-Jihadist forces loosely under Khattab's command. The true significance of Khattab's forces lay not in their size—by spring 1999 he had only about five hundred fighters answering to him—but in his singular long-term influence on the evolution of Chechnya's armed formations and the population at large. Khattab concentrated on the training and indoctrination of would-be terrorists and fighters, as well as the Islamicization of the segments of society they hailed from. Because he was feeding his "graduates" into the ranks of the various private armies, Khattab enjoyed the patronage of other key commanders, who ultimately came to support his efforts to bring permanent change to the character of life in Chechnya.

At the core of Khattab's forces were the roughly seven training centers he established in the Nozhay-Yurtovskiy and Vedenskiy rayons of Chechnya. Most of these camps were located in Soviet-era tourist resorts and several Pioneer summer camps along the Kholkholo River in the Vedeno and Shali, areas that were impounded by the Grozny authorities and handed over to Basayev and Khattab. These facilities, which could accommodate one thousand people at a time, served as much to transform and mold the youth of the Caucasus as to make them expert terrorists.

The six-month course was divided into two stages. The first two months were devoted to Islamist-Jihadist indoctrination—from the tenets of Wahhabi Islam, with the attendant ideological commitment, to lessons in furthering the Islamist-Jihadist cause throughout the Caucasus and the uncompromising jihad against "evil Russia." Bagautdin Magomedov and a team of Caucasian clergy were responsible for this phase. The next four months were devoted to professional training, primarily by Arab experts, in all aspects of guerrilla warfare and terrorism—techniques of mining roads and sabotaging industrial facilities, daytime and nighttime fighting tactics, firing various types of weapons, individual combat techniques, assassination techniques, and ambush tactics, as well as the taking and interrogating of hostages.

The most promising trainees were then given additional advanced training to qualify as commanders. The subjects covered in this phase included how "to instill the terror of death among those who have betrayed Allah" and "how to destabilize the situation, the economy, and finances" in the areas not yet under the Islamists' rule through "bribery, infiltration into power, [and] spreading of rumors," as well as sabotage and terrorism. As part of their graduation process, trainees were sent to neighboring republics on missions of terrorism and subversion in order to prove their dedication and skills. To increase the all-Islamic character of the entire course, the trainees came from the republics of the North Caucasus, the former Soviet Central Asian states, and the former Yugoslavia (mainly Bosnia, Sandjak, and Kosovo).

By the spring of 1999, specialized advanced training programs were being run in each of the seven main camps. The main training center was the Said Bin-Abu-Waqqas Camp near the village of Serzhen-Yurt, where a hundred trainees were groomed for the elite mujahedin units under Khattab's personal supervision. An elite unit of one hundred highly experienced foreign mujahedin was also based in Serzhen-Yurt. The Abu-Jaafar Camp, located in the former Druzhba Pioneer camp, specialized in "partisan warfare tactics," training junior commanders for regular and special forces units. Advanced training in the use of heavy weapons and sabotage operations was conducted in the Yaqub Camp, and advanced demolition and sabotage techniques in the Abu-Bakr Camp; special religious and ideological training, along with tactical skills for commanders of fighting units, were conducted at the Davgat Camp. These, too, were all located at former Druzhba Pioneer camps. Except for the Said Bin-Abu-Waqqas Camp, each of these training camps employed roughly twenty-five foreign instructors and seventy to eighty trainees. Military Imams and specialists in indoctrination, including the tempering of potential terrorist "martyrs," were trained in the al-Dawah Camp. Another religious-military school, whose stated mission was to indoctrinate and prepare Muslims from throughout the Caucasus to create "a unified Imamat from the Caspian to the Black Seas," was the Caucasus Institute, located in the village of Kharashwi. The instructors in the institute were senior guides of the Muslim Brotherhood from numerous Arab countries, as well as Saudi Imams, many with combat experience in Afghanistan and Bosnia. Ibn al-Khattab was a frequent visitor to the Caucasus Institute.

The Islamist-Jihadist leaders were now exploiting their systematic destruction and disruption of the economy in Chechnya and neighboring republics, mainly Dagestan, to help in the recruitment and retention of individuals and to spread their hold over society. Would-be trainees in Khattab's camps were offered a monthly stipend of about $100—well above the average income in the rural and urban slum areas. Graduates from the North Caucasus republics were sent back home to "create bases, select people," and thus begin the process of transforming society. These graduates were given money, weapons, and equipment to start subsidizing whole communities by becoming the primary source of employment, food, and social services. Gradually, these graduates took over segments of the rural population simply by being the primary source of income in the intentionally created economically depressed areas. Within a few months, they transformed villages into bases of the Islamist-Jihadist insurgency—creating a situation where people were fighting in and out of their homes, and any attempt by the Russian forces to hit "guerrilla bases" was bound to inflict civilian casualties.

One especially cynical tactic the jihadists used in furthering Islamicization was exploiting the region's criminal element. After Chechnya's society and economy began to collapse in 1997, an array of heavily armed criminal gangs, most of them ten to thirty strong, rose to prominence in the region. These groups recognized no authority and obeyed no one but their own leaders. They were well armed—with modern small arms, grenade launchers, and machine guns—and highly mobile, using the latest Russian-made and Western-made all-terrain vehicles. By the late 1990s, these gangs were thriving on the fringes of the Islamist state, making money from smuggling goods and drugs on behalf of the Islamist-Jihadist forces, terrorizing the population in the Russian-controlled zones by extorting merchants, kidnapping locals for ransom, and conducting a reign of robbery and murder. By mid-1999, the Islamist-Jihadist authorities were also using criminal gangs for deniable terrorist operations and particularly for kidnapping and assassinations of rivals and their family members throughout the Northern Caucasus. The use of the criminal elements not only achieved practical results—eliminating individuals who were causing the Islamists trouble—but also handed the Islamist propaganda and agitation machine an endless stream of "cases of abuse" of innocent civilians, which they habitually attributed to "cruelty and intrigues of the Russian special services."

To sustain their military and civilian efforts, the Islamist leadership had consolidated, by spring 1999, a formidable financial and logistical support system. They had enough resources at hand to replace the flow of money from the federal authorities in Moscow—thus removing one of the last effective arguments for reaching a negotiated settlement with Russia. "We [in Chechnya] do not need their [Moscow's] money," Salman Raduyev declared at the graduation ceremonies for students of the intelligence school in Grozny. "Money is given to us by several European countries and also Pakistan, Afghanistan, and Iran. We receive money and arms and equipment from them." By now, weapons, funds, and volunteers were arriving in Chechnya and Dagestan via neighboring Georgia and Azerbaijan. Though Azerbaijani leaders tried to disrupt smuggling routes through their territory, routes through Georgia expanded rapidly without any serious interference from Tbilisi. Repeated attempts by the Georgian MVD to seize weapons failed largely because the Muslim population in northern Georgia actively supported the jihad of their brethren in Chechnya and Dagestan.

* * *

In mid-April 1999, the Islamist-Jihadist leadership undertook a thorough survey of the forward deployment of their terrorist forces. Ayman al-Zawahiri convened his aides and senior commanders at a major session in the Tora Bora caves near Jalalabad, Afghanistan. Participants at the meeting resolved to upgrade the security and counterintelligence measures undertaken by the clandestine networks. By the end of the meeting, satisfied with the status of his networks, Zawahiri approved the activation of several operational plans. In a statement issued late that month, Zawahiri's Jihad Islami declared that "the escalating Islamic resistance against the Crusader campaign against the Muslim Nation will not stop," and that conditions were becoming more encouraging for the Islamist-Jihadists worldwide. "The Islamic world in general, in particular the Arab region, is being swept by a wave of Islamic jihadist rejection; the Muslim Nation is vigorously rejecting the policy of humiliation and oppression being pursued against those working to restore Islam's sovereignty over its territory, and is at the same time determined to firmly and resolutely move toward achieving its aim of establishing an Islamic state through preaching, jihad, and exposing suspect actions."

In Pakistan, these sentiments were echoed in a concurrent statement by Qari Saifullah Akhtar, the central amir of the Harakat Jihad Islami, then responsible for the organization's international operations. He declared that Harakat Jihad Islami "would not allow anyone to compromise the blood of the Muslims being spilt in Kashmir and Afghanistan." Harakat Jihad Islami also anticipated a marked escalation and expansion of Islamist jihads worldwide because "the atrocities being committed against the Muslims in Afghanistan, Kashmir, Iraq, Bosnia, Kosovo, and Chechnya have awakened Muslims, and now Muslims would fight against India, the United States, Russia and Britain." Led by the Islamist-Jihadist forces, the Muslim world would ensure that "the eye that looks down upon Muslims as a minority or dreams of reducing them to a minority would be gouged out."

Indeed, by now, the Islamist-Jihadist groundswell in Chechnya was contributing to a surge of terrorist activity all over the West and not just within Russia. Chechen mujahedin were offering support to operations that had nothing to do with the northern Caucasus. Their most substantial contribution was the establishment of the Passport and Visa Service of the Ministry of Shariah National Security, an international center for generating ersatz official documents—mainly high-quality counterfeit and doctored Russian passports. At the behest of Khattab, Basayev, and Yandarbiyev, the ministry issued residence permits and Russian passports that cleared the way for these jihadists to continue from Chechnya to their destinations in Western Europe—usually as immigrants and refugees from North Caucasus republics. Hundreds of terrorists from Algeria, Turkey, Jordan, Pakistan, Saudi Arabia, and other countries passed through Chechnya to pick up their new papers en route to the Western world.

The counterfeiting program was launched by Khizir Khachukayev, chief of the Chechen Republic Passport Desk, after Movladi Udugov, then first vice premier of the Chechen government, who requested that Russian documents be issued to "fighters in the Chechen armed forces who have taken an active part in defending the Chechen Republic of Ichkeria against Russian aggression." It wasn't long before other jihadists began securing similar documents. At first the ministry issued only genuine documents, but soon it was supplying doctored and counterfeit papers to customers like the Algerian instructors at Khattab's camps, who used them to relocate to Western Europe; a Turkmen sabotage instructor named Alisher Saypullayev, who returned to Central Asia with a new identity; a Saudi Arabian named Ali Abdul-Rahman, from Mecca, who became Abdurakhman Timikhanov so that he could operate in the West; and Hamzallah, another Saudi Arabian, who became Khamkhat Ibragimov before vanishing in the West. While some of these "new Russians" were deployed to Moscow and other Russian big cities in order to provide expertise to the local Chechen terrorist networks, most of them used their new identities to pass undetected into Western Europe and Central Asia, where they became instrumental in creating an intelligence and sabotage network that would transform the world of terrorism.

Chapter 12

WAR AGAIN

IN THE EARLY SUMMER OF 1999, ISLAMIST OFFICIALS TREATED THE resolutions of the Second Congress of the Peoples of Ichkeria and Dagestan as a de facto declaration of jihad against Russia. They ramped up their systematic infiltration of Islamist assets into Dagestan, building bases in the countryside, and intimidating the population by stealing cattle, cars, and agricultural machinery from Dagestani civilians, continuing their campaign of ransom kidnappings and other acts of blatant terrorism.

Moscow, too, began drawing up its own war plans. But Boris Yeltsin was anxious to avoid an eruption of war in the region, and he urged restraint. Russian authorities expressed an interest in helping Maskhadov bring "sanity" and calm to Chechnya, sending word through official and unofficial channels in a bid to delay the inevitable. The Russian general staff and the Russian army's senior officers in the North Caucasus were convinced that such overtures would be perceived as signs of weakness and would only embolden the Islamist leaders to challenge Moscow. Yet Yeltsin's interior minister, Sergey Stepashin, remained committed to negotiations with Maskhadov, and by the early summer Russia had begun withdrawing its forces from the Dagestan-Chechnya border area, replacing them with local police and paramilitary forces.

In a move that suggested the Russians' desperation, the Kremlin arranged a highly unusual meeting with Shamil Basayev at a villa belonging to financier Adnan Khashoggi in Beaulieu, near Nice, on the French Riviera. The July 4 meeting was arranged by two Georgia natives, Sultan Sosnaliyev, the defense minister of Abkhazia during the Georgian-Abkhazi war, and Anton Surikov, the assistant of first vice premier Yuriy Maslyukov. Also present was a Turk called "Mehmet"—an adviser to Turkey's premier, Necmettin Erbakan. A friend of both Basayev and Khattab, Mehmet was deeply involved in Ankara's support for the Chechens and was in Nice to guard and support Basayev. Representing the Kremlin was administration chief Aleksandr Voloshin.

In the key discussions, which continued into the morning of July 5, the Kremlin offered what it considered generous concessions. But Basayev insisted not just on unconditional independence for Chechnya, but also on the liberation of Dagestan and all other Muslim republics in the North Caucasus, and the meeting ended in failure. Even Basayev's supporters at the Beaulieu summit felt he had already resolved to go to war and wasn't interested in a last-minute negotiated settlement.

Osama bin Laden was taking a personal interest in the coming conflict. Khattab had been in correspondence with bin Laden and other jihadist leaders, including Abdul Qadir al-Mukhtary (a.k.a. Abu-Maali), commander of the mujahedin forces in Bosnia, Hassan al-Sarikhi (a.k.a. Abu-Abdul-Rakhman), a senior commander and spiritual guide of the Saudi jihadist movement who escaped to Afghanistan, and others in Egypt, Algeria, and Tajikistan, exchanging ideas about launching insurgencies and terrorist campaigns. Then, in the last week of July, a high-level delegation of Islamist-Jihadist senior commanders and terrorism experts, most of them Arabs, arrived from Afghanistan and Pakistan for a weeklong visit to Chechnya. At the time, there were persistent rumors that bin Laden himself led the delegation. (These rumors have never been verified, and they may stem from confusion over the name of another senior Pakistani commander whose nom de guerre, Abu-Abdallah Jafar, was similar to bin Laden's, which was Abu-Abdallah.) The delegation spent most of its time in Khattab's Said Bin-Abu-Waqqas training camp near the village of Serzhen-Yurt, conducting inspections and consulting with Chechen and foreign mujahedin commanders. The delegation reviewed Basayev's and Khattab's war plans, made some last-minute refinements, and agreed on a strategy to flow experienced fighters and trainers, equipment, and funds to the region to help sustain the jihad against Russia.

In early August, the Chechnya-based Islamic Army of Dagestan, under the command of Basayev and Khattab, invaded Dagestan and captured two clusters of villages. The mujahedin immediately imposed a strict Islamist regime and began coercing the population to adopt their version of the "Sharia lifestyle." This time, however, the turf wasn't as hospitable for the Islamists. Authorities in both Moscow and the Dagestani capital of Makhachkala were flooded with demands from the population to expel the Islamist-Jihadist invaders. Moreover, a large number of local volunteers joined the Russian forces, spearheading a campaign for the total eradication of the Islamist-Jihadist menace. By mid-September, the Islamic Army of Dagestan was forced to withdraw into Chechnya. Even in retreat, Khattab claimed success, noting that his forces had killed some two hundred Russians while losing only sixty mujahedin, but Russian security sources disputed his claims.

The actual invasion of Dagestan was relatively limited in scope, but as a signal of the Islamists' growing ambition it rattled the entire region. In mid-August, Ruslan Aushev, the president of Ingushetiya, who had earned the Hero of the Soviet Union medal in Afghanistan, reflected popular opinion when he warned about the long-term ramifications of the invasion. "[These] militants, who intruded into the territory of Dagestan from Chechnya, use religion as a cover." The Islamists' rush to proclaim an Islamic state in Chechnya and Dagestan was evidence "that they wish to expand [that state] at the expense of Ingushetia, Kabardino-Balkaria, and other Muslim states of the region," Aushev warned. "Their main purpose is to create a 'Muslim enclave' from the Caspian Sea to the Black Sea."

The Islamist leaders considered the invasion of Dagestan the first act in their Caucasian jihad, designed to liberate the entire region, punish Russia, and provide a launching pad for their global objectives. Basayev did warn against overconfidence, stressing that these were early battles in a "war that will last twenty to twenty-five years . . . to free Muslims from the Volga to the Don." Indeed, for him the real end to the war would come not merely with a unified Muslim Imamat in the Caucasus but only when the Islamists finally "establish Allah's law in Jerusalem."

Significantly, the influential ideologue Magomed Tagayev emerged during this period as a spokesman for Shamil Basayev. By late August, as the jihadist forces were being driven out of Dagestan, Tagayev continued to stress Basayev's regional ambitions. Despite the military setbacks, Basayev "had not given up plans for establishing an Islamic state in Dagestan," Tagayev noted. The Dagestani military operation, he claimed, had been merely "a demonstration of Islamic force" aimed to excite and mobilize believers throughout the entire region. The jihad in Dagestan, Tagayev claimed, had crossed the point of no return.

Ibn al-Khattab was even more explicit. In early September, he declared that his mujahedin remained determined "to liberate the entire Caucasus," despite the collapse of their Dagestan campaign. The Islamists intended to liberate "all Caucasian regions where Muslims live," and to forge them into "a single state as was under Imam Shamil" in the nineteenth century. Khattab also dismissed any concern over the mujahedin's defeat, maintaining that "the Muslims in the Caucasus have forces to solve any issues by military means." He also explicitly threatened to "strike at targets in Russia" to change the political dynamic in the Caucasus. While he admitted that he didn't know "how long the war in Dagestan will last," Khattab emphasized that his troops, like Shamil's, would be "able to fight for thirty years if needed, until their terms are accepted."

The signals from Chechnya left no doubt that the Islamists would keep striking out once they recovered from the defeat in Dagestan. Moreover, official Grozny had drawn a line in the sand in its negotiations with Russia, no longer entertaining any that meant less than an Islamist and Mafiya-ruled Chechnya, with Maskhadov as a titular head. The Russian government wasn't buying it, and soon Russian forces started moving on the Islamist bases in eastern Chechnya that the mujahedin forces had used to invade Dagestan. The jihadists' response was their most dramatic yet: a wild-card gambit targeting the citizens of Moscow.

* * *

The war in the Caucasus reached Moscow with all its fury on the night of September 8, 1999. A couple of seconds before midnight, a massive bomb exploded in the basement of a multistory apartment building at 19 Guryanova Ulitsa. The explosion brought down the entire building, killing ninety-four and wounding more than two hundred innocent civilians. Then, on September 13, at around five o'clock in the morning, another bomb went off in the basement of a similar building, 6 Korpus 3, Kashirskoye Shosse. Once again, the bomb brought down the entire multistory building, killing 124 and wounding several hundred. The next day, September 14, police found and safely defused a third bomb in the basement of yet another apartment building, this one on Borisovskiye Prudy Ulitsa. The bombing of civilians wasn't limited to Moscow: On September 16, at 5:58 A.M., a truck bomb exploded in front of 35 Octyabrskoye Shosse, an apartment building in Volgodonsk. This time the casualties were limited—"only" 18 civilians were killed and 342 wounded, although 69 required prolonged hospitalization—because some of the families received warning phone calls minutes before the explosion and managed to vacate the building in time.

The magnitude of the carnage shocked not only Moscow, but the rest of the world. At first, even the Islamist-Jihadist leaders were silent after the bombings, although they had issued a little-noticed warning on the eve of the first blast. In an interview with the Czech newspaper *Lidove Noviny,* published on September 6, Ibn al-Khattab made an explicit case for terrorism against Russian civilians. Since the jihadists' objective was "the Russians' departure from the Caucasus," he explained, the fighting must be waged "[n]ot only against the army, but against the entire Russian nation, which will be called to account for what is taking place in [the Dagestani village of] Karamakh, for what they are doing to Muslims. For the suffering of the women and children of Dagestan," he charged, "Russian women and children will answer for the deeds of the generals." Khattab contended that Russia would not withdraw from the Caucasus until its civilians grew personally alarmed by the ramifications of war, and that would come only "when they hear bomb explosions at home, when we begin to kill them. . . . We will stop Russia. Russia will cease to exist."

Shamil Basayev had also scheduled an early September interview with *Lidove Noviny*, but his was held after the first bombing in Moscow, and he tried to disavow the attack. In the interview, published on September 9, Basayev insisted that he "denounce[d] terrorism." He assured the interviewer, "The latest blast in Moscow is not our work, but the work of the Dagestanis." But Basayev also warned that the bombing campaign in Moscow was about to escalate. "Blasts and bombs—all this will go on, of course, because those whose loved ones, whose women and children are being killed for nothing will also try to use [similar] force to eliminate their adversaries." As if to support Basayev's position, on September 15 a group calling itself "the Islamic Liberation Army of Dagestan" claimed responsibility for the bombings, with little elaboration or justification.

Basayev and Khattab were eager to paint the Moscow bombings as a spontaneous grassroots reaction to the recent Russian "aggression" in Dagestan. As later investigations proved, however, the attacks had been in the works at least since late July—more than a week before Basayev and Khattab led their forces across the Dagestani border. There are strong indications that the Islamist delegation to Khattab's Said Bin-Abu-Waqqas camp was briefed on the impending attack, and gave its blessing. After the bombs were set in place, on August 30–31, Khattab and his key aides held an intense marathon of phone calls with key sponsors in Jordan, Syria, Saudi Arabia, and Egypt to coordinate plans with the supreme Islamist-Jihadist leaders, and to call for more resources and volunteers necessary to cope with the inevitable escalation that would follow the attacks.

Russian security services eventually identified the key perpetrators of the Moscow bombing, and established their links to Basayev, Khattab, and their followers. The operation's on-site commander was a Chechen national called Achemez Shagabanovich Gochiyayev, a born-again Wahhabist who had attended one of Khattab's training camps. His chief assistant was Denis Faritovich Saitakov, a graduate of a special industrial-technical school who provided technical know-how. Saitakov had known Gochiyayev since 1992, and had traveled to Chechnya in 1997 under his influence to discover "pure Islam." With half a million dollars supplied by Khattab and fake documents bearing the name Mukhita Laipanov, Gochiyayev had rented the basements of the three apartment buildings for "commercial storage" of sugar and other packaged goods. The bombs were made of ammonia-saltpeter, diesel fuel, and aluminum powder, with an expertly shaped charge of hexogen and RDX high explosives as an accelerator. The bombs were fused with a combination of electronic timers and remote-control detonators to ensure redundancy. The power of each bomb was comparable to three hundred kilograms of TNT. The hexogen and RDX high explosives were packed in sugar sacks in a terrorist base in Urus-Martan, Chechnya, and trucked to Moscow in several rounds, stored at rented commercial storage sites along the way until they reached their destination in late August.

Using his brother-in-law Taukan Frantsuzov's identity documents, Gochiyayev settled in a Moscow hotel and personally supervised the placement of the explosives in the basements of the three buildings. Saitakov also seems to have been in Moscow at the time, but both men escaped to Chechnya after the bombings. Three Chechen terrorists—Adam Dekkushev, Yusuf Krymshamkhalov, and Timur Batchayev—prepared the Volgodonsk bombs from the same stockpiles, transported them to Volgodonsk, and randomly picked the apartment building on Octyabrskoye Shosse because it had an easy parking spot for their van.

The operation may have succeeded, but politically the bombings in Moscow backfired for the Islamists. The public had grown impatient with the growing Chechen menace, and the attacks sparked a movement to address the Chechen problem once and for all. The Kremlin dusted off its war plans, and the Russian security establishment—still smarting over the 1994–1996 war—roared into action, bent on revenge. This time, however, the Russian forces would take a lesson from the jihadists: Slowly but surely, they enlisted loyal Chechen natives in the fight to liberate their country from the Islamist-Jihadists and start rebuilding the country's society and economy.

The Moscow bombing did pay some dividends for the Islamist

jihad, in part by focusing world attention on the crisis in the Caucasus. Even though the Russian forces were prevailing in Chechnya, the Caucasian cause had already become a new Afghanistan for the Muslim world, a rallying cry that inspired radicalization. Meanwhile, driven by a desire to control the region's oil resources and pipelines, the United States was increasing its support for the Chechens—an investment that would ironically help to keep the terrorists' cause alive.

On September 20, 1999, Shamil Basayev upped the ante again, announcing the establishment of a battalion of four to five hundred voluntary "Shahids"—would-be suicide terrorists. "These people will be ready and capable of carrying out the most difficult of tasks," he declared in Grozny. In a subsequent conversation with Western European correspondents later that day, Basayev explicitly threatened "to unleash a wave of Islamic suicide attacks on Russia" unless the Kremlin immediately halted all air strikes on Chechnya. Their specific targets? "Time and circumstances will tell," he said.

Chapter 13

THE INTERNATIONALIZATION OF THE JIHAD

THROUGHOUT THAT FALL OF 1999, AS THEY RECOVERED FROM THE Dagestan defeat, Shamil Basayev and Ibn al-Khattab were conspiring with the Islamist-Jihadist supreme leaders to rethink their strategy. In late September, Basayev assured journalist Mahmud Sadiq of *al-Watan al-Arabi* that "attack and retreat and guerrilla warfare" would "become the fighting strategy against the Russians," and he described the Moscow bombings in this light. "Regardless of who was responsible for the recent explosions in Moscow, they worked in favor of the Dagestanis," Sadiq reported Basayev as arguing. "They gave them self-confidence and increased the number of Muslim separatist supporters. The explosions created fear in the hearts of the Russian forces that the battle was no longer in their favor."

The supreme Islamist leaders continued importing personnel from throughout the Muslim world. From October 3 through October 5, 1999, roughly fifty veteran Arab mujahedin—carrying papers from Kuwait, Saudi Arabia, Yemen, and the Palestinian Authority—arrived at Khattab's main camp at Urus-Martan in central Chechnya. The group was able to enter Georgia legally, having been the first of roughly one hundred Arab mujahedin who received visas from the Georgian consulate in Turkey. Once their tactic was exposed, Georgian authorities in Turkey made it more difficult for walk-in Arabs to secure visas—and the Islamists started looking for new, alternate illegal routes into Chechnya and Dagestan.

The supreme Islamist leaders were exerting increasing control over the jihad movement in the Caucasus, and Osama bin Laden agreed to meet the local jihadists' requests for additional troops and materiel. In mid-October, two senior emissaries—a Basayev representative named Azam Kholboyev, and a Khattab representative known as "Mustafa"—arrived in Qandahar to consult with bin Laden and his senior aides, as well as senior Taliban and Pakistani officials. They discussed a wide array of topics, from support for the coming escalation of the war to the evacuation to Afghanistan of key Chechen and mujahedin commanders if Russian forces should complete their occupation of Chechnya before winter set in. The Taliban promised to support whatever arrangements the emissaries agreed on with bin Laden and his commanders.

At this meeting, the Chechen emissaries also expressed concern about an alarming security breach. Russian authorities had recently confiscated a computer belonging to Khattab's interpreter when he tried to have it fixed in Dagestan. The Russians recovered the entire database—including guidelines for organizing mujahedin training camps, details concerning Islamist ideological training and indoctrination campaigns, and a host of evidence that the commanders in Chechnya were in deep and constant coordination with their counterparts in other parts of the North Caucasus and the Middle East. Khattab and Basayev were afraid that the Russians would use this material in order to hunt down key commanders, including themselves—one reason they wanted a promise of safe haven in Afghanistan. The emissaries insisted that these were only temporary setbacks that need not affect the overall progress of the Caucasian jihad. Bin Laden and his key aides were more somber.

Both Khattab's correspondence with bin Laden and his emissaries' trip to Qandahar were marked by a growing desperation. Khattab implored bin Laden to send mujahedin with extensive combat experience to serve as instructors and/or field commanders. At Qandahar, his emissaries complained that many of the recent Arab mujahedin arrivals had proved a disappointment: With only mediocre military training, they had difficulty operating in mountainous terrain, and morale was low among both the new foreign volunteers and the native Chechen fighters who looked to them for inspiration. The Islamist-Jihadists responded immediately, organizing (with the help of the ISI) a crash training program for about fifteen hundred mujahedin in Afghanistan. These mujahedin included recent recruits from the Caucasus, Russia, and Central Asia, as well as Pakistani, Afghan, Turkish, and Arab volunteers. They would constitute the core of a new force to be committed to battle in spring 2000.

One revelation of the early phases of the Second Chechen War was that the threat of weapons of mass destruction played a large part in the jihadists' strategy. With the direct involvement of high-level leaders in Afghanistan/Pakistan and Sudan, the Chechen Islamists began considering drastic terrorist measures designed to pressure Moscow into abandoning the Caucasus or risk making their civilian population pay an unacceptable price. When the September bombings in Moscow failed to create a grassroots pressure on the Kremlin to withdraw from the Caucasus, the Islamists were anxious to raise the ante—and soon the specter of weapons of mass destruction began appearing in Islamist circles.

Even as Shamil Basayev was making threatening noises about his new battalion of suicide bombers, several senior terrorist commanders began alluding to the use of WMD in future terrorist strikes inside Russian borders. In mid-October, Salman Raduyev vowed to use radioactive material and chemical or bacteriological weapons to carry out several "acts of retribution" within Russia. Later that month, Ibn al-Khattab warned Russia that it would face "a different sort of warfare" after its troops entered Grozny. In communications with Islamist leaders throughout the Middle East and South Asia, Khattab, Basayev, and other senior commanders hinted at a new round of unprecedented attacks against both Russian forces and civilians deep behind the Russian border. And these were not empty threats: Written instructions on "the use of bacteriological weapons" were found on bodies of Chechen and other foreign mujahedin killed in both Dagestan and Chechnya. Vakha Arsanov, the Chechen vice president, hailed the existence of a Chechen sabotage group whose future operations "will force Russia to meet all the conditions set by the Chechen Republic."

In late October, the Chechen military command completed preparations for a new round of fighting they code-named "Jihad-2." The objective of this operation was to "complete the expulsion from the republic's territory" of all Russian forces and presence. One of the cornerstones of Jihad-2 was to be a series of terrorist strikes against transportation, water supplies, food industry combines, and food industry outlets on Ingush and North Ossetian territory—and the operational plans specifically called for "chemical or radiation weapons" to be used in these strikes.

At the same time, Chechen forces were building traps for Russian troops in the northern and northwestern outskirts of Grozny that may have involved WMD. Starting in late October, Russian intelligence observed numerous people dressed in chemical warfare protection suits and gas masks burying barrels and tanks, some as large as railroad cars, in the ground along likely Russian axes of advance. Other barrels were placed near bridges and road junctions on the approaches to Grozny, where men in protective suits and gas masks were observed pouring an unidentified liquid into the barrels. Russian military officials believed that the Chechens planned to blow up these tanks and barrels ahead of the advancing Russian forces, triggering some kind of chemical reaction that would threaten those troops uninjured by the blast itself. The Chechens were also training a small force to work with gas masks, reportedly preparing them to use either yperite (mustard gas) or an unidentified "nerve gas" against Russian forces attacking Grozny.

The Islamist-Jihadist mujahedin, particularly the forces answering to Khattab, were also known to be pursuing radiological weapons. In September 1999, three Chechen mujahedin tried to steal a container with nuclear waste from a Soviet-era storage known as the "112th workshop." The terrorists mishandled the containers, however, exposing themselves to radiation; two of them died immediately, and the third agonized in the hospital for weeks. The storage site in question was filled with similar containers.

Another key potential source was a burial site of cesium and cobalt isotopes located northeast of the Tolstoy-Yurt settlement and south of the village of Vinogradnoye. This so-called "tomb," only fifty-seven percent full, housed solid radioactive waste. In late October 1999, Khattab announced his intent to build a camp atop one of these burial sites, convinced that the Russians would never bomb the camp for fear of an ecological disaster. He was right. "Our aircraft and artillery will not hit these geographical coordinates," Major General Boris Alekseyev noted, although he counseled that an act of terrorism there would generate only minimal exposure to radioactivity, "because the radioactive waste [there] is medium- and low-level." During their occupation of the site, the Islamists removed a substantial number of these containers to unknown locations.

The Chechen jihadists never used any of the WMD they had accumulated against the Russian forces. The collapse of their defensive lines in key cities caught them by surprise, forcing them to leave their heavy equipment and preparations behind as they withdrew to the countryside. In Grozny, the Russians kept their forces in constant motion; an explosion involving WMD would have had only a limited effect on Russian forces, while devastating the unprotected civilian population. It was a risk even the mujahedin commanders weren't willing to take.

Ultimately, the most important outcome of the October meeting in Qandahar was the formal internationalization of the Chechen jihad. The Islamist-Jihadist supreme leadership now took concrete steps to clear the way for a campaign of international terrorism against Russian and Russian-related objectives worldwide.

Among the key points of discussion at the Qandahar meeting was the resumption of terrorist activity within Russia. After reviewing recent intelligence collected by Islamist operatives throughout Dagestan and Russia, the terrorist commanders in Chechnya were convinced that expert terrorists would be able to carry out spectacular operations both in the Caucasus and in Russia's main cities. Their findings were eagerly communicated back to the appropriate authorities in Afghanistan and Pakistan. Impressed with the intelligence reports from the Russian cities, the Islamist leaders agreed to send veteran terrorists for the job. Bin Laden and his commanders also raised the possibility of launching terrorist strikes against Russian objectives in the West and elsewhere, particularly in revenge for a possible Chechen defeat. Starting in late

October 1999, these operations were undertaken by bin Laden's own networks, since the Caucasian jihadists had no such capabilities.

On October 21, 1999, the Islamists' efforts to expand the Chechen jihad into a global cause received a vote of support from the highest Islamist authorities. At the request of the uppermost echelons of the House of al-Saud, a Saudi High Islamic Court chaired by Sheikh Ibn Jibreen declared a fatwa—an official endorsement by the government of Saudi Arabia and the House of al-Saud.

In the fatwa, Sheikh Ibn Jibreen explained that it was both permissible and imperative to provide comprehensive support for the Chechen revolt. He stressed the anti-Islamic character of the war, charging the Russians with "killing, displacement, destruction and oppressive harm which has resulted in the death of many thousands simply because they are Muslims, believers in Allah the Almighty." Jibreen stressed that Muslim brotherhood compelled all Muslims to "help each other and co-operate in that which causes them victory and strength and repelling the deception of their enemies." On the basis of these principles, Sheikh Ibn Jibreen extrapolated the formal Islamic policy toward Chechnya:

> So from these evidences, it is obligatory upon the Muslims to:
>
> First: To supplicate for their Brothers (and Sisters) in that land for victory to overcome their enemies;
>
> Second: To supply them with weapons and real power with which they are able to strive and kill their enemies (on the battlefield);
>
> Third: To strengthen them with financial donations, as they are in desperate need of food, clothing and all that (which assists in becoming) strong and prevents (reduces) the pain of striving and harm, and treating anyone amongst them who has been (struck with) injury or is in pain.

Jibreen went further, decreeing that whoever did not support the Chechen cause was essentially non-Muslim—a major threat in the conservative parts of the Hub of Islam. "So, it is upon the Muslims to take interest in the affairs of their Brothers (and Sisters), since whoever does not do so, then he is not from amongst them (i.e., a Muslim)," he decreed.

Given the sensitivity of Jibreen's ruling, the House of al-Saud began deliberating the ramifications of adopting an actively pro-Islamist policy in the Caucasus, postponing its formal release. Although Riyadh tried to stifle the existence of the fatwa, word of it spread swiftly in Islamist circles all over the Muslim world. By the time of its formal release, on February 25, 2000, the fatwa reflected the essence of the real Chechnya policy of Saudi Arabia.

Not to be upstaged by Riyadh's impending fatwa, the supreme Islamist-Jihadist leaders moved quickly to outperform their Saudi counterparts. On October 31, Sudan's spiritual leader, Hassan al-Turabi—the most important theological authority of the Islamist movement—endorsed an escalation of hostilities in the Caucasus, and issued a Popular Islamic Conference (PIC) resolution clearing the way for worldwide Islamist terrorist attacks as revenge for the oppression of Muslims in Chechnya. In its communiqué, the PIC defined the Russian bombing campaign on Chechnya as "genocide," a campaign of "ethnic cleansing" against the Chechen Muslims. The communiqué deplored "the complete silence" throughout the West, attributing the situation to "a victory by interests over principles." The international community's complacency had "aggravated" the situation in Chechnya, the PIC charged, and should therefore share the blame for the Muslims' plight. PIC ended with a call to all Muslim nations and individuals to pressure Russia to end the strikes and withdraw from the Caucasus.

The first major step in implementing Turabi's decision came in early November, when the Islamist leaders in Western Europe called for a formal declaration of jihad on the following Friday, November 12. Among the participants was Sheikh Omar Bakri Muhammad, one of Osama bin Laden's senior representatives in the West. "The Muslims of Dagestan and Chechnya have made it clear that their objective is to defeat the foreign occupying forces and to defend their lives, honour and wealth in order to put the authority back in the hands of Muslims," the call to jihad declared. "Islamic Movements and Muslims are obliged to support them in their *Jihad,* whether that support be financial, verbal or physical. . . . [T]he fascist rape of Muslim land and people by the oppressive Russian regime has made it vital that Muslims worldwide take immediate action to stop more killings." But the Islamists' language made clear that they were driven by more than humanitarian concern for the plight of civilians in the Caucasus. "We cannot delay supporting the *Mujahideen* and Islamic movements who are working to establish the rule of Allah on Earth i.e. *Al-Khilafah* wherever they are, so Muslims can once more have a shield behind which to fight and a leader who will protect their belief, life, honor, mind and wealth," the invitation concluded.

Though it came from Western Europe, this call for jihad in support of the Chechen cause quickly reverberated in the United States. One of the key groups organizing "real" assistance to the Muslims in the Caucasus was the combined Jamaat al-Muslimeen and al-Gammaa al-Islamiyya, a pair of organizations founded by Ayman al-Zawahiri, bin Laden's key military commander, and Sheikh Omar Abdel Rahman, the spiritual guide of the New York Islamist terrorist network in the early 1990s. Since the 1980s, both groups (in one form or another) had been involved in recruiting volunteers and sending assistance (in the form of satellite radio-telephones, GPS sets, and other not-exactly-humanitarian equipment) to jihadis all over the world.

In early November 1999, Jamaat al-Muslimeen used e-mail to issue a call to action in English. (The phrasing of their communiqué suggests that they were constrained by U.S. laws concerning incitement on the Internet and recruiting Americans for participation in foreign wars.) However oblique its language, the communiqué was clearly an invitation to Muslims in the United States. "Chechens are a brave Islamic nation and will make the Russians pay heavily for their aggression. The question arises: what is the role of the international Muslim community in this struggle?" Surveying the various political and charity options available to Muslim activists, the e-mail urged American Muslims to contact the Jamaat al-Muslimeen to learn what else they could do.

Other American Islamists, apparently less constrained by such laws, urged their followers explicitly to join the jihad. One of these American-originated missives was distributed through one of the key Islamist venues in the United Kingdom, suggesting that it had been approved by the high Islamist leadership. "This is a call to every able member of Muslim *Ummah* to rise for *Jihad* against Genocide and War committed by Russians against your brothers and sisters in Chechnya. . . . Please act upon it and ask other brothers in your community to join this call for *Jihad*. We must all unite together and support our brothers and sisters in Chechnya, both physically, financially, and materially and take lesson from Muslims experiences in Afghanistan during Russian invasion."

And there were other such calls, including some that were issued from the safety of Pakistan and other Muslim countries by organizations with ties to the West. One of the most important was a ruling issued by a key religious leader in Pakistan who was responsible for the recruitment and indoctrination of would-be mujahedin. It was distributed all over Western Europe and the Middle East through Sheikh Omar Bakri Muhammad's U.K.-based al-Muhajiroun network.

This message made the case for a worldwide terrorist campaign in the name of the Chechen cause. The jihadists argued that many Muslim countries were too afraid of the United States to provide tangible—that is, military and financial—assistance to jihadi causes. Therefore, the logic went, Muslims all over the world must take action against both the Western nations and their own pro-Western governments. "The cause of the repeated occupation of Muslim lands like Chechnya, Kashmir, Palestine, etc., is the absence of the *Khilafah* system," the message read. "The present Muslim rulers, in truth, are one of the causes of the occupation. Their presence has removed the shield of the Muslims . . . leaving us exposed to the murderous *Kuffar* dogs. And until the *Khilafah* is established, the Muslim leaders will never act."

On November 12, roughly three hundred Muslim men and one hundred Muslim women and children attended an emotional call-to-arms meeting in London. The meeting began with the showing of a video on the jihad in Chechnya and Dagestan, replete with scenes of dead mujahedin. The commentator on stage suggested that the serene expressions on their faces showed that they were prepared to enter paradise as martyrs to Allah and Islam.

The key speakers were Sheikh Omar Bakri Muhammad, Abu-Hamzah al-Masri (born Mustafa Kamil, a key commander of bin Laden), and an American imam from Chicago. Their speeches mingled hate-filled diatribes against the Russians and Americans with urgent calls to support the jihad in Chechnya. All the speakers sought support of three kinds: "1. Cash, 2. Advocacy, and 3. Men to go to war." They urged their listeners to join "the most experienced and brave fighters in the world," emphasizing the sense of comradeship and inner peace among the mujahedin and the lure of possible martyrdom. The women were exhorted to support and encourage their men to commit to the jihad, because their reward would be in heaven. The crowd interrupted repeatedly with chants of "Russia now, America next!" and cheered as the formal declaration of jihad was read.

The declaration issued at that London gathering was the first official statement of the Islamist movement in a Western country. It left no doubt that the movement advocated terrorism and other forms of violence, warning the Russians against "engaging in this struggle against Muslims," and "remind[ing] them not to forget their fate in Afghanistan and in Chechnya in the past."

On November 14, after the United Nations declared sanctions on Afghanistan because of the Taliban's sheltering of Osama bin Laden, al-Muhajiroun issued a follow-up to the London declaration. This message argued that "international law (UN) is merely a tool for America's foreign policy in the [war] against Islam and the Muslims," whose key target was "Sheikh al-Jihad Osama bin Laden." It concluded with a succinct summary of the state of the jihad against the U.S.-led world order:

> It is about time that we as Muslims realize the cold fact that the U.S. is truly at war with the entire Muslim world, not just Sheikh al-Islam Osama bin Laden or the Taliban, as well as in Iraq, Saudi Arabia, Palestine, Kashmir, Dagestan etc. All are vivid examples of the covert war being waged against the Muslims via international law/organizations. The US (UN) has declared war on the Muslims and we have declared Jihad on them. The mujahedin like Osama, Khattab, Samil, will be supported financially, verbally, and physically by the *Ummah* of Muhammad.

It was another milestone: the first public declaration that tied Osama bin Laden, Khattab, and Shamil Basayev to a single, common, and joint Islamist jihad.

There were immediate reverberations throughout the Muslim world. In the next few weeks, leaders of numerous Islamist movements—particularly militant and terrorist organizations—publicly expressed solidarity with the Chechen jihad and urged the Muslims everywhere to mobilize in support. Their conclusion was nearly unanimous—namely, that only a worldwide armed jihad could save the Chechens from the Russian onslaught and liberate the Caucasus.

Several Islamist religious leaders pointed to the forthcoming holy month of Ramadan as a source of inspiration for believers to commit to the Chechen jihad. This call to arms reached even the most remote communities. "*Alhamdulillah*," wrote an Australian Islamist leader, "the mujahedin have inflicted heavy casualties on the Russian crusaders. . . . With the blessed month of Ramadan approaching, we remind all Muslims to supplicate for the mujahedin in their *salat* and *du'a*, and to invoke *Allah*'s wrath against the Russian army and the Russian people. We pray that He curse them with disease, doubt and destruction and that He punishes and disgraces them at the hands of the mujahedin."

In early December, a number of Islamist groups took a series of concrete steps to focus this international uprising. First to act was the World Islamic Front [al-Jabhah al-Islamiyah and al-Alamiyah], which released a statement from Beirut urging its supporters to attack Russian targets around the world. The World Islamic Front was a name used by a coalition of several Islamist terrorist organizations including HizbAllah, the Sharia Council [al-Majlis al-Shari] of Pakistan, the World Islamic League, the Supporters of the Khilafah Group [Jamaat Ansar al-Khilafah], and the Supporters of Taliban Movement [Harakat Ansar Taliban]. The leaders of the World Islamic Front were fond of flaunting their close affiliation with Osama bin Laden and their support of his jihad cause.

"We call on Muslims everywhere to perform their duty toward their Chechen brothers," their statement declared—a duty that explicitly involved terrorism against Russian targets. "If the aggression on Chechnya is not halted and the Russian forces do not withdraw within 48 hours," the group warned, "we call on Muslims in Russia and throughout the world to deal with Russian targets: airplanes, embassies, and government centers." The group alluded to a fatwa authorizing such actions, calling on Muslims to "retaliate by staging the *Jihad* for the sake of *Allah*'s cause. Therefore, we urge the Russian forces and the Russian government not to go near Muslim countries, whether in Chechnya, Uzbekistan, or elsewhere."

The fatwa in question, issued by bin Laden ally Sheikh Omar Bakri Muhammad in his capacity as the Supreme Judge of the Sharia Court, was read in mosques around the world on December 10, 1999. This fatwa formally obliged all Muslims to answer the call for terrorism against Russian targets worldwide, including the assassination of President Yeltsin and Russian military commanders. "The naked aggression and atrocities committed against the people of Chechnya, such as bombing them, throwing them out of their homelands and threatening to exterminate them all if they do not evacuate their homes by Saturday etc. is a violation of the sanctity of Human Beings and their honor which requires an immediate and comprehensive solution including putting the perpetrators of these crimes on [*Sharia*] trial." Of course, the fatwa itself already constituted the verdict of any such trial.

To participate in this obligatory jihad, the fatwa clarified, Muslims need not go to Chechnya: They could join the anti-Russian terror campaign in their own countries. Sheikh Muhammad did issue an interesting caveat advising against attacking civilian targets: "Russian forces, government, Russian Embassies, military airports and jets etc. . . . are legitimate targets for Muslims wherever they may be. . . . However, it is prohibited to target any non-military or innocent Russians because this would be considered to be murder and therefore aggression against the sanctity of Human Life." The lives of Russian leaders, on the other hand, didn't deserve quite the same protections: "The Islamic verdict regarding President Yeltsin and his commanders is for them to be put on trial in any independent Islamic court for their crimes and for capital punishment to be applied."

Reactions to Sheikh Omar Bakri Muhammad's fatwa came swiftly from throughout the Muslim world. One of the first to express support was Abu-Hamzah al-Masri, the chief of the London-based Supporters/Partisans of the Sharia, which issued a statement entitled "Religion's Ruling on Apostate Russians' Aggression on Muslims" over the signature of its spokesman, Abu-al-Layth. "Muslims are being slaughtered in Chechnya during the month of Ramadan for the second time, merely because of the people's interest in religious matters and their willingness to accept the *Shari'ah* and spread *Allah*'s message to their neighboring states," the statement claimed. "To the Eastern and Western camps, this approach [to being religious] is called terrorism and extremism, and thus all the enemies of the region believe that Muslims' blood should be shed, their property should be destroyed, and their honor should be violated, and that excuses should be fabricated to obstruct the establishment of the region." Such a situtation could not be permitted to continue: "Anything that is associated with Russia should be given the same ultimatum given to the Muslim people of Chechnya, then the Russian infidels should be killed in every part of the earth, and their buildings should be destroyed until the Muslim people conclude a truce with them through a sound legal contract."

From London, Abu-Hamzah al-Masri also commented on the fatwa, urging a global campaign aimed not only at Russia but at all other states that would accord it legitimacy—including Muslim states. "We as Muslims should exert pressure on the states that have relations with the Russians, and if these states fail to support the Chechen Muslims, groups or even individuals should be assigned to join the Muslim fighters' battalions, or instead, they should support them financially and by praying for them." But Abu-Hamzah al-Masri clearly defined the Muslim world's primary objective: "The Russians should be expelled from the Muslims' land, because they cannot be trusted. They should be given the same ultimatum that the Russians have given to the Chechens to leave the capital Grozny." Given abu-Hamzah's serious legal problems with the British government over public incitement for terrorism, this was a most serious threat.

As these statements suggest, by late December 1999 the jihad in Chechnya had become a central facet of the global jihad advocated by Osama bin Laden and the Islamist leadership. Back in late July, bin Laden had declared that "the United States' journey toward destruction will start before the arrival of the 21st century, because this [will be] the century of Islam and therefore the Muslim *Ummah* should declare a *Jihad* against the United States." He identified India and Russia as key U.S. partners in the conspiracy against Islam. On December 24, bin Laden's emissaries read a special message during the Friday sermon urging the escalation of the jihad against "the main enemies of Islam"—specifically identified as the United States, Russia, and India—in Kashmir, Chechnya, and Central Asia. Bin Laden's note was uncompromising: "Anybody disobeying the instructions of the clergy to carry out their decrees against the infidels will be regarded as apostates." This transformation of the Islamist jihad from a voluntary event to a compulsory undertaking was a milestone of paramount importance.

Other Islamist leaders rushed to add their voice to the call for jihad in Chechnya. On December 25, Sheikh Omar Bakri Muhammad delivered a major speech by phone to a conference in Mozang, Pakistan. The address was an attempt to define the significance and context of Islam's ascent and to outline a path toward attaining its objectives. "Islam is not a religion, rather an ideology (complete comprehensive way of life) that ruled most of the world for nearly 1400 years," he declared, and the contemporary jihad should aspire to the same breadth of influence. "*Allah* sent Messenger Muhammad with the *Deen* to dominate the earth and implement the justice of Islam. For this domination the Muslims need a power base from which the other nations can be annexed. This power base is the establishment of the Islamic State, the *Khilafah* which the Muslims must work for wherever they are."

From Sheikh Omar Bakri's perspective, the political and military struggle against America was at the core of the Islamist jihad—but the expansion of the Hub of Islam and the acquisition of additional safe havens for future operations were the most urgent tangible tasks. And, with other Muslim nations slow to heed the call of jihad, the ongoing war in Chechnya was a kind of test of the Muslim population's readiness to meet the challenges ahead. "Muslims in Chechnya are calling Muslims worldwide for help, but it is shameful and a disgrace to the *Deen* of *Allah* that the Muslim armies stand idly by while their brothers and sisters get butchered. When will the Muslims realize and respond to the call of *Allah*? It is time for the sincere Muslims to take a stand, because our corrupt regimes will never have the courage to do anything apart from filling their bellies."

Chapter 14

A "NEW" WAR

WITH THE AIR FULL OF SUCH INCITEMENT, IT WAS ONLY A MATTER of time before the Islamists struck out against a Russian objective overseas.

The decision to do so was signaled by the World Islamic Front's reestablishment as a London-based political front, which was formally announced on January 1, 2000. The front's declared objective was "cooperating in charity and piety and devoting resources and efforts to supporting *Jihad* in Palestine, Kashmir, South Lebanon, the Caucasus, Burma, and Afghanistan." Regardless of its London address, however, the declared signatories of the World Islamic Front were all foreign entities—the Islamic Partisans League (Usbat al-Ansar al-Islamiyah) in South Lebanon, the Kashmiri Partisans Movement, the Islamic Salvation Front in Pakistan, the Taliban Partisans Movement, and the Caucasus Jihad Movement—groups dominated largely by Arab "Afghans." Islamist sources in London noted the absence of British signatories, calling the organization "an information facade which is more concerned with collecting funds and donations for the mujahedin."

Two days later, on the morning of January 3, Ahmad Raja abu-Kharub (a.k.a. Abu-Ubaydah), a lone thirty-year-old Palestinian from the Ayn al-Hilweh refugee camp in southern Lebanon, seized the Safa building in Beirut's Mazraa district. Then, from a window in the building, he fired four rocket-propelled grenades and a burst of small-arms fire at the nearby Russian embassy building, causing slight damage but no casualties. When Lebanese security forces rushed to the Safa building, he seized and briefly held a woman hostage. Kharub was killed after an hourlong battle in which a policeman was killed and seven wounded. Kharub had a note in his pocket reading "I martyred myself for Grozny." His hostage later elaborated on her captor's motives: "He wanted to die a martyr and said it pained him not to have killed a Russian before he was shot."

Kharub's futile attack, it later emerged, had been a remnant of a more complex initiative that had been aborted before completion. He had belonged to the Islamic Partisans League, which controlled the al-Tawari refugee camp, situated between the Palestinian Ayn al-Hilweh camp and the Lebanese-populated Ayn al-Hilweh development district of Sidon. The league's leader—Abd-al-Karim al-Sadi (a.k.a. Abu-Muhjin)—and its key commanders lived in the al-Tawari camp. Under Abu-Muhjin's guidance, the Islamic Partisans League had been preparing plans for strikes against Russian objectives for more than two months, and several teams had been dispatched to Beirut to review their options. Roughly sixteen members of the Islamic Partisans League, including Kharub, were indoctrinated in preparation for the attack. According to an Ayn al-Hilweh-based Palestinian source, these terrorists spent two months "mobilizing the Palestinian street and their supporters against the Russians. They preached that defending the Muslims of Chechnya is a 'divine duty and a major *Jihad*.' They said that 'anyone who kills a Russian goes to paradise and is treated like a martyr.' "

The original plan involved detonating a car bomb in front of the Russian embassy around the first of the year. Shortly before midnight of January 2, however, something went wrong with the device's core detonator, and it exploded while stored in a perfume store belonging to Muhammad Hassan Dahir in the al-Tawari camp. The league rushed to suppress the incident, dismissing the explosion as the result of a commercial dispute. It also managed to keep outsiders—including Lebanese security forces—from accessing the store until all incriminating evidence was removed and destroyed. Kharub, who had prepared himself for martyrdom in the operation, was so agitated by the lost opportunity that he mounted his own one-man jihad the following day.

Kharub's strike was a shadow of the group's original plan, but it was sufficient for al-Muhajirun to release a preplanned statement in London. A representative attributed the attack to the Islamic Partisans League, rather than Kharub as an individual, and called it the first of many such attacks to come. "It is clear that Russian Federation servicemen, buildings and interests cannot be safe in any area of the world until the war against our brothers and sisters in Chechnya is completed," the representative stated. A Palestinian source noted that "a number of Usbat al-Ansar supporters" had already been sent to Beirut and other regions to carry out revenge attacks against the Russians."

Even as the worldwide jihad movement focused on expanding its terrorist ambitions, however, conventional forces were once again squaring off to fight for control of the Chechen nation. The Second Chechen War began in the fall of 1999, when the Russian military mounted an offensive campaign to destroy the terrorist haven that Chechnya had become, and the military phase of the operation was completed successfully by the spring of 2000.

The fighting in Chechnya can be broken down into four phases.

Phase one began with the containment of the Islamist forces after their eviction from Dagestan. In September 1999, Russia deployed forces all around Chechnya, except along its border with Georgia. Late that month the Russians began gradual advances, particularly in northern regions, and by mid-October they had established a "safe zone" on the Terek River, which they would use as a base for future surges.

The primary goal of phase two was to surround Grozny. In the late fall, the Russians dispatched two units on a slow but determined advance toward the city from the east and the west, reaching the city in late December. In the process, the troops cleared several Chechen cities, destroying rebel bases and safe havens along the way. The Islamists deployed mujahedin forces into villages to attract Russian fire onto the civilian population. Mindful of the heavy losses it had suffered in hand-to-hand combat in the first war, the Russian forces ignored the human shields and used heavy fire—tube and rocket artillery as well as aerial bombing—to attack the rebels in their headquarters. Local leaders in many Chechen cities and villages—including Gudermes, Chechnya's second-largest city—responded by turning on the mujahedin forces, evicting them from their midst, and then surrendering to the Russian forces. Meanwhile, Russian special forces conducted several heliborne raids in southern Chechnya, slowing the flow of mujahedin and weapons from Georgia.

Phase three was the siege on Grozny, which lasted from late December 1999 to early February 2000. This phase saw the most intense urban fighting of the war. Already devastated by the 1994–1996 war and never really rebuilt by Maskhadov's government, the city was blasted continuously with artillery and aerial strikes. Under the cover of heavy strikes, five main units tried to penetrate the center of the city; at first they met with well-organized ambushes and sustained heavy casualties, but the aggregate impact of the Russian fire offensive and the resilience of the ground units finally overwhelmed the Chechen resistance.

Then, far more suddenly than it began, phase three ended—and with it the Chechen leaders' ambitions to drive the Russians from their country through conventional warfare.

The end came on February 6, when the Islamist leaders and elite forces attempted to flee on foot in the dead of night using an open corridor to the mujahedin stronghold of Alkhan-Kala. But it was a trap: The mujahedin walked right into a series of dense minefields, suffering heavy casualties as they struggled to get out of the killing zone. Shamil Basayev himself lost a foot that night, and his veteran Chechen units, which had endured the fighting of 1994–1996 and 1999, were devastated beyond recovery. Phase four—which has involved an elaborate combination of mop-up operations, scattered incidents of mujahedin resistance, and Russian "heart-and-mind" operations designed to rebuild this devastated zone—began in earnest after the Chechens' military defeat and effectively continues to this day.

The nerve center of this final phase of the war was in mountainous southern Chechnya, where hundreds of mujahedin retreated during the winter of 1999–2000. The following spring, as the snow melted and the weather improved, Russian troops took special pains to prevent mujahedin and terrorist teams from descending from the mountains to mount new strikes in the Caucasus or Russia itself. One challenge the Russian special forces faced in this phase was the Georgian government's inability, or unwillingness, to stop the flow of Islamist volunteers, weapons, and funds through its borders into Chechnya. The availability of safe havens beyond the reach of Russian forces enabled the Chechen jihadists to persist despite the effectiveness of the Russian military and security services.

In a late March 2000 e-mail, nearly two months after the Chechens' defeat at Grozny, Ibn al-Khattab answered a series of questions from journalist Mahmud Sadiq of *Al-Watan al-Arabi*. "Today, the mujahedin are in a better position than before," he claimed. "Although the Russian forces have captured several major cities, they have become the target of our strikes due to their presence on flat territory, which is open to our attacks. We consider that the fighting has started now," Khattab explained. The terrorist commander put on a bold face, insisting that his forces' defeats had been merely tactical maneuvers required to confuse and subvert the Russian strategy. But even he had a more somber view of the future of the war he and Shamil Basayev had initiated. "What is important, irrespective of the Russian policy and military strategy," he stressed, was that "there is no military solution in Chechnya. . . . We will continue to defend our homeland and religion, and our plans will change as warranted by the situation in the battlefield."

Their sudden and overwhelming losses in early February had put a sudden end to the Chechen jihadists' dreams of prevailing over Russia in a military conflict. With that option closed, the movement turned full attention to its traditional modus operandi: terrorism.

The Chechens' bond with the supreme Islamist-Jihadist leadership was bolstered during these turbulent months after a surprising turn of events: a sudden downturn in Osama bin Laden's health.

Early in that winter of 1999–2000, bin Laden's health collapsed. His kidneys gave him debilitating pain, which evolved into bouts of depression. By January 2000, his overall situation had deteriorated so fully that his closest confidants feared that he might become permanently disabled or even die. Arab doctors were rushed to Qandahar from the Persian Gulf, but the complex treatment they prescribed could not be performed in either Afghanistan or even Pakistan. Faced with this crisis, Dr. Ayman al-Zawahiri—once among Egypt's leading pediatricians—resigned from the leadership of the Egyptian Islamic Jihad to devote himself full-time to the bin Laden organization, and to Osama bin Laden's health.

In January, Zawahiri arranged for an Iraqi doctor to examine bin Laden in Afghanistan. The Iraqi doctor prescribed a treatment involving dialysis, a series of shots, and intravenous infusions, as well as an assortment of rare medications. A thorough search of Afghanistan led to the discovery of a Soviet dialysis machine and related equipment in the basement of a now destroyed Kabul hospital originally built for former president Mohammad Najibullah and the Afghani governmental elite. Zawahiri then called on his longtime connections with the Chechen Mafiya, which dated back to the early 1990s and were cemented by his late 1996–early 1997 clandestine visit to Chechnya and Dagestan, to find and quickly deliver the spare parts for the Soviet dialysis machine and other medical equipment the Iraqi doctor required. The Chechen Mafiya smuggled the equipment from Russia and a number of Central Asian states to Iran, ostensibly to a Tehran institution affiliated with the Iranian HizbAllah. Zawahiri then sent a three-man delegation to Tehran to deliver the dialysis equipment and some essential medications, along with a local specialist who volunteered to join the team treating bin Laden. When the Iraqi doctor returned to Afghanistan in early February, bin Laden was moved to his forward headquarters near Sarobi, Laghman province (between Kabul and Jalalabad), where the medical equipment was installed.

Led by the Iraqi doctor and Zawahiri, the medical team treated bin Laden for more than a month. By late February, he was strong enough to attend brief meetings with guests in the Jalalabad area. His intensive medical treatment continued, and in late March he managed to make a rare public appearance, attending a high-level meeting with the Taliban leadership in Laghman province to discuss President Clinton's upcoming visit to Pakistan. According to witnesses, bin Laden appeared frail, but he was engaged and well-informed throughout the long meeting. A team of medics and more than a hundred armed guards, mostly Arabs, surrounded bin Laden at every moment.

The Chechen leaders were quick to capitalize on the gratitude felt by bin Laden and Zawahiri. In late March, former Chechen president Zelimkhan Yandarbiyev, Maskhadov's emissary to Pakistan and Afghanistan, was named Chechnya's ambassador to Afghanistan and soon traveled to Kabul. The ostensible purpose of the trip was to inaugurate the new Chechen embassy there, but this was a mere excuse; the Taliban's real capital was in Qandahar, after all, and the Chechens already maintained a group of emissaries there. Far more important was a stop Yandarbiyev made in the Sarobi area, where he met with both bin Laden and Mullah Omar to ask for increased support for the Chechen war effort. Yandarbiyev delivered a special message from the uppermost Chechen leadership to bin Laden, congratulating him on his recent recovery (and, not so subtly, reminding him that he "owed" the Chechens).

The Chechens' message was not lost on the Afghans and Arabs who attended the meeting. The Taliban immediately authorized yet another new recruitment and training effort in Afghanistan and Pakistan for mujahedin to fight in the Caucasus—this one massive in scope. They agreed that most of these would-be mujahedin would be trained in the regular camps throughout eastern Afghanistan. However, a cadre of elite terrorists—a mix of Afghan and Pakistani, Central Asian, and Caucasian mujahedin—would receive intense, highly specialized training in a specially converted military camp called Kargha-1, twelve kilometers north of Kabul. A senior ISI intelligence official confirmed that "Chechens . . . and also people from other [former Soviet] countries" were receiving training in the area. Bin Laden's people and the Taliban also arranged for a new flow of expert terrorists, mainly combat-hardened commanders, to replenish Chechen forces depleted in the latest Russian offensives. Finally, bin Laden tapped a number of expert terrorist networks in Russia and the West to plan further strikes against Russian targets, in a new bid to build domestic and international pressure on Moscow to withdraw from the Caucasus.

Little wonder, then, that by early April 2000 the Chechen Islamist-Jihadist leaders were showing a new confidence in their ability to take the offensive against Russian forces, not only in southern Chechnya but throughout the entire Caucasus, and even within Russia. Basayev and Khattab conferred with senior commanders—both Chechens and volunteer "Afghans" and "Bosniaks"—and issued ambitious orders for a major escalation in and beyond Chechnya. On April 9, Shamil Basayev inspected several recently reconstituted bases and mujahedin units in southern and eastern Chechnya. These forces were completing preparations for what Basayev called "the beginning of the spring-summer military campaign." Praising the mujahedin's fighting spirit, Basayev told local commanders that the Chechen forces now had "enough strength and means to deal a decisive blow to the enemy and destroy it." Basayev rejected any notion of a negotiated settlement with Moscow, stressing that "the enemy will be attacked by all available means until the Russian aggressors are expelled from the territory of the independent Chechen Republic of Ichkeria." He also maintained that "the invaders will have to be first evicted from Chechnya [by force]" before the "peace treaty" President Aslan Maskhadov had mentioned might be negotiated with Moscow.

In some sense, these audacious new plans marked "the start of a fundamentally new war—a full-scale third Chechen war," as Oleg Odnokolenko of the Moscow *Segodnya* called it in early April. Yet this new phase also represented a profound change in the character of the Chechen jihad. For one thing, the Chechen forces were now dominated by foreign mujahedin, who were playing an increasing role in the training and commanding of key Chechen forces. For another, popular grassroots support for the war among Chechens themselves was collapsing. Apprehensive that the slowdown in the fighting might be eroding their power base—and goaded by the supreme Islamist leaders to ramp up the anti-Russia jihad—Shamil Basayev and Ibn al-Khattab were spoiling for a fight, whatever the condition of their troops.

The morale of those mujahedin forces was not as rosy as Basayev had recently painted it. Indeed, the testimony of both foreign and Chechen fighters suggested that it was rather grim. One twenty-year-old soldier called Khussian, who had served under Khattab's direct command, represented a growing segment of Chechen youth—once the backbone of the Chechen forces—who were giving up on the war. In late May 2000, Khussian told a European interviewer in Moscow that he had become disillusioned with the war after six months, three of which he had spent in training and three in a few ambushes in remote mountain sites. Khussian left the mujahedin on his own volition—he gave his interview in Moscow en route to stay with relatives in Kazakhstan—but he was never identified or apprehended by the Russian security forces. Khussian readily acknowledged that he felt more comfortable in Moscow than among Khattab's Mujahedin "because I love it here. Moscow has become so beautiful and modern."

Khussian's description of the situation in Chechnya reflected popular sentiment. "Chechen pride pushed me into the war," he explained, but it had not been enough to sustain him. Such disenchantment with the war and its leaders was found even among the elite mujahedin units under Khattab's and Basayev's direct command. Hundreds of Chechen fighters simply disappeared, melting away into the civilian population in the Russian-held parts of Chechnya or the Chechen community in Kazakhstan. This was yet another reason that the foreign mujahedin were assuming a greater role in the war. "Half of us were Arabs, Pakistanis or Afghans. I spoke Russian to them, or used sign language," Khussian noted. Most young Chechen fighters who left the mujahedin did so because they, like the civilians around them, were tired of the war. Furthermore, the Chechens were losing trust in their leaders and cause. "They hate Basayev and Khattab, but they also despise Maskhadov," Khussian explained. "No one listens to him any more." The few operations Khussian participated in during the winter of 1999–2000 convinced him and his comrades of the futility of the war and the urgent need for a solution based on coexistence with the Russians. "We cannot kill all the Russians. All we have to do is to elect a good president who will look after his people," Khussian concluded.

In the summer of 2000, two letters from a Norwegian mujahed using the name "Brother X," who had been fighting in Chechnya under Abu-Jaafar since the winter of 1999–2000, were circulated in Western

Europe for fund-raising purposes. Perhaps inadvertently, these letters portrayed the mujahedin as being constantly on the defensive, frequently hunted down by Russian forces, and largely ineffective.

"During this time I have seen the signs of Allah in us getting out of impossible situations," Brother X wrote in his June 26 letter. He described the mujahedin as "disparately spread throughout Chechnya, always on the move in the woods/mountains (sometimes villages and the plains) and ready to move at a few minutes notice. Occasionally one spends several days living 'on the move'!—that means that you sleep as you move: boots, jacket and all the other gear on! . . . Food is always an issue: after the Shuhadaa (martyrs), I think that food (or the lack of it!) is what I shall remember most about this Jihad." Mujahedin patrols kept narrowly missing Russian dragnets and ambushes, at times barely reaching safety. They were also repeatedly subjected to shelling and bombings, making movement nearly impossible—especially when wounded had to be carried. Brother X complained that the Chechen villages and towns were filled with "informants/hypocrites" who kept betraying them to the Russian authorities. Several mujahedin, including Arabs and Central Asians, were captured or killed in these raids. The mujahedin, including Arab senior commanders, spent most of their time in remote rugged mountain paths—at times forced to keep moving for days on end, relentlessly pursued by Russian patrols and shelling that he described as "extremely accurate considering the difficult terrain… [On] the few occasions when they tried to approach villages to get food, Russian forces immediately sealed the village…and began house-to-house searches." Brother X praised the close shaves he and his fellow mujahedin had managed, but none of his stories suggested any true effectiveness on the mujahedin's part.

In another letter, from July 4, Brother X described a major mujahedin formation being separated into small contingents in a heavily wooded mountain area, with "a lot of artillery and helicopters around us" and Russian positions blocking all nearby roads and paths. Having barely made it into a seemingly inaccessible gorge, the mujahedin were jumped by a small force of SPETSNAZ and had to flee, only to come under intense small-arms and grenade fire from another Russian ambush. "For the rest of the day there was intermittent fighting, shouting and then the most intense pounding by mortars, artillery and helicopters. The intensity was such that sounds of the many explosions became one roar. Field Commander Abu-Waleed, who had arrived some 300m to our rear with more brothers and ammunition, had to go back. It's incredible to behold—the earth shaking as craters are gouged, huge trees shattering like ice, and the air filled with the rush of shrapnel." The small patrols of mujahedin and Russian special forces kept ambushing and attacking each other while moving through the rugged terrain, with both sides suffering casualties, but ultimately the mujahedin were forced to flee with the Russians in pursuit. The experience gave Brother X a healthy respect for his opponents: "These Russians were not your run-of-the mill conscripts, but professional elite troops whose training and expertise dwarfs that of your typical young mujahid."

As with the previous Chechen wars, the Islamist leaders and local senior commanders—particularly Shamil Basayev and Ibn al-Khattab—considered terrorist strikes in Russia, and conceivably the West, to be their secret weapon. On June 1, 2000, in a clandestine meeting of senior commanders in a village in southern Chechnya, Aslan Maskhadov called for his commanders to launch new offensive operations against Russian forces throughout Chechnya. Ibn al-Khattab reported that the mujahedin were "planning to stage major operations against Russian occupation forces in Chechnya." He anticipated that the next round of major attacks would take place "in the next few days" and would soon be coming "on a daily basis, using innovative tactics to ambush enemy troops and to wipe out their reinforcements." An elite mujahedin unit trained in "the techniques of explosive operations [and] terrorist and subversive acts" had already been deployed to Grozny and locations in Russia. Indeed, a pair of bombings in Grozny and Volgograd a few days before were hailed as the first shots in this new campaign.

On June 29—in a declaration that ran counter to the fatwa issued by Sheikh Omar Bakri Muhammad six months earlier—Basayev and Khattab (signing only as mujahedin "commanders" in Chechnya) reaffirmed that they would continue to target the Russian people if "the repeated violations committed by the Russian military against innocent women and children" were not halted immediately. "Our patience has run out in view of these crimes, and we know that the Russians will continue these crimes unless a clear deterrent stands in their way," Basayev and Khattab wrote. Pragmatically, they demanded only that "the Russian government stops its soldiers from committing crimes against the Chechen people," rather than calling for a complete withdrawal from Chechnya. "We warn the Russian government and its military," Basayev and Khattab concluded, that "the war will expand beyond the Chechen borders and will reach the Russian people" unless their demands were met. The decree was formally approved on the next day by the *shura* (Congress) of commanders.

By late July, the Chechen Islamist-Jihadist leaders were ready to strike. On July 24, Movladi Udugov issued a warning in the name of the "High Command" of "the Caucasus Mujahedin," declaring that they considered "all industrial, military and strategic sites" to be legitimate targets. The High Command also vowed to hold the international community responsible for acts of terrorism because the community was not forcing Russia to withdraw from Chechnya. "If the international community does not view our proposals as fair and acceptable, then the Chechen people have the right to force the Russians to comply with these conditions," Udugov's statement concluded.

The intended cornerstone of the Chechen rebels' new campaign was another potentially devastating terrorist strike. On August 18–19, 2000, with support from Osama bin Laden and the Islamic Movement of Uzbekistan (IMU), the Chechens planned to assassinate Russia's new president, Vladimir Putin, during the summit of the presidents of the Commonwealth of Independent States (CIS) at Yalta, Ukraine.

Initially, the supreme Islamist-Jihadist leaders had planned a coordinated strike against three CIS presidents it considered "enemies" of the jihadist movement: Putin, Askar Akayev of Kyrgyzstan, and Islam Karimov of Uzbekistan. The plan was approved by bin Laden and

Zawahiri themselves, and the highest-quality assets were allocated to the operation, which would involve a major explosion during an event attended by all three of the CIS leaders.

A diverse group of expert terrorists spearheaded the operation. Several Arab expert bomb makers (mainly Saudis and Egyptians) prepared the plans and specialized equipment in Afghanistan. The personnel and equipment were smuggled from Afghanistan via Central Asia by operatives of the IMU. To ensure their security, the operatives were smuggled one at a time, the equipment in small batches using avenues frequented by drug smugglers. Meanwhile, relying on the Chechen expatriate community in Ukraine and the Crimean Peninsula, the Chechen Islamist-Jihadists smuggled large quantities of high explosives and other equipment, collected intelligence, and established a comprehensive support base for the operation. About half a dozen highly trained Chechen terrorists were dispatched to Ukraine to supervise the operation.

Things didn't go as planned. In one of its last stages, the IMU smuggling chain was intercepted by Kyrgyz and Uzbek security services in the Fergana Valley. The IMU terrorists revealed their destination under interrogation, and the security officers immediately tipped off their Russian counterparts (who have a strong presence near the borders with Afghanistan as part of the CIS joint antiterrorism effort). In turn, the Russians warned the Ukrainians, and soon an intense chase developed between the Islamist-Jihadist terrorists and the Ukrainian security forces who were trying to prevent their strike. On the eve of the CIS summit, the Ukrainians arrested four Chechens, all of them graduates of the Khattab-Basayev camps, and a few "Middle Eastern citizens" in the Crimea area. All were swiftly expelled from Ukraine. The timing was so close that, on the advice of their respective security services, presidents Akayev and Karimov decided to forgo Yalta altogether, and Putin spent only a few hours there.

As all sides must have recognized, if the plot had succeeded, the Caucasus and perhaps all of Central Asia would have gone up in flames. At first, Ukrainian authorities decided to keep their discovery secret, and the jihadists quietly pulled back their resources. In mid-September, however, the head of the Ukrainian Security Service, Leonid Derkach, revealed the existence of the plot. The next day, the Islamist website Kavkaz-Tsentr issued a communiqué in the name of Basayev and Khattab, along with its own remark that "the Chechen side has so far not commented on Derkach's statement. However, Kavkaz-Tsentr notes that Putin and some other high-ranking Kremlin officials have for a long time been on the Chechen authorities' wanted list for war crimes and criminal actions," and that "the supreme military Majlis ul-Shura [Congress] of the Chechen mujahedin" had offered a reward for "the apprehension or annihilation of military and criminal offenders from the current top military and political leadership of the Russian Federation." The full text of the Edict of the Shura followed, declaring that the Sharia court had "pronounced the death penalty on the following war criminals, political carrion crows and national scum who actively participated in the genocide of the Chechen nation." The list of wanted Russians that accompanied the edict included a price on each leader's head and specific requirements for claiming each reward. The edict was cosigned by "Amir of the supreme military Majlis ul-Shura, Abdullah Shamil Basayev" and "Military Amir of the supreme military Majlis ul-Shura, Khattab."

Chapter 15

MONEY MATTERS

THE SPRING OF 2000 SAW ANOTHER PROFOUND CHANGE IN THE SUPPORT network that undergirded the Chechen war. Between 1996 and 1999, the jihadist campaign had effectively destroyed the Chechen economy; meanwhile, Russian forces had made great strides in denying the mujahedin access to the population at large. As a result, the Chechen insurgency became completely dependent on two external sources of funding. The first source included contributions from Islamist sources, mainly on the Arabian peninsula. The second source was a steady flow of criminal income—mainly from the drug trade, oil smuggling, the counterfeiting of U.S. and other Western currencies, and so on. To sustain the flow of cash from these sources, the Chechen leaders were forced to accede to the demands of their sponsors.

In the early days of the Chechen rebellion, Dzhokar Dudayev's funding network was founded on an expectation of stability in Chechnya and continuing relations with Russia. His three main sources of money included the local population, who could be counted on to provide sustenance for the fighters (from on-site donations of food and shelter to local fund-raising); the Chechen Mafiya in Russia, which supported the rebels as much for nationalistic reasons as for economic; and a few oligarchs who were attracted by Chechnya's unresolved legal status as a tax and financial loophole. By 1999–2000, however, each of these factors had evaporated. The forced Islamicization campaign of 1997–1999, launched to coerce the population into the ranks of the Islamist-Jihadist forces, had destroyed the Chechen economy, forcing ever greater segments of the population into reliance on fighters' salaries and handouts from Islamist charities. Instead of being a source of moral support and financial sustenance for the insurgency, the civilian population became a financial burden for the insurgency.

The Chechen Mafiya was also declining in its support for the rebellion. Russia and the CIS had undertaken persistent efforts to suppress organized crime, but there were other reasons: As a largely nationalistic, secular, and somewhat Russified syndicate, the Mafiya had little interest in the Islamist-Jihadist dimensions of the revolt. They preferred to conduct their activities in Russia and the CIS and send contributions back to the motherland, but the security services' dragnets against the general Caucasian population, and the public hostility and alienation toward Chechens as a result of the Chechen terrorist acts within Russia, complicated the lives of the Chechen organized crime bosses. The Chechen Mafiya had little incentive to sponsor and fund an insurgency that was detrimental to their own ability to survive and do business. The oligarchs, too, had been willing to subsidize the state of affairs in Chechnya for as long as they could do business there—but as the nation descended into chaos, businessmen were regularly kidnapped for both criminal and religious reasons: When Russian troops rolled back through Chechen borders after the summer of 1999, the oligarchs' loophole was closed, and there was no longer any point in contributing to the Chechen insurgency.

And so Chechnya looked more and more toward its most readily available and willing sources of funds: Islamist charities and criminal networks of narcoterrorism and other forms of organized crime. When the Islamist-Jihadists finally focused on this problem, in the late spring of 1999, they responded with an intense new effort to build a funding system sufficient to sustain the war effort. In the process, a huge new influx of cash was committed to the cause of Islamist jihad throughout the Caucasus—not just in Chechnya.

The procurement of substitute funding to sustain the entire insurgency—and the greater population of Chechnya—was a major organizational and financial achievement. The very fact that the supreme Islamist leaders were willing to accept this challenge underscores the importance of the anti-Russian jihad to the international Islamist movement. Moreover, the Islamists' ability to divert resources from so many facets of international organized crime (ironically, to sustain an Islamist movement that was ostensibly fighting crime—that is, crimes against a just Sharia-based society) suggests how fully the broader movement controls the trafficking of drugs, counterfeit currency, women, weapons, and more between Asia and the West.

By the spring of 1999, as the jihadists' rhetoric was escalating, there was a marked increase in the flow of supplies and funds into Chechnya via Georgia and Azerbaijan. It didn't take long for Russian security services to discover this supply line and even capture some of the shipments. The story of these early seizures reveals how effectively Chechnya was able to use the support of neighboring states to bolster the impression, in political and diplomatic circles, that this new Chechen conflict was a war between two independent states—Russia and the Chechen Republic of Ichkeria (ChRI)—a move that helped garner worldwide support for Chechen independence from Russia.

In mid-April, for example, Russian security services captured a Dagestan train car carrying military equipment and uniforms for two thousand soldiers—including camouflage uniforms, tents, waterproof capes, sleeping bags, uniform stripes, and headwear, all of it marked with Chechen insignia, as well as military compasses, electric flashlights, vacuum flasks, camouflage paint, and holsters. The cargo was manufactured in Lithuania and sold to a Baku company using fraudulent end-user documentation (the Lithuanian manufacturer should have thought twice about the order, ostensibly for the Azerbaijani military, when they learned that it was to be emblazoned with Chechen insignia). The cargo was to be sent from Baku to a road metal plant in the North Ossetian town of Beslan, with documents stating that the gear was for the use of workers there, but a later Russian investigation concluded that the Beslan plant had signed no contracts with the companies and was not expecting the cargo. But the shipping documents from Azerbaijan were genuine, suggesting that the shipments were handled professionally on the Chechen side. In Beslan, the cargo would have "disappeared" into Chechnya.

The flow of weapons, military equipment, and volunteers into Chechnya intensified steadily during this period. While the actual smuggling was conducted primarily along a multitude of mountain passes in northern Georgia, and, to a lesser extent, from Azerbaijan into Dagestan and on to Chechnya, the shipments were controlled from a single command center in Azerbaijan. As the fighting escalated and the Chechens suffered military setbacks, there was a marked increase in the flow of funds, weapons, and volunteers into Chechnya, and in the evacuation of wounded and dependents to safe havens across the border.

By mid-October, the Islamists were expanding their infrastructure in Azerbaijan, to speed the transfer of arriving mujahedin to training and operational bases in Chechnya (most of them going to Khattab's primary base near Urus-Martan), and to expedite new operations against Russia and Armenia (including the breakaway republic of Nagorno-Karabakh, formally a part of southwestern Azerbaijan) in case the Russian counterattack should paralyze the Islamist-Jihadist revolt within Chechnya itself. Among the mujahedin handled in Azerbaijan were numerous would-be Shahids who had trained in bin Laden's camps in Afghanistan and had documents permitting legal travel to Azerbaijan.

The Islamist-Jihadist buildup constituted a major expansion of the covert pipeline that had been running since the winter of 1997–1998, initially designed to smuggle weapons, money, and personnel into Chechnya from Pakistan/Afghanistan. For a while, the Islamist-Jihadists exploited Azerbaijan's relatively lax control over its borders, which saw a constant flow of young foreign men working in the region's vast energy industry. The jihadists traveled primarily by truck from the Baku region through the mountains and into Dagestan and Chechnya, and by light aircraft from northern Azerbaijan into the Vedeno Gorge or to Nozhay-Yurtovskiy Rayon in Chechnya. The weapons transported through these channels included shipments from Afghanistan/Pakistan arriving via Central Asia, as well as large consignments of weapons from suppliers in Ukraine.

In early September 1999, fearing that a quick Russian victory in Chechnya would embolden Moscow to increase its support for nemesis Armenia, several officials urged the government of Azerbaijan to increase its support for the Chechen-Dagestani Islamist forces. The policy was embraced publicly on August 20 in a statement by Vafa Guluzade, an adviser to the Baku government on state policy issues. "Chechen and Dagestani fighting should be regarded as a national liberation struggle, not as terrorism as the Russian authorities are trying to present it," Guluzade declared. "Today," he charged, "Russia is actually continuing in the Caucasus the policy of serf Russia, which in [the] 19th century subjugated with fire and the sword the freedom-loving Caucasian nations. . . . Carrying out a military campaign in the Caucasus today, the biggest campaign after the first Caucasian war, Russia is declaring itself a successor of Tsarist Russia." Having gained their independence after hundreds of years of Russian subjugation, Guluzade argued, all Muslim states of the Caucasus should unite their efforts to compel Russia to "change its policy regarding the Caucasus and other national regions before it is too late."

Azerbaijan could no longer feign indifference to the escalating war just across the Russian border; there were both security and humanitarian concerns to consider. On what was supposed to be a secret visit to Georgia and Azerbaijan in late September and early October, Selim Beshayev, vice speaker of the Chechen parliament, solicited the Baku leaders' support for the smooth flow of mujahedin and specialized equipment via Azerbaijan. But Baku rebuffed his overtures, agreeing to discuss only medical support for wounded mujahedin, as well as humanitarian support and shelter for innocent civilians affected by the war. Nevertheless, Beshayev persisted, using both carrots and sticks in his effort to convince Azerbaijani authorities to expand their direct involvement in the Islamist "cause." He promised lavish "unofficial" foreign aid to Azerbaijan—large quantities of cash from diverse sources in Saudi Arabia and other Persian Gulf states—and reiterated the Islamists' promise to assist Baku in "resolving the Karabakh problem" as expeditiously as possible. In a not-so-subtle threat, he also offered to "guarantee" the security of the Baku-Novorossiisk oil pipeline. Baku stuck to its guns, agreeing only to a myriad of humanitarian activities.

* * *

Faced with this rejection, the Islamist-Jihadist leaders decided to cheat Baku, using a number of key facilities in Azerbaijan that were declared as charity and educational organizations formally supported by the Saudi royal family. The Saudi ambassador endorsed them to the government of Azerbaijan. Needless to say, any affiliation with Osama bin Laden's networks was concealed from Baku. The headquarters of these organizations were filled with Arab "teachers" and "managers" from the ranks of such organizations as the International Muslim Brotherhood, the Islamic Salvation Front (FIS) of Algeria, several branches of Islamic Jihad, and the National Islamic Front of Sudan.

In fall 1999, the key Islamist-Jihadist organizations were headquartered in the Baku area. Their key principals were citizens of Saudi Arabia, Yemen, and Somalia. They had dedicated contact men for bringing in funds—usually in cash—from Islamist charities and financial organizations in Saudi Arabia. These Arab principals had huge amounts of cash in hard currency and they were involved in the acquisition of real estate, among other "educational" projects. By now, these charities began setting up several camps near Baku, ostensibly to allow the student body "to study the Koran in a quiet setting." The primary functions of the camps were to indoctrinate the students—largely Arabs, Caucasians, and Central Asians—in the tenets of Wahhabism, and to train them as professional agitators who could "educate" the Muslim population of Dagestan and other parts of the northern Caucasus, as well as Uzbekistan and Kyrgyzstan, into supporting jihadist causes, subversion, and terrorism.

The camps also housed facilities for producing printed, audio, and video propaganda advocating anti-Russian and anti-Western jihad. Significantly, the camps also engaged in agitation against what they perceived as enemy Muslim regimes, particularly the House of al-Saud (even though senior members of the House of al-Saud were among their most generous contributors). On the wall at one camp appeared the slogan: "The fate of the Shah of Iran, who was driven out of his own country by Islamic organizations, awaits the [Saudi] royal family."

The second phase in the expansion of these Islamist facilities began in late September, when a group of Arabs—all carrying legal travel papers from Saudi Arabia, Sudan, Yemen, and Afghanistan—left the Baku area for newly established "religious field camps" in the remote mountains of northeastern Azerbaijan, on the road to Chechnya and Dagestan. All three Islamist "charities" established such camps around this time; among the leaders and commanders of these camps were several individuals identified by Russian intelligence as "proxies of terrorist Osama bin Laden."

In the last week of September, senior officials of the International Muslim Brotherhood, the National Islamic Front, and several branches of Islamic Jihad arrived in Azerbaijan. The following week they contacted local Islamists, looking to establish new routes to move money, weapons, and mujahedin into Dagestan and Chechnya. Among their priorities was the shipment of Stinger missiles from Pakistan. The emissaries acknowledged that they wanted to make Azerbaijan a "springboard for inserting their main forces [into Chechnya and Dagestan]." They also deposited huge quantities of cash in the accounts of the Islamist charities and camps. From late September forward, tens of millions of dollars were transferred into these accounts, from Saudi Arabia via Beirut.

From there, much of this money was transferred to Chechnya by couriers. For example, after one such camp received an electronic transfer of $2 million, an Arab called Bin-Abḍallah (with United Arab Emirate papers) received the money in cash and immediately carried it across the border into Chechnya. On October 5, when Azerbaijani border guards arrested two Arabs (both with Iraqi papers) near the village of Pashbir, they were found to be carrying $300,000 in cash. (The men claimed to be volunteers working for a charity in Chechnya.) All available evidence suggests that these known cases were but a small fraction of the ongoing shipment of funds from Arab countries to Chechnya, as well as the "Koranic camps" in Azerbaijan.

Another indicator of impending Islamist activities in Azerbaijan was a series of reconnaissance trips by Arab terrorist and military experts, surveying the nation's borders with Chechnya and Dagestan. In the first week of October, for example, a team from the Islamic Jihad traveled twice from Baku to the Azerbaijani-Dagestani border, studying the mountain passes and roads near the border. On October 5, a Turkish citizen named Yegid Rejeb was arrested on the Azerbaijani-Dagestani border, heading for Khasavyurt with a Russian passport in the name of Magomed Sattarov. A veteran of the Turkish military, Rejeb had spent time in one of the Baku area camps.

The harsh Caucasian winter temporarily froze most of these cross-border activities, but by early March 2000 they were renewed with vigor. Though the Russian forces had scored some impressive successes on the battlefield, the Chechen war was far from over. The Russians had been unable to resolve one critical problem: the Afghanistan-style internationalization of the conflict. Once they had been fighting a relatively contained rebel movement within Chechnya; now their enemy was the entire Islamist-Jihadist movement, which was pouring thousands of volunteers and millions of dollars into the movement from all over the Muslim world.

The ongoing conflict in the Caucasus was also expediting the revival of militant pan-Turkism as a nationalist movement. This modern-day pan-Turkism—a movement supported by Turkish official and quasi-official voices—was not an Islamist movement, but rather a grassroots outgrowth of an increasing anti-Western, anti-Russian movement that recalled the numerous Turkish-Russian wars between the eighteenth century and World War I. In a sign of this trend, the Turkish military intelligence service in the North Caucasus began supporting the Chechen and other regional rebel movements, and expanded the mission of its training camps in northern Cyprus to include training for fighters headed for the Caucasus. In this context, Turkish leaders started pressuring the leaders of Azerbaijan—its closest ally—to permit the transfer of these volunteers through its territory.

Baku's dilemma was exacerbated in late 1999, when the United States began extending support to yet another anti-Russian jihad. As if reliving the good old days in Afghanistan, Bosnia-Herzegovina, and Kosovo, the Clinton administration was seeking to support and empower another virulently anti-Western Islamist movement in another strategic region. Despite strong misgivings about the policy, the Azerbaijani leaders found it impossible to resist the pressure from both Washington and Ankara. In mid-December 1999, American officials participated in a formal meeting in Azerbaijan to discuss specific programs for the training and equipping of mujahedin from the Caucasus, Central and South Asia, and the Arab world. As a result of this meeting, Washington extended tacit encouragement to both its Muslim allies (including Turkey, Jordan, and Saudi Arabia, with their intelligence services) and American "private security companies" to assist the Chechens and their Islamist allies in their planned surge against Russia in the spring of 2000.

The Clinton administration's motive was to protect its plans for a proposed Baku-Tblisi-Ceyhan oil pipeline. Back in early October 1999, at a meeting with representatives of Russian oligarchs in Europe, U.S. government officials and representatives of several U.S.-based oil companies had offered to share huge dividends from the proposed pipeline if the oligarchs used their influence in Moscow to ensure that Russian forces withdrew from the entire Caucasus, clearing the way for an Islamic state much like that advocated by Basayev and Khattab. Such an anti-Russian Islamic state would inevitably close down Russia's Baku-Tblisi-Novorossiisk oil pipeline—leaving no competition for the Baku-Tblisi-Ceyhan line, and a flood of profits for all concerned. The oligarchs were convinced that the highest levels of the Clinton administration endorsed the proposal, but it fell on deaf ears, and the meeting ended in failure.

After this initiative collapsed, the Clinton administration resolved that Russia was just as likely to lose its alternate pipeline route due both to the violence and terrorism that were escalating along its borders at the time and to the political fallout from charges of war crimes currently being spread in the media. In Washington's judgment, the expanding war in Chechnya would debilitate Russia in the Caucasus. The Clinton administration was also hotly pursuing a "rapprochement" with Iran at the time, and suddenly it appeared that a southern route for the Baku-Tablisi-Ceyhan pipeline through Azerbaijan (with a detour through Iran to sidestep Armenian territory) might be feasible. This route would make the pipeline economically sound and shorten the time required for its completion.

However, this southern route could be cleared only if the Armenians were evicted from a slice of Azerbaijani territory they occupied in the areas between Nagorno-Karabakh and the Iranian border—and so Washington took a sudden interest in finding a solution to the ongoing Nagorno-Karabakh conflict. Despite the growing political violence in both Armenia and Nagorno-Karabakh, however, the Armenians were reluctant to capitulate. Consequently, the escalation of the Chechen war to include a confrontation with Armenia—long a coveted objective of both nationalist Turks and Caucasian Islamists—became a very tempting option. Moreover, plans for such an escalation already existed in Grozny. Back in late July 1999, Movladi Udugov had vowed that Chechnya would "help the Muslim people of Azerbaijan" in the "liberation" of Nagorno-Karabakh. "Islamic military detachments," he explained, "intend to start a march to liberate the Muslim territories of the Caucasus" the following year.

Encouraged by the shift in U.S. policy, Chechen rebel leaders started hinting in public at a wider regional war. In a late March 2000 interview with al-Hayah, former Chechen president Zelimkhan Yandarbiyev noted that "Georgia and Azerbaijan are offering us substantial support and are providing a haven for our refugees. We are therefore trying to safeguard their security." A Russian grand design was the real motive behind its Chechen policy. "Russia is determined to broaden the scope of the war from Chechnya, if it can, because its aim is to control the Caucasus," he warned. "Chechnya is just the beginning." The Chechens would help Azerbaijan and Georgia confront all their enemies, he suggested—not just Russia, but also Armenia and Nagorno-Karabakh. With its increasingly nationalist streak, Turkey would be unable to resist an attack on Armenia, regardless of whether it was Islamist-Jihadist in character.

Thus, by late March 2000, the Chechen Islamist-Jihadist leadership was considering the potential of a confederate Caucasian state including not only southern Russia beyond the northern Caucasus, but also the territories of Azerbaijan, Georgia, Nagorno-Karabakh, and Armenia. It was a vision explicitly stated by Ali Ulukhayev, the self-declared "Ambassador of Chechnya to Azerbaijan," who reported to Zelimkhan Yandarbiyev, the Chechen president's representative in Muslim countries. In a late March 2000 interview with Elkhan Shahinoglu and Agasi Hun of Baku's *Azadlyg* newspaper, Ulukhayev characterized the war in Chechnya as a direct continuation of the Caucasian Muslims' four-hundred-year long struggle against Russia. The Chechen invasion of Dagestan in August 1999, he argued, could not be considered a sudden provocation. "Everything that has been going on here against Moscow since the Russian empire invaded the Caucasus should be regarded as a national liberation struggle [that] will continue until the Caucasus is liberated from occupation. If this event [in Dagestan] did not take place, Russia would have started the war against Chechnya under another pretext," he explained.

Ulukhayev maintained that the unfolding war against Russia must also be viewed in a regional context. "Ichkeria [Chechnya] and Dagestan are the same nation," he declared. "The Caucasian nations were artificially separated from each other." This precept, he explained, determined the strategic objectives of the war. "Chechens will not be satisfied with the liberation of their own territory." Only a regional solution would satisfy the Chechen Islamist leadership. Ulukhayev stressed that "the freedom of Chechens is impossible until all the Caucasian peoples are liberated. We will drive the occupation army up to the [river] Don. We should liberate the territory from the Don to the Volga, from sea to sea [that is, from the Black Sea to the Caspian Sea] and up to Iran and Turkey from Russia and set up a confederate Caucasian state. If we are liberated from the empire, the Abakhazian, Ossetian, and Nagorno-Karabakh conflicts will be resolved by themselves peacefully." Ulukhayev highlighted the urgent importance of resolving the latter conflict, because "Nagorno-Karabakh always was an inalienable part of Azerbaijan."

As for the desires of the region's people—the non-Muslims in particular—Ulukhayev dismissed them as irrelevant, a simple "result of the Russification policy." According to Ulukhayev, the Chechen Islamist-Jihadist leaders and their allies had already earned the right to determine the fate of all other nations and peoples in the Caucasus. "Today, Chechens carry the burden of the Caucasus-Russian war on their shoulders," giving them droit du seigneur over the region. At the same time, he argued, the war would have to be expanded to other fronts if the Islamists were to defeat Russia. "The Caucasian peoples have no possibility of resolving their problems independently," he argued. However, he continued, "if the Caucasian peoples divide this burden equally, then it will be easy to deal with Moscow." After all, he explained, "if, God forbid, the Chechens are defeated, Georgia and Azerbaijan will be the Kremlin's next target." Only by uniting behind the Chechen Islamist-Jihadist leaders would the peoples of the region have the strength to take on Russia successfully.

Under pressure from both the United States and Turkey on the one hand and a growing grassroots radical movement on the other, the leaders of Azerbaijan found it difficult to extricate their nation from the gathering storm. The growing militancy in Azerbaijan had even spread to the country's elite. In late March, a contingent representing the country's active and recently retired military, led by General Zaur Rzayev and former defense minister Tacaddin Mehdiyev, urged President Haidar Aliyev to resolve the Nagorno-Karabakh issue by force. Every day that passed was increasing the world's acceptance of the "Nagorno-Karabakh entity," they argued, reducing Azerbaijan's likelihood of recovering the important region. "The military are confident that it is possible to resolve the conflict and liberate the land only in a military way," the delegation affirmed.

Convinced that a sudden surge of militancy was ill advised, Aliyev resisted. Under his guidance, Baku steadfastly withstood outside pressure to increase its involvement in the Chechen conflict. As Russian forces consolidated their victory, and Islamist-Jihadist threats of further escalation seemed to subside, the Azerbaijani security services moved quickly to counter the Islamist charities' illegitimate activities within their borders. By the early summer of 2000, Aliyev's government had closed down the jihadists' "camps" and smuggling/transportation routes, and established stricter control over the movement of people and funds near the border with Russia. Haidar Aliyev was determined to prevent his nation from becoming either a supporter or a victim of Islamist-Jihadist terrorism.

Chapter 16

THE ROUTINE OF WAR

THE HISTORY BOOKS SUGGEST THAT THE SECOND CHECHEN WAR ended in the first half of 2000. And indeed the Russians made serious gains that year, retaking control of Grozny from the rebels and appointing an interim president, Akhmad Kadyrov, in June.

Yet the terrorist war in Chechnya continued, and by mid-2000 it was almost routine. The Chechen rebels were isolated in the southern part of the country, shielded by the rugged mountains and easy access to the Georgian border and the Pankisi Gorge beyond them. This impasse was punctuated repeatedly by a series of terrorist acts of varying intensity, in Chechnya, Dagestan, and Ingushetiya, and in the heart of Russia. In the first such strike, an August 8 explosion in a pedestrian underpass on Tverskaya Street near Pushkin Square in central Moscow, eight people were killed and seventeen injured. Significantly, the bombing came just days after the Islamists had hinted at a coming "wave of attacks" to mark Chechnya's "Independence Day." In Chechnya, the mujahedin spread leaflets warning civilians to leave their cities and villages that day, "when armed actions will be carried out . . . against Russian installations and soldiers." Besides a few routine clashes, the Moscow bombing—not a military strike, but a terrorist act against civilians—was the only violent action the Islamists mounted in the coming days.

The Chechens' strategy was increasingly focused on terrorism, and particularly on suicide bombings. The month of August 2000 saw the publication of an extensive fatwa titled "The Islamic Ruling on the Permissibility of Martyrdom Operations." This fatwa put forth a laborious case concerning all forms of suicide bombing and other forms of terrorism including the killing of innocent Muslim civilians. It took as its starting point the case of a young female jihadist named Hawa (Khava) Barayeva—a twenty-two-year-old cousin of Chechen warlord Arbi Barayev, and one of the two first female suicide bombers in Chechnya.

On June 7, 2000, Barayeva and a fellow jihadist named Luiza Magomedova drove a truck bomb into the offices of the police commander in Alkhan-Yurt, Chechnya. The truck they were using had been stolen and was packed with high explosives by an expert bomb maker. Having been refused entry into the compound, the two raced the truck through the checkpoint and crashed through a fence—attempting to reach a building occupied by an OMON unit. As it broke through the fence, a team of seven troops ran toward the truck in an effort to block it. When the team reached the truck, the two women blew themselves up, killing two of the troops and wounding the others. The OMON building was damaged by the powerful explosion, and Movladi Udugov later claimed that at least twenty-seven policemen were killed in the explosion.

The English-language service of Kavkaz-Tsentr published a message praising the two women as a source of inspiration for future suicide bombers. "The Russians['] occupation forces in Chechnya have once again suffered a tremendous blow. This time it was at the hands of our sister mujahids [sic] who have unselfishly demonstrated their determination in being part of the struggle against the yolk [sic] of Russian oppression. Praise be to Allah for such women of courage in the Muslim world." Before she set off on her mission, Udugov told foreign journalists, Barayeva had declared that she was "going willingly to my death in the name of Allah and the freedom of the Chechen people."

The fatwa on "martyrdom operations" was a response to a theological question put to the Higher Sharia Court in Chechnya: Did Hawa Barayev commit suicide or achieve martyrdom? A few learned Chechens and Arabs had argued that women could not qualify for martyrdom—that Hawa Barayev had simply committed suicide, which is forbidden by Islam—but their argument had prompted the Higher Sharia Court to reexamine the entire issue of martyrdom terrorism.

The Court addressed the broader issue of terrorism as part of the Islamist jihad, starting with the legal definition used by the Muslim Brotherhood and other Islamist-Jihadist movements. "Allah legislated Jihad for the dignity of this Ummah, knowing that it [i.e., violence] is abhorrent to us," the fatwa advised. "People today have neglected this great duty, and pursued what they love, thinking good lies in what they love, and failing to realize that good lies in that which Allah has legislated." Once again, the Sharia court had affirmed the jihad as an obligatory duty for all Muslims.

The fatwa stressed the role of Chechnya as a major theater in the global jihad. "Allah has blessed us, here in Chechnya, by allowing us to fight unbelief—represented by the Russian army, and we ask Allah to strengthen and assist us. . . . Our martyred brothers have written, with their blood, a history we can be proud of, and their sacrifices only increase us in eagerness for our own martyrdom." The court concluded that Barayev, like the male suicide bombers before her, was "one of the few women whose name will be recorded in history." They praised the "most marvelous example" she set: "The Russians may well await death from every quarter now, and their hearts may appropriately be filled with terror on account of women like her."

The fatwa offered a definition of "legitimate" martyrdom operations—namely, "those performed by one or more people, against enemies far outstripping them in numbers and equipment, with prior knowledge that the operations will almost inevitably lead to death." The most common form of suicide attack, it noted, was "to wire up one's body, or a vehicle or suitcase with explosives," and then to detonate the bomb in a place calculated "to cause the maximum losses in the enemy ranks." But it also noted that martyrdom could be achieved by breaking into an enemy area and firing at close range, "without having prepared any plan of escape, nor having considered escape a possibility."

The Sharia court stressed the effectiveness of such operations in the context of the jihad against Russia: "[T]here is no other technique which strikes as much terror into their hearts, and which shatters their spirit as much. On account of this they refrain from mixing with the population, and from oppressing, harassing and looting them. They have also become occupied with trying to expose such operations before they occur, which has distracted them from other things. . . . Those troops who are not busy trying to foil martyrdom operations are occupied with removal of Russian corpses, healing the wounded, and drawing out plans and policies from beneath the debris." Moreover, the court determined, martyrdom operations were the most cost-effective tactic for the mujahedin. "On the material level, these operations inflict the heaviest losses on the enemy, and are lowest in cost to us. The cost of equipment is negligible in comparison to the assault; in fact the explosives and vehicles [involved have usually been] captured as war booty, such that we returned them to the Russians in our special way! The human casualty is a single life, who is in fact a martyr and hero gone ahead to Gardens of Eternity, inshaa-Allah."

The real impact of suicide bombings, however, would come only if they were part of a prolonged and persistent campaign. "We feel sure that the Russians will not remain long in our land with such operations continuing. Either they will [stay away from crowds], in which case they will become easy targets for attack, or they will gather together to combat the assaults, in which case the martyrdom operations will be sufficient—Allah willing—to disperse them." The fatwa's authors drew lessons from the Palestinians' use of suicide bombings against Israel during the 1990s. "The operations in Palestine . . . were a major factor in convincing the Jews to grant self-rule to the Palestinians, hoping that they could be more easily controlled in that way."

One sticky issue surrounding the use of martyrdom bombings was the potential for civilian casualties among innocent Muslims or Chechens. Using comparable Palestinian fatwas as precedent, the court decreed that since these Muslims did not choose to have the Russians in their midst, they should be considered involuntary human shields. "Killing those used as a shield was permitted by [earlier] scholars out of necessity" on the logic that "the public interest overshadow[ed] the individual interest." The fatwa also decreed that the blood of non-Muslim civilians was permitted in the course of a jihad. "In the case of women and children of the unbelievers, however, they [may] be fired upon [as] an expediency of war even if it is not dire necessity, for war may [require] such action, but the intention should not be specifically to kill the non-combatants." The court did clarify that inflicting Muslim civilian casualties in the course of a terrorist strike was "permissible only if abstaining will lead to a wholesale harm, such as a greater number of Muslims being killed than those being used as a shield, or the Muslims being defeated and their land overrun. In such a case, any Muslims killed as a result will be raised up according to their intentions," the court decreed.

The Higher Sharia Court in Chechnya concluded not only that "martyrdom operations" were permissible, but that a mujahed who perpetrated such an attack was "better than one who is killed fighting in the ranks. . . . One who kills himself because of his strong faith and out of love for Allah and the Prophet, and in the interests of the religion, is praiseworthy."

The publication of the fatwa on "martyrdom operations" had an immediate impact on the public posture of the Chechen Islamist-Jihadist leaders. In the past they had hotly denied involvement in terrorism in Moscow and elsewhere, particularly in Dagestan and Ingushetiya; now, senior leaders such as Shamil Basayev and Ibn al-Khattab openly claimed responsibility for events such as the August 2000 sinking of the submarine *Kursk* (allegedly by a Dagestani suicide bomber) and a fire in Moscow's Ostankino TV tower the same month (allegedly set by a Russian criminal bribed by the Chechens). The Chechen commanders framed these events as a prelude to the next phase in their war against Russia. In a September 1 interview, Khattab boasted that the mujahedin "can strike anywhere that we consider necessary, and any time considered necessary. The future events and attacks that we are preparing will show the Russian unfaithful that their crime will not remain unpunished. Putin and his gang should know that punishment for satanism is inevitable. We shall come to them. And there will be no place for them to hide."

Khattab's warning, however, came at a time when the jihad in the Caucasus was plagued by a growing stream of defections and casualties among Chechen, Dagestani, and Ingush fighters. As Russian forces made advances and the Islamist cause declined in popularity among Caucasian natives, for the first time the mujahedin were unable to recruit enough local volunteers to replenish their losses, let alone increase their numbers. And, as the role of the foreigners grew, their commanders—particularly Khattab—insisted on having a greater say in the conduct of the jihad. In the late summer of 2000, Khattab began taking steps to bolster his position, insisting on nominating several Arabs to senior command positions so that they could both influence and control Chechen field commanders, and launching a campaign to get rid of Chechen commanders who challenged this trend. When the commanders retreated to Georgia and Azerbaijan to regroup and rebuild their forces, a Khattab loyalist accused them of "cowardice and theft," claiming that they had actually absconded from Chechnya with large sums of money.

In these efforts, Khattab was exploiting deep divides between key Chechen commanders on the basis of their tribal affiliations. One of the first commanders Khattab attempted to sack was Ruslan Gelayev, whom Khattab urged the Sharia Court to prosecute for the loss of more than twelve hundred mujahedin in a siege on the village of Komsomolskoye back in February. Khattab's real motive for targeting Gelayev, however, was helping the commander's archrival Arbi Barayev, who insisted that Gelayev was working for Russian intelligence and had lost Komsomolskoye at their order. Yet ultimately Khattab tended not to reward homegrown Chechen commanders of any stripe, even those who supported him; rather, he insisted on nominating either foreign mujahedin or Chechens indoctrinated in Pakistan and Afghanistan for open posts in the "Chechen" forces.

By late August, this trend had become so apparent that Khattab found it necessary to deny publicly that foreign mujahedin were taking over the Chechen army. "There are some tens of Arabs and representatives of other peoples of the East, [but they] never could be decisive neither for the military, nor political conditions in Chechnya," Khattab explained. At the same time, however, Arab and Islamist sources in the Middle East were hailing the role of foreign mujahedin in filling the ranks of "Chechen" fighters. These mujahedin were still mainly

Arab, but there were growing numbers of Afghans, Pakistanis, Turks, and Central Asians among them. By now the Islamist training and support system in Afghanistan and Pakistan was churning out mujahedin units in order to maintain the size of the Chechen forces at roughly two thousand to twenty-five hundred mujahedin (including Chechens and other Caucasians). If the current trend continued, observed Tatyana Gomozova of Gazeta.ru, "it is quite likely that all Chechen rebels will be exterminated. But the number of Arab terrorists will be enough to last for two more military campaigns."

In Pakistan and Afghanistan, Arab and Islamist sources openly acknowledged the growing need to push larger numbers of trained mujahedin to fill the ranks of the Chechen forces. For example, in late August, Abu-Daoud, the nom de guerre of a Yemeni military instructor in Osama bin Laden's Afghan forces, confirmed that earlier in the summer bin Laden had sent four hundred highly trained Arab mujahedin with explosives and weapons to Chechnya. Similarly, one of the leading Kashmiri jihadist leaders, Abu Abdul Aziz—the chief commander of Lashkar-i-Taiyyaba, founder of Markaz al-Dawaat wal-Irshad, and chief of Rabta al-Jihad al-Alami—noted that members of his forces were yearning to go to Chechnya and other worldwide jihad fronts. "As soon as we get a chance, we will fight shoulder to shoulder with our Chechen and Palestinian brothers," he promised.

The continued stalemate in the military operations in Chechnya itself, and the mujahedin's inability to make headway despite a growing number of ambushes and strikes throughout Chechnya, did not seem to worry the Islamist leadership. Instead, the commanders stressed the need for more martyrs in order to stay the course. "Congratulations to the soil of Chechnya for being purified with the blood of these Martyrs," read a mid-October obituary for recently fallen mujahedin commanders distributed as a recruitment tool. "Oh Allah, help us during our difficult times and replace these with better days." To the Islamist leadership, particularly in the Arab world and Western Europe, the Chechen cause was a valuable tool in motivating local youth to volunteer to join the jihad in their own regions. This trend was increasingly visible in the

United Kingdom, France, and other Western European states, where volunteers were being actively sought to train as terrorists for future operations in their own home countries as well as in the United States.

Thus, by the fall of 2000, a new generation of Islamist Mujahedin was taking shape in both the Caucasus and the West—a generation committed more to punishing the United States, Russia, and its Western supporters, than to liberating the Muslims of the Caucasus. This generation showed a growing penchant for terrorist strikes against Russia—regardless of the ramifications of such actions for the local population in Chechnya. In mid-October, a call for financial help from the mujahedin in Chechnya promised their European supporters that "the Russians will soon see" a dramatic change in the situation. "The mujahedin are currently preparing an upcoming series of operations. These operations will be of a different style and magnified destructive force on the enemy," the commanders predicted. "The current operations you read and hear about are the remnants of the old plans and operations," they assured their European supporters. These appeals were reinforced by several warnings by Khattab, Basayev, and other Chechen leaders of new terrorist strikes to come within Russia itself.

As the winter of 2000–2001 set in, the course of the war was dictated as much by the debilitating weather as by the performance of the warring sides. The Russians kept the upper hand, but were unable to force any real change in the Chechens' strategic posture. Both sides continued their relentless ambushes, raids, bombing, and shelling. In late November, the Mujahedin Military Command promised a campaign of "special and fiery" operations that would "wreak havoc in the enemy lines, and completely diminish their spirits." The Command added that "the heat of these planned operations, combined with the freezing winter, will Insha-Allah ensure that the end of winter will bring with it the end of the enemy."

But the mujahedin were incapable of living up to their promises. They attempted repeatedly to escalate their campaign of assassination and terror bombing in Russian-controlled cities and against the oil industry and the railways, but they proved unable to sustain momentum. A few strikes, ambushes, and sabotage operations succeeded, but they were inevitably followed by Russian security sweeps that kept the mujahedin forces on the run, forcing them to withdraw from the civilian population they had once relied on for shelter and sustenance. By mid-December, Chechen commanders were sounding increasingly defensive. "It's already a long time since the front line has disappeared in Chechnya. The mobile groups of Chechen rebels are situated everywhere around Chechnya. This is known both in Moscow and in Chechnya," explained Shirvani Basayev, the head of the Vedeno region and brother of Shamil Basayev. By that time, though, it was the Russian Special Forces who were roaming throughout Chechnya—and on the night of December 20–21, a hit team raided the headquarters of Shirvani Basayev and assassinated him.

Within Islamist-Jihadist leadership circles, there were fewer illusions. A Russian drive to control greater tracts of inhabited lands had forced hundreds of mujahedin back into the mountains in the dead of winter, as the writer of a December 20 plea for help from the West's Islamists acknowledged. "Our conditions are presently extremely severe . . . the temperatures are so bitterly low that we sometimes feel the cold is crushing our bones. And although the enemy is broken, it is still a stubborn beast, gasping for its last breath, but continuing to throw its weight and might around, killing and destroying everything and everyone standing in its path. Our arms are nothing when compared to theirs and our money is but a tiny fraction of the material wealth they possess," he wrote.

Eager to break the impasse, Russia resumed contact with receptive elements in Chechnya in the hope of bolstering anti-Islamist leaders who might ultimately consent to meaningful negotiations. Moscow's primary objective was contacting the Chechen leaders and commanders who had been attacked by Khattab. At first, the jihadist leaders attempted to subvert these contacts by joining in. On November 11, Khattab leaked to Qoqaz.net that "high-level secret negotiations are currently taking place, aimed at withdrawing Russian troops from Chechnya. If these negotiations prove successful, they will result in either an immediate withdrawal of all Russian troops from Chechnya, or the partial withdrawal of the troops, followed by a full one at a later stage." When the Kremlin repudiated the claim, the Islamist-Jihadists dropped the issue. But a few Chechen commanders attempted to sustain contact with Moscow, keeping their options open against the possibility of a future purge by Basayev and Khattab. In late November, Viktor Kazantsev, the presidential envoy in southern Russia and the former top commander of Russian forces in Chechnya, acknowledged that his aides were indeed in contact with Ruslan Gelayev—Khattab's main nemesis. "We are carrying on talks with those who have normal minds," Chief of the General Staff Anatoliy Kvashnin told the Russian Duma. "All those who want to return to peaceful life."

By mid-December, having been relegated to secondary position in a jihad dominated by Basayev, Khattab, and their lieutenants, Aslan Maskhadov—still the titular president of Chechnya—tried to regain prominence by asserting a role in a new cycle of negotiations. First he sent several emissaries to renew contact with the antiwar factions in the Duma and among Moscow's intellectual elite. Some of these contacts expressed interest in the offer of negotiation, but set conditions for their resumption—a viable cease-fire chief among them. Maskhadov and his emissaries knew he couldn't deliver such a condition, so instead they sought to increase pressure on Moscow by highlighting the plight of the civilian population and hinting that the war was no longer in their interest. This theme dominated Maskhadov's message to Muslim Chechens to mark the end of the Ramadan and Eid al-Fitr festivities. "Today, the Chechen people, *hostages of a bloody war*, celebrate this holiday in conditions of humanitarian catastrophe," Maskhadov said [emphasis added]. "Hundreds of thousands of people are without shelter, tens of other thousands have been killed, more than one hundred thousand people have been wounded and twenty thousand disappeared." At the same time, Maskhadov stressed that the Chechens were "not defeated but determined to be victorious." When Maskhadov's intermediaries refused to withdraw their demand for a cease-fire as a precondition for political negotiations, Maskhadov withdrew his offer. "Maskhadov is the president of an independent state and it is inconceivable to impose such conditions on him," Maskhadov's aide Said-Hassan Abumuslimov told Western reporters on December 27.

Ultimately, the jihadist leaders prevailed. The Chechen commanders and leaders who wanted an end to hostilities had no real power base. In his Eid al-Fitr message, Shamil Basayev committed Chechnya to another century of war, claiming that victory was a matter of time. "Our tactic is to have no tactics. We are nowhere to be seen. We have chosen a method of warfare that the Russians cannot predict," Basayev said. "Whether this war lasts ten, twenty, or one hundred years, we will fight on." Alluding to Khattab's recent efforts to tighten control over the jihad, he promised, "We have . . . put aside our personal ambitions and disputes for the sake of our struggle and Allah the Great." As for the Russians? "The best thing they can do is fire their generals and get out of here as quickly as possible."

Chapter 17

CHECHNYA AND THE PALESTINIAN PROBLEM

IN LATE SEPTEMBER 2000, DESPITE UNPRECEDENTED CONCESSIONS by Israel, Palestinian leader Yasser Arafat rejected the U.S.-brokered peace plan and unleashed a new wave of fighting against Israel—a new Palestinian Intifadah. Although the Intifadah was declared by the nationalist leadership, the bulk of the Palestinian forces fighting Israel were Islamist—albeit with varying degrees of dependence on the international Islamist-Jihadist movement. Nevertheless, the Intifadah captivated the attention of the entire Muslim world, including the Caucasus. Before long, the Palestinian Intifadah had become a new rallying cry for the entire Islamist-Jihadist movement worldwide. In many cases, Islamist groups put the struggle for the liberation of al-Aqsa ahead of their own jihad. Even the Islamist-Jihadist leadership in Chechnya seemed to be more preoccupied with the Intifadah in Israel and the territories than with the deteriorating situation in their own backyard. Indicative of this trend was the growing involvement of the Chechen mujahedin—the local Islamists and the international jihadist volunteers—in support of the Palestinian jihad.

The Chechen leaders' focus on the Palestinian jihad was an inevitable outgrowth of the burgeoning anti-Semitism within the Chechen Islamist movement. Since the mid-1990s, and especially since the summer of 2000, the Chechen leadership had resorted to blaming the "international Jewish conspiracy" for major setbacks in the jihad. The now-dominant foreign mujahedin in Chechnya promoted a number of extremist Islamist organizations with highly anti-Semitic ideology, and banned the activities and even the existence of Jewish organizations.

The virulent anti-Semitism of the Chechen leaders was most troubling precisely because it did *not* constitute the reawakening of any traditional enmity against Jews or their religion or culture. As a purely cynical instrument used by the Chechen Islamist-Jihadist leaders and their Arab allies to gain support and preserve the endorsement of the international Islamist-Jihadist leadership, the fanning of anti-Semitic flames in the Caucasus was one of the most disturbing by-products of the jihad.

By mid-2000, the most prominent proponent of anti-Jewish policy among the Chechen jihadists was "Abu-Omar"—the nom de guerre of a Chechen commander close to Khattab. Yet anti-Semitic incitement was prevalent among most Chechen commanders. Several of them told their fighters in public addresses that "the Jews are the cause of all the misfortunes the Chechen people are facing." Early that summer, another commander, Arbi Barayev, asserted on Grozny TV that "the Jews are our enemies, and we will kill them, kidnap and rob them everywhere we meet them." Predictably, this summer also saw an increase in the kidnapping and torture of Jews, including Israeli citizens, in Chechnya. Some, including young children and the elderly, were cruelly maimed in order to extract higher ransoms from their relatives and/or the Jewish community.

The outbreak of the Intifadah in the fall of 2000 brought the anti-Semitic sentiments of the Chechen jihad into the spotlight, and the Islamist leadership was soon ready to turn its words into action. On October 8, the Chechen leadership formally committed itself to active support of the Palestinian jihad. Shamil Basayev chaired an extraordinary session of the Majlis in his capacity as Amir of the Supreme Military Majlis ul-Shura of the Mujahedin of the Caucasus. As a number of Chechen websites reported, the Majlis "expressed its concern over the situation in Al-Quds (Jerusalem) and sharply condemned Israel for the mass killings of Muslims, calling the Israeli leadership the main ringleader of the new wave of violence on Palestinian land." The Majlis then declared that it had readied a detachment of 153 Chechen mujahedin to travel to Jerusalem "to help the Palestinian Muslims," and established an ad hoc leadership group "to devise a route to move the detachment to Palestine." Promising that the detachment would be "fully equipped with everything necessary and is capable of acting absolutely autonomously" for an extended period, Basayev added that this was only the Chechens' first contribution to the Palestinian jihad. "Chechnya is ready to supply 1,500 mujahedin to help the Muslims of Palestine," he claimed, despite the Chechen forces' struggles at home. "No matter what difficulties we experience, the main problem for all Muslims is the liberation of Al-Quds. Fighting in Chechnya does not relieve us of responsibility for this holy town," Basayev stressed.

Basayev's threat may have seemed outlandish, given the Chechens' straitened circumstances, but it was serious. Basayev had indeed committed 153 mujahedin to send to Jerusalem, although he misrepresented their origins: Basayev had diverted more than 150 mujahedin from training camps in Jordan, Turkey, and Afghanistan/Pakistan, not Chechnya itself. They included both Chechens and foreign (predominantly Arab) volunteers from numerous countries, some of whom could travel to Israel on their own passports to launch urban terrorism missions there. The mujahedin had been preparing for deployment to Chechnya, and the decision to send them to Israel instead was a considerable sacrifice, given the Chechens' manpower shortages at the time. It was yet another sign that Islamist-Jihadists' assimilation of the Chechen movement was increasingly at odds with the Chechens' national interest.

A few days later, on October 16, the Chechen leadership formally reiterated its commitment to assisting the Palestinian jihad in a long and detailed document headed "Urgent Letter to Our Brothers in Palestine." The unusually emotional letter started with a commitment to solidarity with the Palestinian Intifadah. "Even if our homes are thousands of miles apart, our hearts could not be closer to you," the letter affirmed. "We continue to grieve because our beloved Aqsa is in the hands of lawless occupiers, the Jews. Without doubt, they are the worst of all nations Allah has created." Given the disparity in military capabilities between the Palestinians and the Israelis, the letter stressed, the Intifadah—like the Chechens' jihad against the Russians—might take generations to triumph. "It is true that your weapons are nothing but mere stones. But in your land and in your hands, there is a difference. With you, every stone throws horror in the hearts of the Jews. With you, your stones become like weapons of mass destruction." The Chechens stressed the crucial importance of enduring losses and sacrifices to the sustenance of the jihad, and they urged the Palestinians to follow their example. "Our nation is a ship but unlike other ships, it only floats in the blood of its sons. The less blood that is sacrificed for the sake of Allah, the lower our ship sinks. And should the blood run dry, then the ship will be destroyed. We ask you by Allah, do not deny us your blood. There is no honor to us except with the blood of martyrs." The Chechen document concluded with a specific commitment: "We promise you that soon insha-Allah, you will hear of what we do with the brothers of the Jews, the Russians, in retaliation for Al-Aqsa and for you. Wait for a forthcoming operation which we have named 'Operation al-Aqsa.' That is all that we can do to comfort you and heal your wounds."

It didn't take long for the Chechen Islamist-Jihadists to attempt to make good on their promise. In mid-November, they hijacked a Russian Tu-154 aircraft en route from Makhachkala, Dagestan, to Moscow, diverting it to Israel—an event intended to be the catalyst for a much wider Operation al-Aqsa. The hijacking itself was performed by several terrorists, all Russian speakers from the Caucasus. The hijackers made no political statement while onboard, but they left several clues as to their intentions, mentioning al-Aqsa to the Azeri ground crew in the context of their destination in Israel. Significantly, in their communications with air traffic control, one of the hijackers demanded to fly to Israel's Ben-Gurion Airport rather than just Israel, as they had demanded at the start of the hijacking. Instead, a pair of Israeli fighter jets forced the Tu-154 to land at the remote Negev air base in Ovda; by then the Israeli security system was at full alert, and Prime Minister Ehud Barak, who was on a flight to Washington, had decided to turn back.

Once the Russian airliner landed, the hijacking fizzled. When the Israeli counterterrorist teams reached the plane, a lone, seemingly mentally unbalanced individual claimed to have hijacked the aircraft "to warn the white race against the rise of the yellow peril." Yet there were many unanswered questions about the hijacking: A close examination of the passenger lists and documents concluded that there were five people with false passports onboard, and both the Russian captain and the Azeri refueling team had referred to several hijackers, who appeared coherent and determined.

Later, after interrogating a number of captured Palestinian terrorists and reviewing documents seized from Arafat's compound in Ramallah and other Palestinian locales, the Israeli intelligence services concluded that the hijacking had been intended as only the first step in a larger Operation al-Aqsa. The hijackers had expected to coordinate with accomplices awaiting them at the Ben-Gurion Airport and the surrounding area, precipitating a lengthy crisis. With the Israeli security forces swarmed in to defuse the hostage situation, the Palestinian forces—including a number of "Chechens"—would have shelled or launched rockets at the Ben Gurion compound from nearby hills, hoping to hit the assembled Israeli forces. The Palestinians expected the hijackers to approach the siege as a martyrdom operation, taking as many passengers and Israelis with them as possible. But the diversion to Ovda forced the hijackers and their Palestinian cohorts to call off the operation—and the Caucasian terrorists returned to Russia as innocent passengers. In the fall of 2000, as both Moscow and Jerusalem were preoccupied with more pressing crises, there was little incentive to get to the bottom of the hijacking plot.

Meanwhile, a number of mujahedin dispatched from the Caucasus started assuming prominent roles in the fighting against Israel, both in Arafat's forces in the territories and in HizbAllah-led operations along the Israeli-Lebanese border. These "Chechen" mujahedin included some Chechen fighters, but also Dagestanis, Ingush, and so on, as well as Central Asians from Kazakhstan and Uzbekistan who were part of the Chechen contingent in Afghanistan and Pakistan. There were also Circassians (descendants of Caucasian Muslims exiled by the Ottomans to the Arab Middle East in the nineteenth century), many of them highly trained veterans of the Jordanian special forces and security services, who had contributed to the revolt in Chechnya before moving on to the Palestinian Intifadah. Finally, there were numerous "Chechen" Arabs—Palestinian, Syrian, Egyptian, and Saudi Arabian veterans of the war in Chechnya, who now rallied to the Palestinian cause.

The first decisive "Chechen" participation in an operation took place in Har Dov, or Shaba Farms, on November 16, 2000. That day, a detachment of expert terrorists crossed the Israeli-Lebanese border in the central zone—overlooking the Upper Galilee—and traveled unhindered along a civilian road into the northern parts of the Golan Heights. The detachment then reached the Israeli patrol road from the Israeli side of the border, planted a big bomb on the side of the road, and safely withdrew into southern Lebanon. The bomb was activated by remote control, but slightly ahead of time, causing damage but no Israeli casualties.

Subsequent investigation confirmed that the bomb was a "Chechen" operation. The remote-control mechanism was new to the theater and therefore not covered by the myriad of electronic jammers used by the Israeli Defense Forces (IDF). The main charge was concealed in a new fashion in trees and branches—not the "rocks" used by the HizbAllah and their Iranian instructors—and essentially missed by the troops. However, the remote-control fuse, the overall bomb structure, and the form of camouflage were identical to bombs used against Russian forces in Chechnya. Moreover, after the bombing a report was sent to the headquarters of the Secret Islamic Revolutionary Army—the joint headquarters of Imad Mughanniyah and Osama bin Laden in Lebanon.

By late 2000, there were two types of "Chechens" operating in the territories: commanders, many of them the Chechens, including veterans of the Russian/Soviet armed forces with extensive combat experience, and operators, most of them Arab mujahedin with extensive operational experience. The commanders, experts in urban warfare and fortification, were brought in to teach the Palestinians lessons they had learned during the siege of Grozny and other cities, where lightly armed Chechen forces had engaged Russian heavy armored columns, causing extensive casualties to the Russians. The Palestinians knew the IDF was planning on using a similar tactical approach in the Intifadah, and the Chechen commanders offered them tactical instruction, training, and help in building fortifications in most Arab towns. The operators were mainly snipers and bomb makers, who operated on the battlefield with Arafat's elite units (among them the units known as Preventive Security, Tanzim, and Shadows). These Arab mujahedin went beyond mere training, actually participating in operations. The sniper who killed Israeli major Sharon Arma while he was walking inside an Israeli divisional headquarters in the Gaza Strip on November 24, for example, was one of these Arab mujahedin.

Equally important has been the training and support infrastructure established in Jordan—mainly in the Palestinian refugee camps in the Amman and Zarqa areas. The trainers were predominantly wounded Chechen mujahedin sent to Jordan for medical treatment and rehabilitation, along with numerous Jordanian Circassian veterans of the jihad in Chechnya. On instruction from Basayev and Khattab, they opened training centers modeled after Khattab's camps in order to teach advanced bomb-making techniques (using both military equipment and off-the-shelf materials easily available in Israel), recruitment and running of martyrdom-seeking networks, and espionage/counterespionage trade craft. Depending on the needs of the various Palestinian networks, courses ranged from a few weeks to about six months. Local commanders, particularly in the West Bank, were encouraged to send promising young fighters to attend these camps.

These training initiatives would have a major impact because until then the bulk of the expertise within Palestinian ranks rested with the jihadist networks in the Gaza Strip. The Intifadah, and the work of the Israeli security services, had made it virtually impossible to move fighters between the Gaza Strip and the West Bank, so the Chechen-run training camps filled a strategic void. The early graduates of the Chechen training program made an impact very quickly. One of the first trainees was a twenty-seven-year-old Jordanian religious teacher named Fuaz Badran, who was teaching at the time in Tul-Karm in the West Bank. Soon after the outbreak of the Intifadah, he traveled to Jordan and attended one of the first courses for a few months. He learned how to make martyrdom belts as well as identify, recruit, and indoctrinate would-be martyrs. He returned to Tul-Karm and was put to the test in the form of running a deniable martyrdom strike in March 2001. Badran was then recruited by the local HAMAS leader, forty-one-year-old Abbas al-Sayyid. For the next year, Badran provided a stream of bombs and would-be martyr-bombers, and taught a number of local would-be bomb makers himself.

The operations of the al-Sayyid and Badran network peaked on Passover Eve, the night of March 27, 2002. A suicide bomber they had equipped walked into a hotel in Natanyah and blew himself up in the middle of a Seder crowd, killing some thirty and wounding hundreds. Among the victims were relatives of the victims of the recent Dolphinarium bombing in Tel Aviv, who had been invited for a communal holiday Seder in order to help alleviate their pain. To this day, this remains the most lethal terrorist strike in the history of Israel. HAMAS was quick to take responsibility. Soon afterward, Israel launched a massive crackdown—Operation Defensive Shield—in which this and other key networks were destroyed. Yet the Chechen-run training system remained the dominant source of know-how among the jihadists in the West Bank. Meanwhile, starting in the summer of 2002—anticipating a U.S. invasion of Iraq—the Jordan-based, Chechen-run training system expanded its activities and began churning out mujahedin—predominantly Jordanians and Saudis—for operations in Iraq. Many of the early would-be martyr-bombers in Iraq were of these Chechen- and Circassian-run programs.

Ultimately, this influx of advisers and terrorists from the Caucasus had a profound impact on the overall character of the Intifadah. Although they operated within the context of Arafat's own forces—mainly the Al-Aqsa Martyrs' Brigades—these "Chechens" were fierce Islamist-Jihadists, many of whom had volunteered for previous foreign jihads. Most of them had been handled via the bin Laden establishment—commonly known as al Qaeda—and were committed to his brand of uncompromising global jihad. Ideologically, they were also very close to the HAMAS: A few of the Palestinian "Chechens" had actually been members of the HAMAS Izz-al-Din Brigades before leaving Gaza and Nablus for advanced training in Afghanistan/

Pakistan. The prominent role of these "Chechens" in Palestinian special operations was bound to draw HAMAS into closer cooperation with the Fatah (Arafat's own militant organization), further inflecting the Intifadah with an Islamic cast. Any such radicalization was bound to increase the likelihood of terrorism aimed at the heart of Israel and a rapid slide toward regional war.

Chapter 18

THE CHECHEN JIHAD AFTER 9/11

IN 2001–2002, THE ISLAMIST-JIHADIST SUPREME LEADERSHIP FINALLY launched its global jihad. This decision would affect Islamist-Jihadist activities worldwide—from training in Afghanistan and Pakistan to building a new logistical infrastructure for evacuees from Afghanistan through Central Asia to Chechnya and onward to the Balkans. The supreme leadership also established a sophisticated new communications system designed to withstand the inevitable onslaught of Western intelligence and security services in the aftermath of the spectacular September 11, 2001, strikes in New York and Washington, D.C.

In early 2001, the center of Islamist-Jihadist activity was still in Afghanistan and Pakistan. In Afghanistan, the jihadists launched a major escalation in the northeast, designed to consolidate the hold of the Taliban and their patrons, the Pakistani intelligence service (the ISI), over the country. Chechens and other Muslims from the Caucasus played a major role in the ranks of the 55th Brigade—the Taliban's and Osama bin Laden's "foreign legion"—that was trained, led, and sponsored by the ISI. Chechen mujahedin were prominently involved in fighting the Northern Alliance in the summer of 2001. Concurrently, the jihadists intensified their efforts to graduate as many fighters as possible from their training programs before the coming strike on the United States. Among these foreign mujahedin were several hundred trainees from the northern Caucasus and southern Russia—many of whom were identified by the ISI as Chechens.

Maintaining secure lines of communication between Afghanistan and the Caucasus, along the historic Silk Road, was a major strategic objective of the Islamist-Jihadist supreme leadership, particularly after the spring of 2000. The leadership relied on the Afghanistan-to-Chechnya axis for both operational and economic reasons. The flow of expert terrorists, specialized equipment, and written communiqués between Europe and the Afghanistan/Pakistan hub was increasingly conducted along this axis. As confrontation with the U.S.-led West drew near, the supreme leaders relied more heavily on this clandestine channel rather than on commercial flights in and out of Pakistan, which were being monitored by hostile intelligence and security services with growing frequency. The Afghan/Pakistani drug trade in Western Europe and the United States was also dependent on these transportation routes along the Silk Road and through the Caucasus, which were dominated by the Chechen Mafiya and its Islamist allies. With a major escalation of the Islamist jihad looming, protecting this flow of drugs and funds was more crucial than ever.

Since the late 1990s, Osama bin Laden himself had been preoccupied with his network's reach into Central Asia and the Caucasus. After the 1998 U.S. cruise missile strikes, he launched a crash program to build a new headquarters far from the Pakistani border. His new headquarters and communication center was constructed in a natural cave system in the Pamir Mountains in Kunduz province, near the Tajikistan border. Special care was taken to ensure that the facility was concealed and protected against both espionage and bombing. Pakistani experts considered it "impervious" to all known forms of espionage and attack. Bin Laden chose the site in part because Russian forces were stationed nearby—he felt sure that Washington would never dare bomb the location for fear of a Russian response—and in part because Iranian intelligence, which had a strong presence in Tajikistan, could be counted on to offer a safe escape out of Afghanistan and into the Caucasus should the need arise. This new compound became operational in the spring of 2000.

Even as all these military preparations were ongoing, Pakistan was also trying to reach a compromise with Afghan rebel Ahmad Shah Massud, leader of the United Islamic Front for the Salvation of Afghanistan (also known as the Northern Alliance). Mindful that Massud had secretly cooperated with the KGB during the 1980s, the ISI feared that Massud might revive his cooperation with Moscow to interdict the flow of mujahedin, drugs, and goods from Afghanistan and Pakistan into Central Asia and on to the Caucasus. Hence, in mid-February 2000, bin Laden took a personal interest in reaching a compromise with Massud, dispatching two senior veteran Arab "Afghans" with impeccable credentials to northern Afghanistan. They were provided with conciliatory letters from the Taliban's leader, Mullah Muhammad Omar, and defense minister Mowlawi Obaidollah. Flying first to Kunduz, bin Laden's emissaries then took a special car to Takhar, where they met two longtime Afghan compatriots, Abdorrab Rasul Sayyaf and Borhanoddin Rabbani.

Rabbani and Sayyaf arranged for the two Arab "Afghans" to meet Massud, but the meeting went very badly. Massud would hear nothing about cooperation with the Taliban. Instead, as one observer reported, "Massud called Osama a terrorist, and described his presence in Afghanistan as harmful for the country." Massud also refused to provide assistance for bin Laden's planned "activities in Central Asia and Chechnya." The Arabs left the meeting "in anguish and despair," reported a number of anti-Taliban Afghan sources, among them Mokhles of the Northern Alliance. To Islamabad, the meeting only confirmed the ISI's warning that "Massud has built very close relations with the Russians and the government of Tajikistan" and that "Massud wants to defend the borders of the former Soviet Union against Islamic 'fundamentalism.' " The ISI urged the Taliban leaders to take quick action against Massud, and in March a bin Laden protégé known as Jonaidollah organized an ambush of Massud's convoy while on a working visit to Tajikistan. (The assassination attempt was a failure: Massud was not in his car at the time of the attack.)

In the summer of 2000, Osama bin Laden moved his headquarters to Kunduz, northern Afghanistan, in preparation for a new attack on Ahmad Shah Massud and the Northern Alliance. Mokhles warned Massud that bin Laden "has recently increased his activities in the northern regions of Afghanistan. He wants to extend the scale of his activities in Central Asia and to take part in the war against the Russians in Chechnya."

Encouraged by the Russians (through intermediaries in Tajikistan), Massud launched a preventive offensive in mid-June, but it was haphazard and indecisive. Despite the flow of Russian supplies and promise of aid, Massud was reluctant to commit his forces to anything more than a series of ineffective skirmishes in the general direction of Kabul. In mid-July, the Taliban launched their major offensive against the Northern Alliance, gaining large amounts of territory in the north in early August. But the Russians rushed additional weaponry to the region, enabling the Northern Alliance to put pressure on the Taliban forces and then launch a counteroffensive that was threatening a key Taliban supply route within a few days. Now it was Pakistan's turn to worry, and the Taliban tried to ensure victory by committing forces to the battlefield. These Pakistani forces quickly facilitated the Taliban's reinvigorated push northward in Afghanistan, consolidating its control over segments of the Amu-Dariya (Oxus) riverbanks. The Taliban's fall 2000 victories were made possible by the active participation of at least two Pakistani brigades from the Gilgit area, but they were supplemented by an increasingly cohesive strike force of Central Asian and Caucasian Islamist-Jihadist forces organized and sustained by bin Laden's commanders. This force would soon be further consolidated into the 55th Brigade.

The Pakistani/Taliban forces, supported by the Islamist-Jihadist leaders, made considerable advances during September and October, capturing key Northern Alliance strongholds throughout northeastern Afghanistan. They managed to surround Massud's main supply points with Tajikistan, but the Alliance prevailed after the arrival of Tajik and Uzbek special forces, as well as help from Iran. Despite such spasmodic episodes, however, Afghanistan was sliding firmly into the hands of the Taliban and their Pakistani overlords, and Afghan commanders were increasingly changing sides. In early November 2000, as the harsh winter set in, fighting came to a halt—leaving Islamabad and bin Laden satisfied that the southern routes into Central Asia and the Caucasus were firmly in the hands of the Pakistanis and their Islamist-Jihadist protégés. Conditions were finally right for the Islamist-Jihadists to launch their grand strategy for the Silk Road.

The training and equipment of Islamist-Jihadist forces continued, readying mujahedin for dispatch to the Taliban front, mainly as part of the 55th Brigade, and for infiltration into Central Asia and the northern Caucasus. The network of camps in Afghanistan were devoted mainly to general training, while those in Pakistan handled more specialized advanced training; at one time or another, most of the camps hosted Chechens and other Muslims from the Caucasus and southern Russia. The main Afghan camp, where many Chechens received military training from Pakistani experts, was Kargha-1, about 7.5 miles (12 kilometers) north of Kabul. Chechen mujahedin also received training in the large al-Badr I and al-Badr II complexes near Khowst, Paktia Province, including a force of roughly 350 Chechen, Caucasian, and scattered Arab fighters earmarked for dispatch to Chechnya who trained there in early 2001. The Chechen and Arab commanders and expert terrorists—mainly bomb makers—were trained by Egyptian devotées of Ayman al-Zawahiri in the nearby Abu-Bakr camp. Veteran foreign mujahedin were trained in the Omar, al-Khuldan, and Farouk camps along the Afghan-Pakistani border in preparation for deployment to Chechnya to bolster the forces of Khattab and Basayev.

Another training camp, in Darunta, Nangarhar Province, was the primary center for training predominantly non-Arab expert terrorists for special terror strikes. In early 2001, the camp was commanded by an Egyptian known as Abu-Abdallah and an unnamed Algerian expert bomb maker with experience in clandestine operations in France and Western Europe. By the spring of 2001, roughly three hundred trainees—mainly from the Philippines, Malaysia, Turkey, and Egypt—were undergoing final training there before being "dispatched to help the Islamic Legion in Bosnia, Chechnya and Azerbaijan," as a Pakistani intelligence source reported.

In Pakistan, Caucasian mujahedin were integrated into the ISI-sponsored training system, managed by several Islamist-Jihadist organizations, most of which were active in the anti-India jihad in Kashmir. The ISI-sponsored system provided advanced training not only in clandestine and terrorist activities, but also in sustaining networks. The key Islamist-Jihadist groups in Pakistan with close ties to the jihads in Kashmir and Afghanistan—Harkat-ul-Mujahideen, Hizb-ul-Mujahideen (the military arm of the Jamaat-i-Islami political party, also active in fund-raising for the Chechen jihad), Al-Badr, the Wahhabist group Sipahe Sahaba Pakistan, and Lashkar-i-Taiyyaba, the military wing of the theologically significant Markaz al-Dawat wal-Irshad—one of the leading Islamist-Jihadist schools—all harbored Chechen mujahedin.

Most significant, however, was the presence of Chechens in the key theological schools and camps in Pakistan, which were then busy training the future leaders of the Islamist-Jihadist movement. The Islamist leaders were clearly making a long-term investment in the Chechen jihad and its ground troops. Dozens of Chechens targeted for senior-level leadership training were sent to the Haqqaniya Mosque and madrassa (Islamic religious school) at Akora Khattak near the Afghan border, joining some three thousand students from all over the world who had come for at least a year of advanced studies. The Chechens attended a comprehensive religious-leadership program ranging from Koranic and Sharia studies to the use of state-of-the-art, high-tech communication facilities and computers for propaganda and educational work. Chechen candidates for military and security leadership attended a special program run by recently retired ISI officers under the cover of the Lashkar-i-Taiyyaba. By 2000–2001, most of the Chechen mujahedin were attending a special program for would-be commanders from Chechnya and Central Asia run in Russian by a mix of Pakistani ISI veterans and Soviet-educated Afghan KhAD/WAD veterans. By then, a comparable program was being run for mujahedin from Western Europe and the United States. After U.S. and European diplomats complained to Islamabad about the program, the parts of the program involving the use of heavy weapons and special sabotage techniques were moved across the Afghan border, to the Kargha-1 camp north of Kabul. All foreign trainees, including the Chechens, were driven to Kargah-1 for training as required.

* * *

In the summer of 2001, as fighting between Islamist-Jihadist forces and the Northern Alliance escalated in northeastern Afghanistan, many of the foreign trainees in both Afghanistan and Pakistan, including the Caucasian mujahedin, were dispatched to the front, most of them to bolster the ranks of the 55th Brigade. By now, there were between eight hundred and a thousand mujahedin from Chechnya and the northern Caucasus in Afghanistan (some estimates put the number as high as fifteen hundred). The Northern Alliance quickly learned to respect their Chechen opponents. In October 2001, when the first CIA officers reached northern Afghanistan to prepare for the coming U.S. invasion, General Bismullah Khan of the Northern Alliance singled out the Chechen and Uzbek mujahedin for special comment. As the CIA's Gary Schroen recalled, Bismullah reported that "the Arab units maintain themselves separately from the Taliban forces, taking up key positions within the overall front lines. . . . They are well equipped and well trained. They fight like devils, especially those Muslim fighters from Chechnya and those from Uzbekistan. Unlike Afghans, these devils aim when they shoot. . . . And they do not like to surrender. They fight hard and they fight bravely." These Chechen and Uzbek mujahedin, Bismullah warned, "will be the glue that will hold the Taliban together when we attack."

From the start, the Chechen fighters lived up to their billing across the breadth of the northern front, holding the line against Afghan forces that were supported by U.S. special operations forces. One key clash involving Chechen mujahedin took place in Kunduz close to the Tajik border, near the ostensibly impervious headquarters and communication center bin Laden had built in a cave system in the Pamir Mountains between 1998 and 2000. Now, as fighting escalated in northern Afghanistan, the Kunduz area became the center of the Islamist-Jihadist resistance under bin Laden's personal guidance.

By late October 2001, the core of the Islamist-Jihadist defenders including a few thousand Pakistani regular troops (originally from the Gilgit area) and an elite force of about a thousand foreign mujahedin—including about three hundred Chechens—were under the command of a Saudi called Omar al-Khattab. The name similarity gave rise to rumors that Ibn al-Khattab had come from Chechnya to join the fight in Afghanistan. Several thousand Taliban fighters and their families were also in the area. As the U.S.-led Afghan forces tightened the siege over the Kunduz perimeter, the Chechen and Uzbek mujahedin marshaled a tenacious rear-guard defense, covering the withdrawal of the Pakistani and Afghan forces. On-site Northern Alliance commanders decried the "cruelty" of the "Chechen gunmen" they were facing, as well as their willingness "to resist without consideration of their losses."

The siege of Kunduz was completed in the first days of November. U.S. airpower hammered the jihadists day and night, but the advance of the U.S.-led Afghan forces was slowed repeatedly by pockets of fierce resistance from what Northern Alliance commanders identified as "groups of Arab, Chechen, Uzbek, and Pakistani fighters." With many of their relatives fighting in the Taliban ranks, the Northern Alliance commanders opened negotiations with the enemy, offering safe passage to the besieged Afghans in return for the surrender of their Pakistanis and other foreign prisoners (who should have included bin Laden himself).

However, on the eve of the negotiations, the highest levels of Pervez Musharraf's Pakistani government interceded with their counterparts in the Bush White House, asking for a clandestine air corridor to Kunduz in order to evacuate the Pakistani military personnel. To avoid humiliating their newfound ally, Washington consented to Islamabad's request and permitted a number of Pakistani Air Force C-130s to make nocturnal flights to a designated airstrip in the Kunduz area. According to Indian intelligence officials, however, the Pakistanis used the corridor to fly not only more C-130 sorties than agreed upon, but also several sorties by ex-Afghan Air Force An-24s and An-26s—evacuating not only their own personnel, but also Osama bin Laden, several senior Islamist-Jihadist commanders, and most of the foreign mujahedin, out of fear that their capture would lead to damaging revelations about Pakistan's behavior. By this account, a total of four to five thousand fighters were evacuated by the Pakistani airlift.

Having just been declared "allies" by the Bush administration, Islamabad was loath to have any of the ISI-sponsored jihadists—including bin Laden and Zawahiri—captured and interrogated by the United States, for such interrogation would immediately have led to the discovery of the extent of Islamabad's support for anti-U.S. jihadist terrorism. It was therefore imperative for Islamabad to get all the incriminating "evidence" out of harm's way. At the same time, however, Islamabad was apprehensive about the ability of U.S. intelligence to track events in Pakistan, so Musharraf decided not to take a chance by shielding bin Laden and other high-profile senior commanders on Pakistani soil. Instead, the ISI helped them return to Afghanistan and continue the fight. Ultimately, however, following the collapse of the Taliban and the discovery of the sad state of U.S. intelligence, Musharraf provided bin Laden, Zawahiri, and their cohorts with shelter in Pakistan—rather than have them be captured by the United States—which they enjoy to this very day.

In Pakistan, senior ISI officers pressured bin Laden and other high-profile commanders to return to Afghanistan immediately and continue their jihad without implicating Pakistan. Some of the foreign mujahedin evacuated in the airlift—including Sheraly Akbotoyev, an Uzbek senior commander from southern Kyrgyzstan—also volunteered to return to Afghanistan. The Islamist-Jihadist leadership had other plans for the majority of the foreign mujahedin. Bin Laden instructed Juma Namangani, an Uzbek senior commander then responsible for the jihad in Central Asia, to press forward with strategic plans that had been in effect since the spring of 2001, using the majority of the foreign mujahedin just evacuated from Kunduz.

Akbotoyev, who was captured in Afghanistan in mid-2002, revealed details of the Central Asian and Caucasian surge during his interrogation by Kyrgyz security authorities. Back in May 2001, he disclosed, Taliban leader Mullah Muhammad Omar and Osama bin Laden had established a new international organization known as Livo designed to expedite the liberation of Central Asia. Akbotoyev defined the declared goal of Livo as "the creation of an Islamic State in all of Central Asia, which would engulf all of Kazakhstan, Kyrgyzstan, Tajikistan, Turkmenistan, Uzbekistan, and China's Xinjiang province." Juma Namangani—the founder of the Islamic Movement of Uzbekistan (IMU) and a close confidant of bin Laden—was appointed as the chief of Livo, and he would be advised by a supreme council consisting of Mullah Omar, Osama bin Laden, Tahir Yuldashev (another IMU leader now in Waziristan, Pakistan), Khassan the Uyghur (a senior leader of the Xinjiang separatists), and two Taliban commanders known only as Ubaidullo and Aimani. Their role was to facilitate the training and deployment of Central Asian and non-Pushtun Afghan mujahedin earmarked for the Central Asian surge. At first, Akbotoyev revealed, Livo was composed largely of mujahedin from Uzbekistan, but they were soon joined by fighters from all the Central Asian nationalities and ethnic groups, as well as a cadre of Arab and Chechen "Afghans." By the time of Akbotoyev's capture in summer 2002, Livo had already established a wide network of underground cells "in different parts of Central Asia."

After the fall of Kunduz in late 2001, one of Livo's immediate goals was to establish clandestine corridors to smuggle key commanders and mujahedin—mostly Arab and West European "Afghans"—across Central Asia to Chechnya, which was chosen as their new safe haven and springboard for future operations. According to plans set by bin Laden and Namangani, the majority of the Chechen mujahedin who survived the fighting in northern Afghanistan were to escort several Arab senior commanders across Central Asia to Chechnya, along the way staying in safe houses previously arranged by Khattab and his deputies. Oybek Rakhimov (a.k.a. the Uzbek), a longtime senior commander and aide to Aslan Maskhadov, was put in charge of ensuring logistical and financial support for the arriving Arab commanders. (Rakhimov was killed on January 16, 2002, by Russian special forces, who retrieved documents concerning the financial system for the Arab "Afghans" and confirming Maskhadov's involvement.) The first group of Chechen and Central Asian mujahedin left Pakistan in mid-November to survey the proposed routes to the Caucasus, and to activate the support system and safe houses along the road. The first group of mujahedin that included prominent Arab commanders left for Chechnya in late November, and by late December all the Arab and West European "Afghan" commanders were safe in Chechnya and ready to resume operations.

The two most important jihadist commanders who reached Chechnya in the Livo-run evacuation were Muhammad Shawqi al-Islambuli (a.k.a. Abu-Khalid), one of the leaders of Ayman al-Zawahiri's inner circle, and Mustafa Hamzah (a.k.a. Abu-Hazim). A longtime affiliate of HizbAllah International, and al Qaeda's chief intelligence representative in Iran since the late 1990s, Islambuli had been evacuated to Afghanistan in the summer of 2001 to avoid embarrassing Tehran with his presence.

Both Islambuli and Hamzah were still in Khattab's camps in mid-February 2002 when they chaired a major conference of senior Islamist-Jihadist commanders about the resumption of their jihad. The conference was attended by Khattab's deputy Abu-al-Walid, a number of key Chechen commanders—including the Akhmadov brothers and Movsar Suleimenov (a.k.a. Movsar Barayev, who represented the forces of his late uncle Arbi Barayev)—and commanders from Great Britain, Sweden, and Germany (including both recent arrivals from Afghanistan and Pakistan and newcomers who traveled secretly from Western Europe). The attendees discussed rejuvenating and expanding the Islamist-Jihadist networks in Western Europe, replacing their lost strategic base in Afghanistan with a new support infrastructure in Chechnya and the northern Caucasus, and using West European and Caucasian mujahedin for new terrorist operations in Europe and Israel. The participants also agreed on a plan to smuggle the West European commanders, Arab "Afghans," and other mujahedin from Chechnya to Europe.

Soon thereafter, Islambuli and Mustafa Hamzah returned to their posts in Iran (traveling via Georgia and Armenia). There they helped activate two way stations—a special Revolutionary Guards guesthouse in the Namak Abroud area, north of Tehran close to the Caspian Sea, and a safe house in Torbat-e Heydariyeh, south of Mashhad in northeast Iran—as bases for a growing operational group of about twenty commanders. Some of these leaders arrived in Iran via Chechnya; others, including Abu-Musab al-Zarqawi, crossed directly from western Afghanistan. By the spring of 2003, this Iran-based Islamist-Jihadist command cell was spearheading the launch of the Islamist jihad against U.S. forces in Iraq. Islambuli supervised the conduct of terrorist strikes in Baghdad and other parts of Iraq; Hamzah replaced him as al Qaeda's chief intelligence officer in Iran. Their key aides were Egyptian "Afghans" who had arrived with them from Afghanistan and Pakistan via Chechnya.

Chapter 19

CHECHENIZATION IN AFGHANISTAN

BY EARLY NOVEMBER 2001, U.S. AND ALLIED SPECIAL OPERATIONS forces, supported by a myriad of Afghan tribal forces, were closing in on the Taliban and al Qaeda strongholds in eastern Afghanistan. As the U.S. bombing raids intensified and the arrival of U.S. and allied ground forces were expected momentarily, Osama bin Laden crossed back into Afghanistan to join the local Islamist-Jihadist forces. On the night of November 10, he convened the key tribal chieftains and commanders in the Jalalabad area to discuss the coming battle. Though the main goal of the meeting was to embolden the attending leaders for the coming fight in Afghanistan, bin Laden stressed that this was only one off-stage battle in the wider worldwide jihad, and that other jihadist fronts should not be sacrificed in order to help Afghanistan. "The Americans have a plan to invade, but if we are united and believe in Allah, we'll teach them a lesson, the same one we taught the Russians," bin Laden assured his audience. The jihadists' duty, he stressed, was "to fight in Palestine, Chechnya, Kashmir, and everywhere that Muslims are being oppressed and tortured by the infidel."

Over the next few days, the foreign mujahedin withdrew with their Afghan and Arab allies into the mountain strongholds of Tora Bora, near the Pakistani border, to prepare for their first major battle with U.S. and allied forces. Given their proven performance in the fighting in northern Afghanistan, many of the foreign mujahedin were allocated specific tasks. By mid-November, an elite force of a few hundred Chechens, Yemenis, and

Algerians were handling security as the Arab and Taliban forces, including bin Laden's own convoy, evacuated the Jalalabad area and repaired to the Tora Bora compound. At first, this elite force was slated to return to Jalalabad and repel the U.S. and allied forces in fierce house-to-house combat. Bin Laden was convinced that inflicting heavy Mogadishu-style losses on the American forces, entangling them in protracted urban battle with serious civilian casualties, would force the Americans to withdraw. With their experience in urban warfare in Grozny, the Chechens were assigned the key roles in this plan. Yet bin Laden ultimately discarded this plan—among other reasons, because of repeated pleas from local notables to spare Jalalabad from relentless U.S. bombing—and the Chechen, Yemeni, and Algerian mujahedin withdrew to the heart of the Tora Bora compound to form bin Laden's second ring of defense. (The first ring—the core—was made up of bin Laden's own bodyguards.)

In late November, local Pushtun forces (funded by the United States) laid siege to Tora Bora. But their efforts were tenuous and largely ineffective, and they dragged on as American bombers intensified their attacks on both real and decoy targets. By now, there were between sixteen hundred and two thousand al Qaeda fighters in the Tora Bora compound, including both Afghans and foreigners—some four hundred Chechen mujahedin among them. With no U.S. forces to engage in Tora Bora, bin Laden decided that holding onto the compound was unnecessary, and that he should withdraw to a safe haven from where he could continue the worldwide jihad. Over the next few days, bin Laden and about five hundred Arabs quietly withdrew into Pakistan. Meanwhile, the predominantly Pushtun Afghan mujahedin melted into the countryside. Bin Laden left a small decoy force behind to conceal his escape and lure the Americans and their allies into protracted skirmishes. This force was put under the command of Abdallah Tabarak, a Moroccan transit worker who had served as bin Laden's bodyguard since his days in Sudan. The four hundred Chechen mujahedin, known for their skills in mountain warfare, were deployed around the perimeter of the Tora Bora complex, prolonging the impression that key commanders still lived there. Chechen and Yemeni mujahedin also maintained a number of decoy tent camps, which the U.S. forces bombed repeatedly.

By early December, with bin Laden and the key Arab mujahedin forces already safely in Pakistan, only a hard core of Chechen, Algerian, and Yemeni mujahedin remained in Tora Bora. Given the ineffective performance of the U.S.-funded Afghan forces, the United States and their allies steadily increased the intensity of their bombing campaign. Certain that U.S. ground forces would soon be deployed to Tora Bora to hunt them down, bin Laden and his key commanders dispatched the Chechen, Algerian, and Yemeni mujahedin to hole up in the most inaccessible mountain ridges and create the illusion that bin Laden and his forces were still hiding in the Tora Bora compound. The Chechens were instructed to simulate a far larger force, to mislead U.S. intelligence and distract allied attention from the steady flow of actual al Qaeda soldiers across the Pakistani border.

In mid-December, bin Laden's son Salakh-Uddin arrived in Tora Bora with a host of communication equipment—including bin Laden's own satellite phone—to bolster the deception. Chechen and Algerian mujahedin were still the core of the defense. They resisted all efforts by Afghan emissaries to convince them to surrender. By now, there were 250 Yemenis, 180 Algerians, and 350 to 400 Chechens still in the Tora Bora area. Abdullah Tabarak tried to induce the Americans to invade the Tora Bora compound by establishing—with the help of a few Chechen mujahedin—a bogus tactical communication network that frequently featured bin Laden's own voice. The al Qaeda forces knew that the United States depended heavily on technical intelligence, and they used the ploy to keep the allies focused on the Tora Bora seige.

Once the American troops finally lost interest, Tabarak turned his attention to the safe withdrawal of the surviving mujahedin. To shield their escape, he sacrificed his own security, joining a few Yemenis and Algerians and heading toward the Pakistani border while still using bin Laden's satellite phone. U.S. technical intelligence finally homed in on Tabarak, and a few days later he was ambushed and captured by U.S. special forces. By then, however, the roughly three hundred surviving Chechens, along with the Yemenis and Algerians, had already safely withdrawn to join the main al Qaeda forces in Parachinar, Pakistan.

From there, using the LIVO escape routes, they made their way across Central Asia to Chechnya.

For the Chechens, the battle for Tora Bora constituted the end of their organized presence in Afghanistan. A few dozen mujahedin, including Chechens, withdrew southward toward the Qandahar area, where Taliban and al Qaeda units were still operating. Along the way, they skirmished frequently with the Afghan forces run by the CIA and special operations forces (SOF). In his book *First In,* the CIA's Gary Schroen described one of these battles. A large force of Afghans, led by a U.S. special operations team, ambushed three foreign fighters running in open terrain toward a hill. The Afghans opened fire, but they missed the three men, who kept running as if they hadn't noticed the shots. "Then, from down the line, one of the Afghans watching the three men steadily cross the open ground shouted, 'Chechnya! Chechnya!'" Schroen reported. "The cry was picked up by others. 'Chechnya!' A wave of panic and fear, so intense that [the American SOF trooper with them] could feel it physically, swept through the line of men on the hilltop." It was a memorable moment: Under a hail of fire, the rallying cry of three Chechens far from home caused sixty of the CIA's Afghan fighters to abandon their positions and flee in terror. It took a two-thousand-pound GBU-31 bomb, dropped from a B-52 and guided by the U.S. SOF team, to kill what Schroen termed "the Chechen[s] standing so proudly" on the hilltop overlooking the Afghans' positions.

As Schroen recalled, this incident conformed with many other battles the CIA's Afghans were conducting at the time:

> In every battle they had fought . . . there had been rumors and reports that a group of Chechens was fighting with the Taliban. They were reported to be fanatical, fierce fighters, well trained and experts with their weapons. After one particularly tough engagement a few days earlier, a number of the dead among the Karzai forces had been found to have been killed by a single shot to the head. This was incredible to the Afghans, none of whom actually aimed their weapons but rather trusted Allah to guide their bullets. They thought that such accurate fire had to be the work of the Chechens.

* * *

In March 2002, the U.S. forces launched Operation Anaconda in the Shah-i-Kot valley. Anaconda was to be the largest operation involving U.S. ground forces, intended to destroy what was perceived to be the last major al Qaeda stronghold in Afghanistan. All together, about a thousand foreign mujahedin—Arabs, Chechens, Pakistanis, and Uzbeks—gravitated to peaks and caves of Shah-i-Kot valley. Individual Chechens were identified in key positions by Afghan militiamen working for U.S. intelligence. There were reports of "Chechens" setting up and manning heavy DShK machine guns and rocket launchers on the high controlling grounds overlooking routes the U.S. and allied forces would have to take in order to penetrate the valley. Significantly, the Arab mujahedin in the area were generally deployed to the rear, controlling longer-range heavy weapons such as mortars and sniper rifles. The Chechens, on the other hand, were deployed to the most dangerous positions, where they were most likely to engage U.S. ground forces as well as helicopters. The Chechens set up sophisticated ambush positions, arranging to be covered by fields of fire. As U.S. aerial activity intensified on the eve of the battle, the Chechens assumed greater prominence among the air defense crews as well. The members of one of the most proficient and threatening DShK mountaintop units were Russian speakers of Caucasian (i.e., white) appearance. The locals called them "Chechens."

After the battle, U.S. intelligence would retrieve from these mountaintop positions several documents handwritten in Cyrillic script—including highly professional range cards for artillery, mortar, and machine-gun fire, prepared for the coming clash with the Americans—as well as a notebook featuring sketches and instructions on how to build homemade bombs and sabotage techniques for bridges, buildings, buses, and cars, apparently written earlier at an Afghan or Pakistani training camp.

From the start of the battle on March 2, 2002, Operation Anaconda did not go according to plan. The order of battle was scuttled almost immediately in the predawn hours by the ambush and shooting down of an MH-47 helicopter carrying special operations forces. The ambush was a professional operation, spearheaded by a coordinated barrage of DShK and RPG fire from positions known to be manned by Chechens. In the first barrage, Petty Officer 1st Class Neil Roberts of the Navy SEALS was shot out of an MH-47 that ultimately crash-landed. From a distance, U.S. SOF teams watched as mujahedin from the nearby positions reacted. As Malcolm MacPherson reported in his book *Roberts Ridge*:

> A Chechen approached him who seemed to be in charge. He pointed the muzzle of an AK-47 at Roberts' head. And if Roberts was still alive, the Chechen executed him with a single bullet. If he wasn't alive, then the shot was meant as a 'security round' to guarantee that he was indeed dead. The local time was 0427. . . . Now, the Chechen who had shot Roberts from close range bent down and straddled his body. He drew a blade and tried to decapitate him, cutting his throat to the bone. He bent over his body for two minutes, searching him. He found the strobe, which he handed to other fighters gathering around. Then he disappeared into bunker #2.

Although subjected to relentless aerial bombing soon afterward, the Chechens held their positions tenaciously. Later that day, they shot down another SOF MH-47 on a rescue-and-reinforcement mission. Elsewhere in the area, the most effective fire directed at the U.S. and allied forces also came from positions held by Chechens, who seemed able to fight regardless of losses and damage to their side.

The basic goal of Operation Anaconda had been to send elements of the Afghan National Army—commanded by General Zia Lodin as well as local tribesmen, all of them guided by U.S. CIA and SOF—into the valley, supported by heavy U.S. and allied bombing. These troops would push the local Taliban and al Qaeda forces into confrontation with U.S. forces, composed mainly of units from the 10th Mountain and 101st Airborne divisions, which would use their vastly superior firepower to destroy the remaining mujahedin.

Yet the United States had overestimated the capabilities and resolve of their Afghan "allies" and underestimated the military skills and resolve of the mujahedin forces entrenched in Shah-i-Kot. In the ensuing combat operations, the Chechen mujahedin—now fighting as individuals in mixed formations—stood out for their dogged determination, as well as for carrying their dead and wounded with them. The U.S. forces—both the special operations forces fighting with the Afghans and the U.S. military units—soon discovered that virtually all the tactically advantageous terrain was covered with mujahedin positions, and that the most demanding positions—that is, those most likely to be subjected to heavy bombardment—were, in the words of an SOF participant, manned by "Chechens, and they were superb mountain fighters." As the fighting continued, the Chechen mujahedin linked up with the die-hard Pushtun Taliban, who were intimately familiar with the terrain. They consolidated their forces on the peak of Takur Ghar, where bunkers from the 1980s still harbored large caches of weapons and ammunition. The Chechens quickly established a command-and-control network, using radios to control and coordinate an unbroken defensive perimeter. These hardcore mujahedin, including the Chechens, held their ground for more than two weeks, despite overwhelming firepower and serious losses.

By mid-March, Operation Anaconda was over. The combination of the U.S.-led Afghan forces' disappointing performance and the U.S. forces' inability to gain a territorial advantage in the Shah-i-Kot valley, left a number of corridors open for the mujahedin to use in withdrawing once the Islamist-Jihadist supreme command had decided it was no longer useful to hold the ground. Having inflicted serious losses on the allied and Afghan forces (who repeatedly fled when encountering stiff resistance), the vast majority of the mujahedin withdrew to Pakistan in an orderly fashion. Led by Uzbek commander Tahir Yuldashev (who took over after the Afghan commander Saif Rahman Mansour was killed in a U.S. bombing on the first day), several hundred Arab, Uzbek, and Chechen mujahedin safely reached Pakistan. Most of the Uzbeks elected to remain in Pakistan's North-West Frontier Province (NWFP). But most of the Chechens were eager to look for new jihad battlegrounds, and by the early summer of 2002 most of them had left Afghanistan. Though a few individuals remained, joining local Taliban and Pushtun forces to rejuvenate the grassroots Afghan resistance against the U.S. invasion, it would be another two years before Chechen fighters would once again play an important role in Afghanistan and Pakistan.

The Islamist-Jihadist supreme leadership—including Osama bin Laden and Ayman al-Zawahiri—recognized the Chechen contribution to the struggle against the United States. On December 2, 2001, the Saudi-owned London-based newspaper *Al-Sharq al-Awsat*, one of the most authoritative Arab newspapers, began publishing an eleven-part serialization of Zawahiri's new book, *Knights Under the Prophet's Banner*. The series' editor called the book Zawahiri's "last will," and Zawahiri himself explained that he wrote the book "to fulfill the duty entrusted to me toward our generation and future generations. Perhaps I will not be able to write afterward in the midst of these worrying circumstances and changing conditions." Much of Zawahiri's book was devoted to articulating the strategy and logic of the coming stages in the global jihad—and both the Chechen mujahedin and their Caucasus safe haven played a large part in his vision.

Zawahiri's main thesis was that the U.S.-led war on terrorism, then in a formative phase, was not merely a reaction to the terrorist strikes of September 11, 2001, but rather the inevitable collision of two megatrends defined during the 1990s: the jihadist movement and the American lust for energy resources. He stressed that the Arab "Afghans" were the founding fathers and standard bearers of the new global jihad, and that their presence in remote jihad fronts worldwide throughout the 1990s had facilitated the radicalization and Islamicization of these crises. "These young men have revived a religious duty of which the nation had long been deprived, by fighting in Afghanistan, Kashmir, Bosnia-Herzegovina, and Chechnya."

The contribution of these Arab "Afghans" went far beyond the radicalization of localized conflicts. Through these efforts, Zawahiri stressed, the jihadist movement was capable of consolidating the forces of two regions that were of global strategic significance—regions that had already proven their ability to stave off American forces. He celebrated "the emergence of two Islamic states that liberated their territory under the slogan of jihad in the cause of God against the infidel occupiers of Muslim lands. Those two countries were Afghanistan and Chechnya." These two nations, he stressed, had become "the safe haven and destination of emigrants and mujahidin from various parts of the world, or what the United States describes as Arab Afghans, fundamentalists, terrorists, and so on." It is this global shift, Zawahiri explained, that was driving the U.S.-led West to pursue its so-called war on terrorism.

Zawahiri also elaborated on the role of Chechnya in the still-unfolding struggle that was rooted deeply in its heritage. Zawahiri emphasized that "the Chechen mujahedin's defiance of Russia, their insistence on liberating the Muslim Caucasus, and their determination to complete the jihad begun by Imam Shamil, may he rest in peace, against Czarist Russia posed a great threat to the influence and interests of the United States, for the Caucasus floats on a sea of petroleum whose estimated reserves are no less than the oil reserves in the Arabian Gulf, especially as the U.S. influence in Central Asia is increasing and taking the form of military bases, spy stations, oil companies, and joint maneuvers." In short, Zawahiri argued, it was more imperative for the jihadist movement to triumph in the Caucasus than in any other jihadist front then active.

Zawahiri analyzed in great detail the strategic ramifications of the success of the Chechens' jihad for the Caucasus as a whole, and predicted a jihadist tidal wave of historic significance originating from Chechnya.

> The liberation of the Caucasus would constitute a hotbed of Jihad (or fundamentalism, as the United States describes it) and that region would become the shelter of thousands of Muslim mujahidin from various parts of the Islamic world, particularly Arab parts. This poses a direct threat to the United States, represented by the growing support for the jihadist movement everywhere in the Islamic world. If the Chechens and other Caucasian mujahidin reach the shores of the oil-rich Caspian Sea, the only thing that will separate them from Afghanistan will be the neutral state of Turkmenistan. This will form

> a mujahid Islamic belt to the south of Russia that will be connected in the east to Pakistan, which is brimming with mujahedin movements in Kashmir. The belt will be linked to the south with Iran and Turkey that are sympathetic to the Muslims of Central Asia. This will break the cordon that is struck around the Muslim Caucasus and allow it to communicate with the Islamic world in general, but particularly with the mujahidin movement.

Zawahiri did not neglect the Chechen people's more immediate goal for the jihad—namely, the eradication of Russian presence in the Caucasus. He explained that "the liberation of the Muslim Caucasus will lead to the fragmentation of the Russian Federation and will help escalate the jihad movements that already exist in the republics of Uzbekistan and Tajikistan, whose governments get Russian backing against those jihadist movements." But he explained that this development would also further the interests of the anti-U.S. jihad: "The fragmentation of the Russian Federation on the rock of the fundamentalist movement, and at the hands of the Muslims of the Caucasus and Central Asia, will topple a basic ally of the United States in its battle against the Islamic jihadist reawakening," he argued. He acknowledged that Washington was aware of this strategic threat: "For this reason the United States chose to begin by crushing the Chechens by providing Western financing for the Russian Army, so that when this brutal campaign against the Chechen mujahedin is completed, the campaign can move southward to Afghanistan, either by the action of former Soviet republics [of Central Asia] that are U.S. agents, or with the participation of U.S. troops under the guise of combating terrorism, drug trafficking, and the claims about liberating that region's women," Zawahiri elaborated. "In this way, the United States will have destroyed the two last remaining hotbeds of resistance to it in the Islamic world."

In late 2001, Zawahiri stressed, this strategic posture was already crucial for the unfolding jihad against the United States. Global alliances and blocks were already being formed. To combat the U.S.-led war on terrorism, he warned, "a fundamentalist coalition is taking shape . . . made up of the jihad movements in the various lands of Islam as well as the two countries that have been liberated in the name of jihad for the sake of God (Afghanistan and Chechnya). If this coalition is still at an early stage, its growth is increasingly and steadily accelerating," Zawahiri predicted. "It is ready for revenge against the heads of the world's gathering of infidels, the United States, Russia, and Israel. It is anxious to seek retribution for the blood of the martyrs, the grief of the mothers, the deprivation of the orphans, the suffering of the detainees, and the sores of the tortured people throughout the land of Islam, from Eastern Turkestan to Andalusia [the Islamic state in medieval Spain]." Zawahiri added that these jihadist movements would also play a crucial role in the Islamicization of society, laying a foundation for the rise of Islamist-Jihadist societies and peoples. "A single look at the history of the mujahidin in Afghanistan, Palestine, and Chechnya will show that the jihad movement has moved to the center of the leadership of the nation when it adopted the slogan of liberating the nation from its external enemies and when it portrayed it as a battle of Islam against infidelity and infidels." It would be hard to find a more direct expression of the significance of the jihadist movements in Chechnya and Afghanistan.

Zawahiri concluded by predicting that both Chechnya and Afghanistan would remain pillars of the jihadist movement worldwide and foundations for the jihadist hold on the Hub of Islam. "Backing and supporting Afghanistan and Chechnya and defending them with the heart, the hand, and the word represent a current duty, for these are the assets of Islam in this age. The Jewish-Crusade campaign is united to crush them. Therefore, we must not be content with safeguarding them only. We must seek to move the battlefront to the heart of the Islamic world, which represents the true arena of the battle and the theater of the major battles in defense of Islam."

As if heeding Zawahiri's words, many of the Chechen mujahedin in Afghanistan were sent westward to Lebanon, where active preparations for a new wave of worldwide terrorist strikes were taking place. Starting in the late fall of 2001, a growing number of al Qaeda mujahedin swarmed into Lebanon and Syria, smuggled from Afghanistan via Pakistan, Iran, and Iraq. The Islamic Revolutionary Guards Corps organized the clandestine transfer of these terrorists and fighters, with Saudi Arabian money covering transportation costs. By early February 2002, a few hundred fighters were stationed in Lebanon—mainly in the Ein Hilweh Palestinian refugee camp in southern Lebanon, the HizbAllah-controlled Biqaa, and other Palestinian refugee camps. Though most of these were Arabs, there were also dozens of both Chechens and Kurds in their ranks, as well as smaller numbers of Central Asians, Kashmiris, and West Europeans.

Soon after their arrival, most of the Chechens were deployed to the HizbAllah bases in the Biqaa, where they absorbed local tactics while sharing their expertise in sophisticated explosives, anti-aircraft expertise, and urban guerrilla warfare. As additional battle-hardened Chechens arrived in Lebanon from the Caucasus, via Turkey and Syria, their next role was being deliberated in the the bin Laden-Moughaniyah forward headquarters in southern Beirut under the command of Muhammad Abu-Zubaidah—the third-highest commander in al Qaeda. In early 2002, the United States had identified Abu-Zubaidah as the driving force behind the resurrection of the al Qaeda networks in the West, a program he ran from numerous headquarters in Lebanon and Syria in cooperation with Syrian and Iranian intelligence services, the HizbAllah, and the Palestinians.

In early March, Abu-Zubaidah traveled to Yemen to attend a major terrorist conference in the settlement called Damatj in the northern Yemeni city of Saidy. The conference, which opened on the night of March 10, 2002, was attended by commanders and/or their representatives from Algeria, Saudi Arabia, Libya, Egypt, Jordan, Iraq, Turkey, Lebanon (both the HizbAllah and Sunni mujahedin), and the Palestinian HAMAS. The conference resolved to designate two theaters of operations as priorities: the European NATO countries and the nations of the Caspian Sea basin. In the latter theater, the main objectives would be to target American and Israeli diplomatic corps in the region, particularly in the South Caucasian countries, as well as oil companies and facilities in Kazakhstan, Turkmenistan, and Azerbaijan. The campaign was to be spearheaded by the Chechen and Central Asian mujahedin then stationed in Lebanon.

From Yemen, Muhammad Abu-Zubaidah traveled to Pakistan to brief bin Laden and Zawahiri. In late March, however, he was shot and captured in Karachi, Pakistan. In the face of this loss, the Islamist-Jihadist command called a halt to the many operations of which he had firsthand knowedge, and instructed the key mujahedin and operatives to go underground until further notice.

The main exception was the rapidly expanding al Qaeda network in Western Europe.

Chapter 20

MOSCOW STRIKES BACK IN CHECHNYA

STARTING IN THE FALL OF 2001, THE DEMANDS ON THE ISLAMIST-Jihadist forces in Afghanistan had a direct impact on the situation in Chechnya and the northern Caucasus. For a brief moment, the world, particularly the U.S.-led West, was willing to acknowledge that the ongoing insurrection in Chechnya was an integral part of the Islamist-Jihadist assault on the West. Emboldened by this recognition, Moscow seized the moment to strike out against the leadership of the Chechen jihad—and soon the jihad as a whole began to falter.

Moscow had actually begun taking an assertive new strategy toward the Chechen rebels in the spring of 2001. At first the Chechens were able to retaliate, increasing their campaign of terrorist bombing throughout the cities of Chechnya—using bombs to blow up buildings, cars, and other infrastructure. Despite the damage and casualties these strikes caused, however, the Chechens proved incapable of gaining the upper hand from Moscow.

The Russian crackdown came after a series of intelligence breakthroughs. In mid-June, for example, FSB and MVD troops exploited intelligence to conduct a search operation in the villages of Alkhan-Kala, Yermolovka, and Kulary, about ten miles southwest of Grozny. The operation resulted in the targeted killing of Arbi Alaa-Eddin Barayev, the senior commander of the Special Purpose Islamic Regiment (the Chechens' special forces unit), and eighteen of his commanders.

That same month, Russian intelligence picked up the trail of

Barayev and his command group, and the Chechen commanders were unable to shake them. On June 15, Barayev and his group sought shelter in his home village of Alkhan-Kala, hoping to disappear into the civilian population, but the Russians were right behind them. According to the Kavkaz-Tsentr website, Barayev reported to his superiors that "twenty minutes after his arrival, Russian aircraft bombed the house." Hearing the roar of the jets, Barayev and his guard left the house "literally a few seconds before the bombing." By now Russian troops had effectively made it impossible for Barayev and his bodyguards to leave Alkhan-Kala for the mountains; the mujahedin were forced to disperse into neighboring villages.

Between June 21 and 24, Russian special forces conducted a major mop-up operation in Alkhan-Kala and its environs. Within a few hours, acting on a well-timed intelligence tip, they identified the house where Barayev was hiding and laid siege. A senior Arab commander, the Amir of the Special al-Bara Subunit, and numerous Chechen and Arab commanders and bodyguards were with Barayev, but Russian mobile forces were able to block the roads connecting the various houses and villages where the fighters were hiding, and the mujahedin were incapable of assisting their commander. An intense firefight erupted between the besieged mujahedin and the Russian special forces surrounding them. By the time the fighting ended a few days later, Barayev, the amir, and at least seventeen other mujahedin were killed, and all others in the house were wounded and incapacitated.

Even as this siege was ongoing, Russian mobile forces were launching raids on nearby villages harboring mujahedin, with extensive fire support from artillery and strike aircraft. Roughly fifty of Barayev's troops were killed, including ten other field commanders who had accompanied him to Alkhan-Kala. Several mujahedin were captured, many of them betrayed to the Russian forces by the civilian population. At least thirty houses were destroyed in the heavy fighting, and there were local civilian casualties, mainly in Alkhan-Kala. Civilian casualties were relatively low because most of the local population had escaped as the Russian forces were moving in (except for those in the tightly sealed Alkhan-Kala), and the Russians kept them at a safe distance until the fighting ended. One Russian serviceman was killed and six were wounded in the operation.

In the coming months, the Russians would make even better use of such real-time battlefield intelligence. Most of the tips came from the local population, long since alienated by the Islamist-Jihadist forces, or from a growing number of fighters defecting to the Russian side. The Russian forces used the information to mount effective new strikes and raids on mujahedin sanctuaries. Both Ibn al-Khattab and Shamil Basayev survived close calls: Khattab was wounded in the shoulder and leg in one early winter strike in the Vedenskiy region (three of his bodyguards were killed), and Basayev was severely wounded in a similar raid in December, forcing him to be evacuated across the border for medical treatment.

By now, Khattab had assumed overall command of Islamist-Jihadist forces in Chechnya. He soon set about remaking his command structure, elevating fellow Arabs at the expense of Chechen commanders—particularly those who didn't follow the neo-Salafite/Wahhabi teachings he favored. But the forces he commanded were beginning to fracture. Facing pressure after the collapse of the strategic headquarters in Afghanistan and cut off from their flow of funds and instructions, the Chechen jihad was beginning to turn on itself in self-doubt. A growing number of Chechen and Caucasian fighters and commanders had begun cooperating with Russian intelligence—resulting in a growing number of close calls for Khattab and members of his inner circle. Late in December, the Russians launched several raids that came very close to killing Khattab and succeeded in killing Abu-Sayak, his deputy, and six other foreign mujahedin.

Convinced, not without reason, that these attacks were the result of leaks from his immediate entourage, Khattab ordered a thorough investigation, conducted by a small team of ruthless Arab and Pakistani mujahedin with extensive counterintelligence experience. At Khattab's instructions, they concentrated on the few native Chechen commanders who remained in top positions.

The investigators soon focused on two veteran field commanders—

the Sadayev brothers—who had been close to Shamil Basayev since the early days of the Chechen revolt. Khattab had long suspected them of colluding with the enemy, especially after a late September incident. A Saudi mujahed called Abu-Yakub had long been one of Khattab's closest confidants and one of his main money men, using a network run by family members to move funds from Saudi Arabia to Chechnya. After 9/11, Khattab instructed Abu-Yakub that funds previously earmarked for the Chechen jihad should be diverted for the rejuvenation of the global Islamist jihad—a diversion that required misleading the Chechen commanders about the flow of funds into the northern Caucasus. The Sadayev brothers were instrumental in forging the financial paperwork and evidence presented to the Chechen leaders, which suggested that Saudi Arabia had cut back its funding for the jihad in Chechnya. In reality, millions of Saudi dollars were being diverted to al Qaeda's coffers for use elsewhere in the world.

In late September, Russian security services cornered Abu-Yakub and the Sadayev brothers in a village called Staryye Atagi. The Russian troops surrounded the house where Abu-Yakub and the Sadayev brothers were hiding in a well-camouflaged cellar, and the swiftness of the raid was clear evidence of a timely intelligence tip. Abu-Yakub and his mujahedin escort held out for a few days but were eventually killed by the Russians; they were found in the cellar with an entire month's worth of food. Amazingly, however, both Sadayev brothers managed to evade the Russian dragnet and reach Basayev's camp. Khattab was furious about the killing of Abu-Yakub, but until the winter of 2001–2002, the Sadayev brothers were shielded by Basayev, who vouched for their loyalty. The minute Basayev was wounded and taken away to recuperate, Khattab unilaterally ordered the brothers' execution.

The security investigation within the jihadists' ranks only further alienated the Arab mujahedin from the Chechen civilian population. Increasingly on the run from the Russian security forces, by now Arab mujahedin were threatening and terrorizing village elders in order to prevent leaks and cooperation with the Russians. Their efforts were counterproductive: Such fratricidal infighting engendered animosity among key commanders, which Russian intelligence exploited by launching a new offensive in the spring of 2002. With security slackening as a result of the flow of mujahedin from Afghanistan and Pakistan via Central Asia, Russian security forces were soon able to target-kill a growing number of key commanders—both Chechen and Arab.

The Russian surge peaked in spring 2002, when the Russian intelligence services assassinated Ibn al-Khattab. It was an impressive achievement: By then, Khattab was hiding in a labyrinth of tunnels and bunkers in a remote ridge near the border with Georgia. Very few people knew the exact location of the isolated site, and access was strictly controlled.

The key to the operation was one Ibragim Magomedov—a friend and confidant whom Khattab used to run sensitive errands. A thirty-year-old Avar from the village of Gimry, Dagestan, Magomedov used at least one false foreign passport (bearing the name "Ibragim Alauri") to maintain Khattab's contacts with several sources of funding in the Arabian Peninsula as well as Khattab's own family in Saudi Arabia. Every few weeks, Magomedov illegally crossed the border from Dagestan to Azerbaijan, from where he would fly to Turkey and the Middle East using the fraudlent passport. On his way back, he would bring large sums of money and letters and parcels for Khattab.

The Dagestani branch of the FSB security service first noticed Ibragim's brother Gazi Magomed, and used him to entrap Ibragim. Their first step was using an FSB officer to gain Magomedov's confidence by helping him several times with the crossing of the Russian-Azeri border. When the FSB then confronted Magomedov with what they knew, they compelled him to cooperate.

In early March 2002, the FSB learned that Magomedov was going to Saudi Arabia to pick up a package for Khattab from his mother (including letters, a Sony video camera, and a wristwatch). The FSB intercepted the package while Magomedov was in Baku and laced the letter with poison—a neurotoxin that was absorbed through the skin, rapidly causing a heart attack or suffocation. On March 19, Magomedov delivered the package to Khattab. Half an hour later, Khattab emerged from his bunker where he was reading the letters, white and disoriented. He fainted periodically, but regained enough clarity to have Magomedov, who was immediately grabbed by Khattab's bodyguards, released. Minutes later, Khattab was dead.

Moscow needed proof of Khattab's death to present to the public, and when they received a tip that Khattab's chief bodyguard Ilyas Isayev (a.k.a. Elsi the Red) was arranging to have a videotape of Khattab's burial sent to Khattab's family, Russian special forces tracked him down. "We waited for Elsi and ambushed him in the mountains," an FSB officer recounted. "We killed him and retrieved the tape." The tape was soon broadcast on Russian TV.

Ultimately, the key to the FSB's success was the precise intelligence that enabled the Russians to identify and get to the Magomedov brothers in the first place. In principle, the FSB was doing its best to take advantage of divisions within Khattab's camp in order to recruit, in the words of the FSB officer, "people who could get so close to him as to be able to kill him." But Khattab was obsessed with his own personal security—particularly the loyalty of the people who knew his whereabouts—and getting the first clue was not a simple task.

Several Chechen former commanders and fighters insist that Khattab was betrayed by none other than Aslan Maskhadov, with the help of Abu-al-Walid. Khalid Yamadayev, a former Chechen commander who joined the Russians, claimed that Maskhadov tipped the Russians about Khattab's routines (which is possible) and even that he may have bribed a member of Khattab's immediate entourage to let the package through without inspection (which is unlikely). According to Yamadayev, Maskhadov feared that the Chechen cause and revolt were being lost to other Islamist-Jihadist priorities under Khattab's command.

Maskhadov was certainly aware that he was losing what power and authority he had to the Basayev-Khattab camp. Moreover, Khalil Ganiyev, one of the few Chechen mujahedin in Khattab's immediate circle, has provided a detailed explanation of the motives and roles of Maskhadov and Abu-al-Walid in the elimination of Khattab. "Khattab was in both their ways," Ganiyev argued. "He was actually the most senior military leader of the mojahedin; in comparison with him Maskhadov was nobody. True, he had officially been elected president. But he did not have real power. He never went beyond Kurchaloi and

Shali, areas where his relatives live: No one wanted to receive him anywhere. Abu-al-Walid wanted to take the place of the chief representative of the Muslim Brotherhood in Chechnya, whom Khattab was at the time."

According to Ganiyev, Maskhadov organized Khattab's assassination because he was afraid of his growing power, while Abu-al-Walid exploited the assassination in order to become "the main representative of the Muslim Brotherhood organization in Chechnya." By then, the tensions between Maskhadov and Khattab had been growing for some time. "Maskhadov and Abu-al-Walid were ready to eliminate Khattab back in the summer of 2001," Ganiyev explained. "Khattab suspected this. He realized that he was in Maskhadov's way and that Maskhadov wanted to eliminate him but he did nothing. He procrastinated." In Ganiyev's view, Khattab was sure "they would not dare to raise a finger against him. He counted on his closeness to Shamil Basayev, although he depended less and less on him. It is possible that in the end Khattab would have done something, but Maskhadov and Abu-al-Walid got there first."

Ganiyev claimed that "Maskhadov provided a figure who could be put forward as the person who committed the murder on the special services' orders." Maskhadov identified "the Dagestani Ibragim Alauri" (that is, Magomedov) as an appropriate connection to Khattab because he was able to implicate Magomedov/Alauri in actions against Khattab—most likely his embezzlement of funds from the Arab world. "He [Magomedov] refused but did not tell Khattab about this in time," Ganiyev noted. It seems clear, then, that Maskhadov knew that Magomedov could be turned, and had his name leaked to the FSB.

According to Ganiyev, it was Abu-al-Walid who needed Khattab's death confirmed on videotape so that he could quickly consolidate his position as Khattab's successor. "To this end, al-Walid deliberately sent one of Khattab's aides, Elsi, on a hazardous journey" with the tape. "As 'luck' would have it, on 10 April Elsi was killed and the federal forces got hold of the tape," Ganiyev explained.

Shamil Basayev also had to move fast to prevent Maskhadov from capitalizing on Khattab's death. Within days after the assassination,

Basayev ordered from his hiding place that Magomedov be hunted down, interrogated, and executed. Magomedov was feeling confident about his fate—after all, Khattab had ordered his release minutes before his death—and he assumed that the FSB would no longer bother him. So when Abu-Ziyad and Abu-Rabia—two Arab emissaries close to Basayev and Abu-al-Walid—invited him to Baku to discuss his next mission to the Middle East, he accepted. Basayev's Chechen hit squad followed Magomedov to Baku, where he was interrogated in a safe house and then shot in the head at close range. His body was found in a ditch near Baku with five bullet holes to the head. Islamist-Jihadist sources later reported that "the traitor had been gunned down for treason and cooperation with the Dagestani secret services."

Even as Magomedov was being eliminated, Abu-al-Walid, then thirty-four years old, was moving quickly to consolidate his position as Khattab's undisputed successor. Walid, whose real name was Abd al-Aziz Bin Ali Bin Said al-Said al-Ghamdi, came from the Governorate of Baljrashi in southern Saudi Arabia. A cousin of two of the 9/11 hijackers, he (like Khattab) had spent formative years (1986–1987) with Abdallah Azzam in Afghanistan/Pakistan. He fought in the jihads in Afghanistan, Tajikistan, and Bosnia-Herzegovina during the late 1980s and early 1990s, and his performance in Bosnia brought him to the attention of the Islamist-Jihadist supreme leaders. Abu-Maali, then the commander of the mujahedin in Bosnia-Herzegovina, recommended Walid to his friend Khattab when he organized the first group to go to Chechnya. Although he settled in Bosnia and married a local woman, Abu-al-Walid readily agreed to go to Chechnya, and in 1995 he was part of the first group of expert terrorists who escorted Khattab to Chechnya. Since then, he had been fighting there as a member of the Basayev/Khattab inner circle. Abu-al-Walid's ascent to power and his quick consolidation of his own headquarters confirmed that the Islamist-Jihadist leaders had anticipated the assassination of their commanders and had prepared succession plans well in advance.

After Khattab's assassination, Basayev and his high command went underground. They began spreading rumors that Basayev himself had been assassinated by Russian secret services, hoping to divert the

FSB's attention and dissuade would-be traitors from betraying him by cooperating with Maskhadov.

A few weeks later, once he felt more secure, Basayev unleashed a series of terrorist strikes in the northern Caucasus. The first was a bomb in a market in Vladikavkaz, Northern Ossetia, that killed eight and wounded forty-five on April 29. Basayev decided to target Vladikavkaz solely because he was hiding nearby and he was confident that his communications with the bombers would not be intercepted by the FSB. Then Basayev ordered a massive bombing in Kaspiysk, Dagestan, on May 9. The remote-controlled detonation of a MON-90 mine during the victory day parade caused forty-five civilian fatalities and wounded more than one hundred—mainly World War II veterans and their grandchildren who were marching together. Chechen spokesmen characterized the terrorist bombing as "retribution" for Khattab's assassination, but Basayev ordered these strikes less to avenge Khattab's blood than to remind the world of his relevance.

And Basayev had reasons to worry. After Khattab's death, Maskhadov had hastened to regain control over the Chechen movement, convening a council of leading Chechen commanders and establishing a new Shura that excluded Basayev, Abu-al-Walid, and their allies. Addressing these commanders, Maskhadov vowed to revitalize the Chechen national liberation struggle in Chechnya. It was a direct challenge to Basayev's leadership and to the character of the jihad, and the strikes in Vladikavkaz and Kaspiysk were Basayev's response. He also used messengers to send threats to the key Chechen commanders who had rallied to Maskhadov's cause and began issuing public communiqués and news stories reacting to key issues of the day, demonstrating that he still had his finger on the Chechen pulse.

Basayev used a mid-May interview with the Prima News Agency to articulate his strategy for the next phase of the jihad. Basayev argued that the stalemate between the Chechens and their Russian adversaries was all but irreversible. "Mujahedin know what to do, and neither I nor other leaders are needed very much now," he conceded. The future lay in further terror strikes within Russia, which could generate enough political fallout to force a decision in the Chechens' favor. As far as the

Islamist-Jihadist leaders were concerned, he stressed, "all of Russia is at war." Addressing the Russian people directly, he invoked the same justification bin Laden and other Islamist leaders had used to justify terrorist strikes against civilians. "Do not assume that we consider you a peaceful population," Basayev told the Russians. "To us you are unarmed military men, because those who by a majority approve the genocide of the Chechen people cannot be peaceful citizens. According to the Shariah, mere verbal approval of war put peaceful citizens in the ranks of the enemy. You are just the unarmed enemy."

As the weather improved in spring and early summer 2002, routine terrorism and sporadic fighting resumed in Chechnya. Interestingly, this escalation of violence raised fears among the Islamist-Jihadist leadership that the Chechens' separatist struggle might revert to its old nationalist character, losing its jihadist mission. The media's ongoing preoccupation with the global reverberations of 9/11—and particularly with the spectacle of Russia's alliance with the United States and the West in the war on terrorism—was raising public doubts about the viability of the Islamist-Jihadist movement. As Basayev himself had conceded, at the grassroots level the Chechen insurrection had no real leadership. And Maskhadov's efforts to reestablish himself as a leader and commander were faring no better than Basayev's: When he issued a call to professional fighters—mainly Chechen but also Caucasian and Arab—to return from their safe haven in Georgia to resume the fighting, his summons was largely ignored.

By now, the grassroots Islamicization movement throughout the northern Caucasus was virtually irreversible. At the same time, however, there was a growing popular backlash against the sluggish pace of the revolt; many everyday Chechens had begun to pine for a return to normal life, even if it meant returning to Russian domination. And this appetite for normalcy and a better quality of life now led a growing segment of society to cross over and cooperate with the Russian authorities. Increasingly, the real support for the anti-Russian insurrection came from the Islamist-Jihadists rather than the population at large. Any would-be leader of the anti-Russian struggle would have to gain the Islamists' support—to convince them that he was fighting the jihad. Maskhadov and his inner circle realized this: Unable to build a power base on their own, they knew they would have to reach a deal with the Islamist-Jihadist leadership in order to remain relevant.

Abandoning his attempt to marginalize his terrorist rival, Maskhadov now consented to a formal new power-sharing arrangement with Basayev. On June 6, 2002, Maskhadov formally convened a gathering of commanders, ostensibly intended to review restructuring options for the Chechen government. Basayev and a number of Islamist imams (both local and Arab) became the dominant voices in the conference, which evolved into a three-week summit. Before it was over, Maskhadov, Basayev, and their respective inner circles had formulated a new strategy for the northern Caucasus, as well as a legal structure for the leadership of the ChRI—mandating a new distribution of power between the president and the Islamist-dominated Majlis ul-Shura. Giving the shura this newly expanded role was the Islamist-Jihadists' attempt to counterbalance and restrain Maskhadov's ambitions. Observers in Moscow considered the conference a "Maskhadov defeat," a watershed marking the Islamist-Jihadists' formal domination of the Chechen revolt. Maskhadov had resigned himself to figurehead status. "In the final analysis," Paul Murphy observed in his book *The Wolves of Islam*, "Maskhadov may have simply had little choice if he wished to go on fighting." Yet in real terms such subtle shifts were increasingly symbolic—and would remain that way as long as the Russians retained the upper hand in Chechnya.

Under the government's new command structure, based on the principles of the Islamist-Jihadist shura, the newly dominant Basayev and his Islamist-Jihadist allies resolved to "increase guerrilla warfare on the entire territory" of Chechnya and to undertake "certain other important military measures" outside of Chechnya. The conference empowered the new command to concentrate on escalating their jihad in the form of new attacks within Russia and in the West. Maskhadov also endorsed the Islamist-Jihadist goal of "creating an Islamic state and Shariah rule" throughout the entire northern Caucasus. In a message laced with Islamist-Jihadist terminology, he predicted that the aggregate impact of these measures would "force the Russian occupiers to leave our long-suffering land."

In practical terms, Basayev emerged from the conference in undisputed control over the next wave of fighting, installing his Islamist-Jihadist loyalists—both Caucasian and Arab—in key command positions. Abu-al-Walid was named commander of the eastern front, and another follower, Dokka Umarov, commander of the southwestern front. Basayev loyalists were also nominated to key political positions that had been held by Maskhadov's people: Movladi Udugov was named head of the External Subcommittee of the Information Committee—that is, official spokesman for the Chechen jihad—and Zelimkhan Yandarbiyev was named special representative of the president of Chechnya to the Muslim countries, a job that combined political and fund-raising responsibilities—putting both now completely out of Maskhadov's control.

A key goal of the agreement between Basayev and Maskhadov, reached in the opening days of the conference, was to rejuvenate the liberation jihad in Chechnya itself. Maskhadov was convinced that success in this arena would strengthen his hand, and he and Basayev agreed on a new plan that would essentially reprise the Chechens' most successful previous offensives—namely, the surprise capture of Grozny, Gudermes, and Argun.

Toward this end, the Islamist-Jihadists used considerable new funding from the Middle East to acquire and smuggle large quantities of weapons and ammunition to the region. In mid-June, Georgian security authorities seized a cache of munitions, missiles, and other equipment worth roughly half a million dollars—just one of many such shipments. Basayev also ordered a huge number of elite forces to infiltrate the three cities and to take up positions in the villages and hills overlooking the cities. The Chechen forces launched an intense intelligence and reconnaissance operation in and around Russian facilities in the cities—but the sudden influx of young males hanging around in the slums was quickly noticed by locals.

By this time, a growing number of Chechen mujahedin were cooperating with Russian authorities. Most of them were Chechen nationalists who had lost confidence in Maskhadov's ability to rejuvenate the liberation movement and who resented Basayev's Islamicization campaign. The ranks of the mujahedin suffered both a seepage of defectors and deeper penetration by Russian intelligence. On June 24, the eve of the intended offensive, Russian special forces raided Maskhadov's forward headquarters, missing him by minutes, but discovering and seizing elaborate plans for the forthcoming offensive. The most impressive of these was an operational plan for a joint offensive by more than a thousand fighters of Basayev's and Maskhadov's, intended to generate a serious flare-up in Grozny. The discovery of the plans deprived the Chechens of the element of surprise, and they hastily withdrew most of the mujahedin from the cities to safe havens in the remote mountains; many of those who didn't leave were arrested by Russian security forces. As the Russians launched their rounds of arrests, several hundred nationalist fighters—including a number of midranking commanders—changed sides, recognizing the futility of their fight and Maskhadov's failing grasp on power. The Chechens' plan to storm Grozny, Gudermes, and Argun was altogether scuttled.

The only real escalation in the summer and fall of 2002 came in rural areas dominated by jihadist forces, buoyed by the unbroken flow of mujahedin and resources from Afghanistan-Pakistan. The mujahedin launched a few successful raids in the late summer, using surplus SFSAMs to shoot down two Russian helicopters—a gigantic Mi-26T in mid-August and an Mi-24 gunship in early September. The missiles used in these attacks came from secret stashes near key Russian cities, such as one cache hidden in a cemetery overlooking the runway of Moscow's Vnukovo airport, which was discovered by Russian security services in early June.

Yet these strikes were hardly sufficient to turn the tide and reignite the popular struggle in Chechnya. With Maskhadov and the nationalist leadership effectively neutralized, the Islamist-Jihadist leaders and commanders turned back to the bigger picture—and to new targets in Moscow and Western Europe.

Chapter 21

STRIKES IN MOSCOW AND IN WESTERN EUROPE

IN LATE 2002 AND EARLY 2003, THE SUPREME ISLAMIST-JIHADIST leaders attempted to launch a spate of spectacular terrorist strikes throughout Europe—from Moscow in the east to Paris and London in the west—from their strategic safe haven in a cluster of terrorist camps in southeastern Chechnya and just across the border in Georgia's Pankisi Gorge. While the Moscow strikes succeeded, those planned for Western Europe were narrowly averted at the last minute.

These terrorist strikes had been conceived during a commanders' conference in mid-February 2002. Chaired by two Egyptian senior Islamist-Jihadist commanders, Muhammad Shawqi al-Islambuli and Mustafa Hamzah, the meeting included the senior Chechen commanders Abu-al-Walid, Movsar Suleimenov (a.k.a. Movsar Barayev), and the Akhmadov brothers, as well as several commanders from Great Britain, Sweden, and Germany. The participants resolved to replace the lost strategic base in Afghanistan with a new support infrastructure in Chechnya and Georgia, to use that safe haven to expand the Islamist-Jihadist networks in Western Europe and launch new operations from Moscow to London.

In the wake of the conference, Basayev established a new elite "martyrdom battalion," formally named the Riyad-us-Salikhin [Fields of Righteousness] Reconnaissance and Sabotage Battalion of Shakhids [Martyrs]—that was dedicated to the conduct of new terrorist strikes in Moscow and other Russian cities. Both the Akhmadov brothers and

Movsar Barayev would play central roles in the battalion's operations. In a stark deviation from comparable entities in the Middle East and Asia, nearly half the would-be martyrs in Basayev's battalion were young women. The other half was composed of veteran fighters, most of them graduates of Khattab's training camps.

At around this time, the Chechens also launched a new effort to secure supply and transportation lines between the northern Caucasus and the Moscow area, dispatching a number of long-term sleeper agents to Moscow. Two of Basayev's key operatives—Aslambek Khaskhanov (a.k.a. Zaurbek) and Ruslan Elmurzaev (a.k.a. Abubaker or Yassir)—had already approached Akhyad and Alikhan Mezhiev, a pair of Chechen brothers living in Moscow, asking for their help in establishing a clandestine support system in Moscow. The brothers were to rent three apartments to be used as safe houses and to acquire vehicles for their travel and cargo transport; they were given money, including $2,500 in cash, and told to start immediately. In April a longtime Basayev operative named Khampash Sobraliev acquired a small house in the village of Chernoe, in the Moscow Oblast, paying for the $20,000 house in cash. A Chechen family moved in, quickly building a high fence around the property, and soon a regular stream of visitors began appearing at the house, many of them driving expensive foreign cars and SUVs. Most of the visitors were suspected of having connections to the Chechen Mafiya.

Yet the sleepers made little actual headway in Moscow until mid-2002, mainly because the Chechen Islamist-Jihadist leaders were on the defensive—preoccupied at first by the Russians' assassination of Khattab and then by the power struggle that culminated in the Basayev-Maskhadov summit in June.

By now the specter of a U.S. confrontation with Iraq was in the air, and the resulting tension in the Middle East had a direct impact on the Islamist-Jihadist infrastructure in the northern Caucasus. That summer, a new forward control center and several advanced training bases were activated in the Pankisi Gorge. Key centers were established in Omalo and in two other hamlets near the central settlement of Duisi—villages that had been frequented by Arab volunteers traveling to and from

Chechnya since late 1999. The Chechen Islamist leadership assigned Abu-Hafs al-Urduni as commander of the Pankisi Gorge infrastructure. A Saudi-Jordanian, Abu-Hafs had been long-term operations chief for both Khattab and Abu-al-Walid; his nomination reflected the importance of this new center. The other key commander in Pankisi Gorge was Abu-Atiya (born Adnan Muhammad Sadiq), a Jordanian now married to a Chechen woman. Abu-Atiya and another Jordanian known as Abu-Taysir (a.k.a. Sheikh Taysir Abdallah or Abu-Khadishakh) were part of the group of expert terrorists and commanders that Osama bin Laden had sent to help Khattab and Basayev in Chechnya in the spring of 1999. Both Abu-Atiya and Abu-Taysir were among the closest long-term associates of another rising Jordanian commander—Ahmad Fadil Nazzal al-Khalayleh (a.k.a. Abu-Musab al-Zarqawi). In early 2002, having traveled from Afghanistan via Iran to Iraq to assume command of the fledgling Islamist-Jihadist forces there, al-Zarqawi nominated Abu-Atiya as his representative in the northern Caucasus. In August 2003, Azerbaijani security forces captured Abu-Atiya as he was trying to travel to the Middle East via Baku and extradited him to Jordan with the help of the CIA.

The primary purpose of these new camps was to serve as a home base for the launching of new strikes on U.S., Russian, and Western installations in Russia and Western Europe. Toward this end, the Islamist-Jihadists procured large quantities of high explosives and stored them in local hideaways. Since Abu-Atiya was considered knowledgeable in the use of toxic gases, he was put in charge of the preparations for the spectacular strikes. Under him, in an isolated installation, a team of six chemists worked on developing poisons and chemical and biological weapons, initially intended for use on smaller-scale Western targets in Central Asia, and then in more spectacular strikes after their effectiveness was proven.

The Pankisi Gorge center became operational in July 2002 with an attempted biological weapon attack in Turkey. On July 10, Turkish authorities learned of an envelope containing a "highly toxic substance" being couriered from Abu-Atiya in Georgia to one "Musab"—a sleeper operative in Turkey (most likely in Istanbul). Musab's cell was instructed to use the product as the basis for a "biological poison" within twenty days (before the substance lost its potency), launching a biological attack on the U.S. and Russian embassies in Ankara. The plan was thwarted when those who prepared the package in the Pankisi Gorge mishandled the toxic substance: By the time Musab and his network received the material it was completely useless, and with Turkish security authorities on to their plot, Musab and the would-be terrorists escaped back to Georgia.

The supreme Islamist-Jihadist leaders took the failure in Turkey very seriously. It confirmed recent warnings from Abu-al-Walid that Russian pressure had made it impossible for jihadist leaders in Chechnya to provide the comprehensive training and long-term preparations required to launch effective strikes within Russia—or anywhere else, for that matter. The one service of which the jihadist camps were capable was providing short-term safe havens for pre-mission briefings and mission launches. Yet the supreme Islamist leaders still considered the anti-Russian jihad critically important, as Islambuli and Mustafa Hamzah conveyed when they returned to Iran from their trip via Chechnya.

Both Osama bin Laden and Ayman al-Zawahiri wanted to stage a dramatic breakout from their failing war in Afghanistan, in the form of a new series of strikes that would "bring the war to the heart of the enemy." They targeted Western Europe for these new strikes, hoping to incite and mobilize local Muslim communities there. At the same time, Saddam Hussein was looking for a way to divert global attention away from the gathering storm around Iraq, convinced that a deniable terrorist strike in Western Europe would shock and mobilize the Europeans to stand in the way of America's plans to attack Iraq. With the collapse of their power base in Afghanistan, the Islamist-Jihadist leaders needed a new base for training and preparations, and at the request of emissaries of bin Laden and Zawahiri, Saddam Hussein agreed to allow Islamist-Jihadist training operations to be moved to Iraq and allocated resources in Iraqi training centers—especially Unit 999 of Iraqi Military Intelligence in Salman Pak near Baghdad. It was a quid pro quo: In exchange for Saddam's assistance, the jihadists would launch a series of spectacular strikes in Western Europe that would benefit both parties.

During the summer, a stream of Islamist-Jihadist terrorists from the Middle East and Western Europe arrived at Salman Pak for a specialized training program run by members of Unit 999. According to Palestinian terorrists who were being trained at the time at a nearby facility, this select group of jihadist terrorists were specifically identified as members of al Qaeda. In addition to the many special operations techniques taught at Salman Pak, the jihadists also received elaborate training on chemical weapons and poison, ricin in particular.

In September, the first graduates of this program were dispatched to the Caucasus. On their way to their deployment zones, the jihadists were taken to a derelict complex of houses near Halabja, in Iraqi Kurdistan, where they conducted experiments to familiarize themselves with chemical weapons and poisons. The area where this training took place was nominally under the control of Ansar-al-Islam—Osama bin Laden's Kurdish offshoot, led by Abu-Musab al-Zarqawi. From there, a few Islamist-Jihadist detachments were to travel to Turkey to prepare for striking U.S. bases with chemical weapons once the war started. Most of the Salman Pak graduates were dispatched to the Pankisi Gorge to assist their Chechen counterparts in launching terrorist operations against Russia, but some were sent to train Islamist teams arriving clandestinely from Western Europe in sophisticated terrorism techniques, including the use of chemical weapons and ricin.

The role of Chechnya and Georgia as a springboard for operations in Western Europe was facilitated by Abu-Doha (born Rashid Bukhalfa). An Algerian who arrived in Afghanistan in the late 1980s, Abu-Doha grew close to Abu-Musab al-Zarqawi during their joint training and operating in Afghanistan in the 1990s. By mid-2002, Abu-Doha was operating in London, recruiting operatives from Muslim communities in Western Europe. He also communicated with Zarqawi, requesting assistance in organizing expert trainers and special weapons for his operatives. They agreed that this training would take place in Chechnya and Georgia, given the ease of travel from Europe and the safety provided by the local Islamist-Jihadist infrastructure. Abu-Doha was arrested in the United Kingdom in 2004, where he now is fighting extradition to the United States for his role in the abortive Millennium Plot of 1999–2000.

Among the first jihadists from Western Europe to be trained in the Caucasus were Menad Benchellali and Mourad Merabet, two French Algerians from Vénissieux, a low-income, predominantly Muslim suburb of Lyon. The son of a radical preacher and Islamist activist, Benchellali had trained in Syria and Sudan in 1995–1996, in Afghanistan in 1998–1999 (where he stayed well over a year), and again in the summer of 2001. Merabet, a chemist by education, arrived in Afghanistan in 2001 with Benchellali, and the two men experimented in chemical weapons and poisons—including testing on animals. Another French Algerian from Vénissieux in the group that traveled from Afghanistan to Georgia—who was subsequently caught in France—told French interrogators that Benchellali was trained in producing ricin and making "chemical or bacteriological products to commit an attack." In the fall of 2001, after the United States invaded Afghanistan, Benchellali and Merabet were among the expert terrorists dispatched to the Chechen safe havens in the Pankisi Gorge. There, the group expanded their knowledge on chemical weapons and other WMD, working with Arab trainers (including Salman Pak graduates) dispatched by Zarqawi. The training took several months, floated by funds from France sent through Aslanbek Bagakhashvili, another Georgia-based associate of Zarqawi.

Anxious to develop some forward momentum, Shamil Basayev and Abu-al-Walid hoped to mount a new strike against Russia that would have the impact of the Budennovsk maternity ward seizure or the Moscow apartment bombing—shocking Russia and the world, and reestablishing the Islamist-Jihadists' dominance in the Caucasus. The Chechen leaders convinced the supreme Islamist leaders that a spectacular strike in Moscow could be a key to the dramatic breakout they were seeking. In the summer of 2002, with a growing number of graduates arriving from Salman Pak, the supreme leaders agreed to assign top-quality resources and experts to facilitate the Moscow strikes.

In August, Aslambek Khaskhanov and Ruslan Elmurzaev made a clandestine visit to Moscow to inspect the support system, securing two of the vehicles and the cell phones that would be used in the forthcoming strikes. To reduce suspicion among the Russian security services, they were joined by forty-two-year-old Yassira Vataliyeva, a veteran female fighter of Basayev's. Vataliyeva helped rent the safe-house apartments using fake documents and reconnoitered the anticipated targets, studying security procedures and traffic patterns and videotaping the sites.

The Islamist-Jihadist commanders were simultaneously developing two alternative plans. The first was to attack the two houses of the Russian Parliament—the State Duma and the Council of the Russian Federation—using chemical weapons to maximize fatalities among Russia's parliamentarians. The alternative was to target a major cultural institution in Moscow, an option that led to prolonged debate over the prospect of a multitude of civilian hostages. The jihadists studied three major theaters in Moscow—the Moskovsky Dvorets Molodyozhi, the Estrada, and the Dubrovka. At Basayev's behest, they also considered the Kurchatov Institute, Russia's preeminent nuclear research institute, but the idea proved impractical and was abandoned at an early stage.

To supervise the potential strike on the Parliament, Basayev selected a fiercely loyal commander known only as "Abdul," a veteran of previous terrorist strikes including the Moscow attacks organized by Raduyev. Abdul visited Moscow several times that summer to inspect the preparations, but in the early fall he was killed in a Russian raid in Chechnya, and Basayev abandoned the Parliament option. Khaskhanov and Elmurzaev, who were in charge of actual preparations for all the Moscow strikes, were instructed after Abdul's death to concentrate on the three theaters, and Elmurzaev chose the Dubrovka after personally visiting the three theaters. In an interview given during the siege, he explained, "We chose the [Dubrovka] theater because it is in the center of the city and there were a lot of people there."

As the date of the strike neared, Basayev decided that Movsar Barayev—a participant in the original commanders' meeting in February—would assume command of the theater strike, which would involve taking civilian hostages to enhance the operation's publicity impact. Moreover, the Islamist-Jihadist leaders wanted someone like Barayev, who was close to the jihadists' Arab supporters associated with the attack, to stress that this was a jihadist strike. Basayev would later acknowledge that Barayev was a last-minute choice: In a late April 2003 interview with Kavkaz-Tsentr, Basayev recalled that "I included Barayev in the group only in late September. I had only two hours to talk to him and give him instructions."

Basayev didn't give final authorization for the strike until mid-October, once he was satisfied that preparations had been successfully completed. Shortly thereafter, on October 18, Basayev, Barayev, and Elmurzaev met with Maskhadov and Sheikh Abu-Omar, the Saudi emissary of the International Muslim Brotherhood in Chechnya and a functionary of the major Saudi Islamist foundations (which would be closed and outlawed by the U.S. and Saudi governments in the aftermath of 9/11). Alerting them to the coming strikes, Basayev predicted that its strategic-political impact would match that of his own 1995 attack on Budennovsk, known as Operation Jihad. Maskhadov agreed, stressing that he was fully aware of the strategic significance of the forthcoming strike. "We have practically accomplished a transition from guerrilla warfare to offensive combat operations," Maskhadov told the group. "I am convinced, and I do not have a shadow of a doubt on this, that during the concluding stage of our struggle, we will definitely hold an even more unique operation, similar to [Operation] Jihad. And with this operation, we will liberate our land from Russian aggressors."

At this stage, except for Basayev himself, only Elmurzaev was familiar with the details of the intended strike—including its initial target date of November 7, the anniversary of the Bolshevik revolution, now celebrated as the Day of Accord and Reconciliation. In mid-October, he briefed Khaskhanov about the specifics of the forthcoming operation, reporting that Shamil Basayev had personally ordered him to prepare "a very large action" involving the seizure of hostages. As Elmurzaev explained it, Basayev's plan called for an escalatory campaign of bombing in Moscow—using both car bombs and suicide bombers in public places—in order to create an escalating atmosphere of fear in the city and across Russia. Each bomb would have greater power and lethality than the last, and they would culminate with the attack on a crowded theater.

But this escalating-bombings plan had a built-in flaw: It put the Russians on high alert before the terrorists' most vicious attacks could be triggered. The first attack, a small bomb in a car parked near a McDonald's in southwest Moscow, took place on October 19. One person was killed in the attack, and a few were wounded. Another, more powerful car bomb was parked near the Tchaikovsky Theater Hall in central Moscow, but it failed to explode; on the night of October 19, Elmurzaev clandestinely removed the failed bomb from the car. By this time, several car bombs and "women's martyr belts" (vest-bombs) were stored in Moscow awaiting the next round of bombing. But the Russian security services had swung into action after the McDonald's blast, and by the morning of October 20 Aslambek Khaskhanov was feeling the heat. Convinced that it was no longer safe for him to stay in Moscow, he flew to Nazran, Ingushetiya, on false papers later that day. His fears were justified on October 22 with the arrest of Aslan Muradov, the owner of the Tavriya car that exploded near the McDonald's.

Losing Muradov was a major blow to the jihad: Operating as "Artur Kashinskiy," he had been a Chechen sleeper in Moscow for at least ten years, acquainted with many of the safe houses, weapons and explosives storage sites, and other jihadist operatives in the greater Moscow area. In reaction, Basayev decided to launch his next strike immediately, before Russian security landed any more blows to the network.

Since mid-October, the jihadists tapped to participate in this next attack had been making their way to Moscow—individually or in pairs—using various means of public transportation. They dispersed to numerous safe houses and thereafter stayed away from each other: Only a few more senior members knew how to reach their comrades in order to activate the entire ring. At least some of them were summoned by Emurzaev for a meeting near the Luzhniki sports stadium on October 21—a mere two days before the attack. At the meeting they were informed of the date and location of the attack, given specific instructions for each cell member in the initial assault (including the time and place where their cells would be picked up by minibus and driven to the strike), and instructed to make sure that all participants memorized their instructions. It was a tight and secure network, making leaks virtually impossible.

Still, Basayev and his fellow leaders were concerned that the Russian security services might close in on the Moscow cells before the operation could be launched. At the last minute—on October 21, the day before the Dubrovka bombing—Elmurzaev was instructed to create yet another major diversion to distract security forces. He gave two of the female would-be martyrs suicide belts/vests from existing stockpiles, and early on the evening of October 23, just hours before the Dubrovka attack was scheduled to begin, he sent Akhyad Mezhiev to drop the two women off in crowded places at the center of Moscow so that they could blow themselves up along with a maximum number of casualties. Mezhiev headed for the crowded and trendy Pyramid Café at Pushkin Square, but traffic was heavy—and by the time they reached the center of Moscow, they heard on the news that the Dubrovka attack was already in progress. Their window of opportunity had closed.

With security heightened throughout the city, Mezhiev removed the suicide belts/vests from the two women, disarmed them, and drove them to the train station, where he bought them tickets to Nazran, Ingushetiya. After sending them off that night, he handed the belts/vests to his brother Alikhan, who transferred them to members of another Chechen cell. The next day, Mezhiev called Elmurzaev—now already inside the theater—to report the events of the previous night. The security services, who intercepted the call, identified Mezhiev and arrested the two brothers shortly thereafter.

The main event—the assault on the Dubrovka Theater—was launched a little bit after 9:00 P.M. on that night of October 23. At least twenty-two men in camouflage fatigues and nineteen women in long back robes with their faces veiled arrived at the theater in three minibuses, storming through the entrance and scattering through the concert hall, where the musical *Nord-Ost* was playing to an audience of roughly one thousand. Following their instructions, the terrorists swiftly took their designated positions, blocking all doors and placing a number of bombs in key positions. The terrorists had brought a veritable arsenal into the theater: twenty-five bomb belts/vests, two 88-pound bombs in metal cylinders, thirty-nine mines and booby traps, 114 hand grenades, fifteen AKSU-74 assault rifles, and eleven handguns.

One of the terrorists stepped on the stage and fired a few shots into the ceiling. "You are hostages," Movsar Barayev declared. "We've come from Chechnya. This is no joke. We are at war." Elmurzaev, who was first introduced as "Yassir," quickly emerged as the man in charge; Barayev later told a journalist that his "aunt Zura"—Arbi Barayev's sixth wife and widow—was the leader of the women's group. The Chechen women all declared their desire to die as martyrs because they had lost their male family members in the war and therefore had no reason to live. Yet at least two female terrorists later contradicted that assertion in conversation with the hostages, claiming to be married to (and pregnant by) male terrorists who were also at the theater.

At first, the operation went according to plan. Members of the terrorist unit placed bombs in key structural positions, where they would bring down the hall on top of the hostages if detonated. Elmurzaev had brought the bomb he had removed from the car in the aborted second bombing, having fixed its faulty timer. Other terrorists ordered the audience to show their hands and drop their bags and cell phones on the ground. Even the mysterious appearance, some two hours into the siege, of a screaming young woman in the middle of the hall did not disrupt the operation: She was caught and summarily executed by the terrorists. Shortly after the woman was killed, the terrorists let a few hostages leave—foreigners (mainly Muslims and Georgians) and a pregnant woman—sending them out to meet the Russians with a list of their demands. A few more escaped on their own, but the terrorists hardly complained; the spectacle of several terrified civilians who could convey the sense of horror inside only served to put further pressure on the Kremlin.

As the freed hostages reported, Barayev had announced that the terrorists were on "a martyrdom mission" to stop the war in Chechnya and would "go all the way" unless Putin met their three demands: (1) to allow a rally against the Chechen war to be staged for the media now gathering near the Dubrovka Theater; (2) to organize a second and larger rally on Red Square, where Russia would announce that it was ending the war; and (3) to have all Russian troops withdraw from Chechnya by the morning of October 26. Elmurzaev interrupted Barayev's monologue to state calmly that if their demands were not met, the terrorists would do to Moscow "what Hitler couldn't," elaborating that widespread carnage would be inflicted by numerous "suicide squads all over the city" that were awaiting his signal to strike out.

The next two days—October 24 and 25—degenerated into a hostage-taking and media festival. Credible and amateur mediators, media personalities known for their sympathy toward—and even activism on behalf of—the Chechen cause were sporadically permitted into the theater, where they were provided exclusive interviews with Barayev and some of the female terrorists. The reporters got firsthand glimpses of the traps the Chechens had set, and of the hostages' conditions. Visitors were forced to trade statements of sympathy for the Chechen cause in return for drinking liquids and bathroom privileges for the hostages (in an improvised facility in the orchestra pit) as signs of the terrorists' humanitarian concerns. Throughout, the hostage takers and their de facto accomplices were building media pressure on the Kremlin to accede to the terrorists' demands—an absurd campaign, given the terrorists' own repeated declarations that they had come to Moscow to die with their hostages in order to demonstrate to all Moscovites the pain and horrors of the war in Chechnya. Senior Kremlin officials offered free passage to the terrorists in return for the lives of the hostages, but the idea was rebuffed. As mediator Giorgiy Yavlinskiy later recounted, Barayev and Elmurzaev "were unable to come up with any coherent negotiating position," because they never intended to negotiate anything. But the terrorists' campaign worked, perhaps better than expected: The lure of participating in negotiations drew an endless flow of political and media personalities in and out of the theater and a resulting stream of photo ops with the awaiting media outside.

Some of the most unmediated coverage of the terrorists and their demands came from the Middle East. On October 23, immediately after the theater was stormed, Al-Jazeera TV broadcast a prerecorded message in which the terrorists announced that their attack had been launched "on orders from the Chechen Republic's military commander"—that is, Shamil Basayev. The terrorists announced their three demands and declared that the operation was intended to convey the horrors of Chechnya to the people of Moscow and the Russian government.

From exile in Qatar, Movladi Udugov emerged as the terrorists' spokesman, planting a flag for the international Islamist-Jihadists as players in the event. In BBC interviews and later postings on Chechen and Islamist-Jihadist websites, Udugov declared that the "mujahedin" were "not going to surrender. The building is mined." The terrorists were "demanding the withdrawal of [all Russian] troops from Chechnya," Udugov stressed, and would accept nothing less. Throughout the crisis, Barayev and Elmurzaev were in constant touch with Udugov by cell phone, and Barayev also exchanged e-mails using his laptop. Udugov recast the content of some of these calls as "exclusive Kavkaz-Tsentr interviews," which he posted on the Web and distributed to Arab media. Udugov hammered home Barayev's assertion that the mujahedin were holding more than a thousand hostages in a hall surrounded by bombs, and that all would surely die if any rescue attempt was made.

On October 25, Barayev had at least two conversations with Zelimkhan Yandarbiyev, then also living in Qatar. In these lengthy calls, Yandarbiyev tried to get Barayev to acknowledge that Maskhadov had known nothing about the forthcoming strike (even though by then there was ample evidence that Maskhadov had blessed the idea of launching major strikes in Moscow as early as his June talks with Basayev). Yandarbiyev suggested to Barayev that the operation was losing its effectiveness with passage of time as public sympathy for the plight of the hostages increased—but instead of calling for an end to the operation, he urged that the strike be brought to a dramatic climax. Shortly after their second conversation, Barayev told journalist and activist Anna Politkovskaya (who was mediating at the time) that he would "wait only a little while longer" for Putin to declare the end of the war and withdraw all Russian troops from "one district of Chechnya" as a prelude to a total withdrawal. Barayev set 6 A.M. on October 26 as the deadline. Some of the terrorists informed the hostages that they would begin their execution around that time, and a few hostages managed to make frantic phone calls to their relatives, who informed the authorities. (On May 8, 2003, on the basis of these conversations, the Russian government issued an international arrest warrant against Yandarbiyev, formally requesting his extradition from Qatar. When Qatar refused, Russian intelligence assassinated Yandarbiyev in Qatar in February 2004 using a car bomb.)

Unwilling to call Barayev's or Elmurzaev's bluff, the Kremlin ordered an audacious rescue operation before dawn on October 26. The raid had been in the works for days: On the night of October 23, Russian security services planted numerous sensors inside and around the theater and began collecting data about the hostage takers. Elite units, mainly the FSB's Alfa and Vympel special forces units, immediately started preparing for a rescue raid, refining its plans as they accumulated new information.

As the hours passed, their intelligence suggested that the hostage takers were on the verge of psychological collapse. By October 24, some were taking stimulant drugs and medications to stay awake—albeit with reduced capacity. The following day, Barayev and some other terrorists raided the theater's bar and started drinking hard liquor to calm their nerves. The drinking had a devastating effect on their psyches: As devout Muslims, they had never had alcohol, and the drink had an immediate effect on their senses, coordination, and reactions. Yet the Russians monitoring the situation weren't convinced that any of the hostage takers were so impaired that they wouldn't be able to detonate their bombs or fire into the crowd of hostages if so instructed. On the contrary, by the evening of October 25, the Russians began to fear that the cumulative stress might make the terrorists increasingly prone to extreme violence and erratic behavior.

That afternoon and evening, Vladimir Putin was thoroughly briefed on the special forces' rescue plan. The security services informed him that numerous sources (including the Barayev-Yandarbiyev conversations and testimony from some of the mediators) suggested that the terrorists were exasperated, and their threats to begin the execution of the hostages early next morning must be taken seriously. That evening Putin authorized the rescue operation, prudently leaving it to

Aleksandr Machevsky, the senior Russian official on site, to decide the time of the raid.

Shortly after midnight, the hostages began to crack. When a young man darted from out of the crowd, screaming, a female terrorist opened fire, missing him but hitting two other hostages. Barayev rushed to the scene—visibly distraught and indecisive. Some forty-five minutes later, he called for emergency services to evacuate the wounded (one of whom later died). Barayev then addressed the hostages—shifting between death threats and optimism about negotiations with a Kremlin representative and a "green corridor" out of Moscow all the way to Chechnya. Yet the hostage takers were increasingly edgy and trigger-happy. Some of the female terrorists were nervously playing with the wires of the bomb vests they were wearing. Between two and three o'clock in the morning, the terrorists fired a few bursts inside the theater, as well as out of the windows—ostensibly at suspected Russian special forces. By three o'clock there were hostages screaming inside, the pace of gunfire was growing, and the Russian monitors discerned the thud of at least one grenade going off. The intelligence experts monitoring the situation warned Machevsky that the hostage takers were on the verge of explosion, likely to start slaughtering hostages out of sheer frustration and rage.

At 3:08 A.M., having confirmed that the special forces commanders were ready to go, Machevsky gave the signal. In less than fifteen minutes it was all over.

The special forces pumped a psychochemical gas—also known as "knock-out" gas or BZ—through the theater's ventilation system. As the strong, sleep-inducing gas took effect, some two hundred special forces from the Alfa and Vympel units broke into the building. Twelve strike teams rushed into the main hall and nearby stairways and corridors, swiftly engaging the terrorists and shooting them at close range before they were able to activate their bombs. Meanwhile, snipers took out terrorists standing or moving near windows. A number of terrorists in remote hallways were not affected by the gas and put up resistance. They had to be chased around by Russian strike teams until they were gunned down. All the terrorists were eliminated swiftly, many with a single or two bullets to the forehead, and none of the bombs and grenades went off. A special strike team rushed upstairs and found Barayev sitting in the room he used for media meetings, a bottle of brandy in his hand. They put three bullets in his forehead before he was able to make a move.

Only after special forces had secured the building, announcing the death of the terrorists and the neutralization of the bombs, were medical and emergency personnel permitted to treat the hostages. By now, more than one hundred of the hostages had succumbed to side effects from the gas. Several weaker and older people died of heart attacks induced by the gas, quite a few choked on their own vomit, and others broke their necks as they lost consciousness and fell. The medical plight of the hostages was further complicated by the fact that area hospitals were given no advance warning about either the impending rescue or the medical characteristics of the gas, and medical help was slow to arrive, chaotic, and inefficient. Several hostages who came out of the theater alive died within hours or days due to improper medical care—bringing the total fatalities to more than 130.

Still, the rescue raid was a success. The Russian forces had confronted an unprecedented challenge—a theater with a thousand hostages—and managed to eliminate the terrorists and free the vast majority of hostages. The casualty rate among the hostages was about fifteen percent—well below average in comparable operations. Amazingly, the Russian forces suffered no casualties in the operation.

There was another reason behind the terrorists' failure to detonate their bombs: Even before they were carried into the Dubrovka Theater, they had been rendered inactive by a Russian agent in the ranks of the Moscow Chechen network. Back in the spring of 2002, as the network in the Moscow area was beginning to organize, they were joined by Arman Menkeev. A half-Chechen (Kazakh father, Chechen mother), Menkeev was a former major in the GRU SPETSNAZ who claimed to have left the service in disgust in 1999 over the launch of the Second Chechen War. A specialist in bomb making and other clandestine skills, he offered the Chechens badly needed expertise—and a secret operative with all the necessary papers to move and stay freely in Moscow. He managed to make the Dubrovka bombs inoperative by attaching empty batteries to the fuses and supplying them with insufficient quantities of accelerator explosives, so that any detonations would fizzle.

At the time when Machevsky ordered the strike, he could not have been certain that Menkeev had succeeded in neutralizing the bombs. Moreover, the Chechens had two other bomb makers—one Chechen and one Arab expert—in the Moscow area, and the bombs inside the theater could have been built and/or repaired by them. So the special forces could not have taken a chance with the bombs. Only later would the Russians learn that most of the bombs inside the theater were indeed Menkeev's defective weapons.

After the Dubrovka siege ended, Menkeev was arrested by the FSB as part of the dragnet in the greater Moscow area, and then quietly released. His role would have remained secret—as it should have been—had it not been for the bureaucratic rigidity of the Russian legal system. On November 22, 2002, when asked by a Russian magistrate why Menkeev was not brought to court after his documented arrest, the FSB representative acknowledged that Menkeev was actually "loyal to the government" and that he had proven that he "knows how to keep military and state secrets." (It wasn't the first time Menkeev had proven his value in this way: During the 1990s, Menkeev was decorated for bravery several times while fighting in Chechnya. His official residence remains in Ryazan—the headquarters of Russia's elite Airborne Troops and the GRU SPETSNAZ.)

In later interviews, Shamil Basayev would stress that the terrorists were unable to detonate their bombs because the detonators were "non-functional," attributing their failure to a traitor in the Chechens' ranks and vowing revenge. On November 1, he issued a statement through Udugov's Kavkaz-Tsentr acknowledging that the hostage seizure had been carried out by members of the Riyad-us-Salikhin under his direct command, and vowing that this was the first of a wave of coming attacks throughout Russia. In the self-contradictory statement, Basayev first pronounced the raid "a successful military operation in the very lair of the enemy, in his heart—Moscow," but then went on to say that "the operation was aimed at stopping the war and halting the genocide of

Chechen people, and if it failed as such it should at least have shown to the entire world that the Russian leadership can destroy without hesitation and regret its own citizens in the center of Moscow by using the most brutal methods." Basayev conceded that "we unfortunately did not attain our main goal—an end to the war and genocide of the Chechen people," but defended the Chechens' right to seize a thousand innocent theatergoers by comparing their number with the number of innocent civilians allegedly killed by the Russians in Chechnya during the war. He lambasted Western critiques of the attack as hypocritical, complaining that they made no mention of the victims in Chechnya.

Basayev stressed that the main lesson of the Dubrovka Theater crisis was that there was no point attempting to negotiate with the Kremlin through terrorism or any other means. "Our martyrs came to Moscow not to seek support, public opinion or sympathy. They came to end the war or become martyrs," he explained. This choice would not be offered again, he vowed, for the next wave of strikes would be far bloodier. "People without any demands, who will not be taking anyone hostage, will come next time . . . and their main goal will be to destroy the enemy and deal the most severe blows onto him," Basayev promised. Under his command, the Chechens would continue inflicting civilian casualties throughout Russia until the suffering became unbearable for both the public and the Kremlin. "[S]ooner or later, like it or not, the Russian people and leadership will have to end this bloody slaughter. They will have to stop this war, agree to peace and get off our land. Sooner or later we will achieve victory, but as long as there is one single Russian soldier on Chechen land, this war will go on, and from now on it will take place both here, on Chechen territory, and throughout Russia—the aggressor country." Basayev concluded by appealing to "all shahids to join the ranks of the Riyad-us-Salikhin reconnaissance and sabotage battalion of shahids," for it had long been a Chechen principle that "war can only be stopped by war."

Even as Basayev was making his pronouncements, a new group of West European mujahedin were training in the Chechen camps in the Pankisi Gorge. Islamist-Jihadists had been arriving at Abu-Atiya's camps from Western Europe for training in bomb making and the use of chemical weapons. The graduates were to return to their home countries with practical knowledge in the conduct of spectacular and most lethal strikes against the civilian population. One of Abu-Atiya's favored proposals for future operations in Western Europe was to cause widespread poisoning by dispersing cyanide derivatives through the water mains of European cities. French security forces would later retrieve methylene blue, an antidote to cyanide, originating from the Pankisi Gorge camps. Other trainees were taught how to use SFSAMs against civilian aircraft in Western Europe and were told that such missiles would soon be smuggled from Chechnya. In the coming years, those trained in the Pankisi Gorge would come to populate the Islamist-Jihadists' new web of European cells, operating under Abu-Musab al-Zarqawi.

Starting in the early summer of 2002, high-quality French-Algerian operatives, including both foreign mujahedin already fighting in Chechnya and roughly two dozen recent arrivals from Western Europe (mainly France, the United Kingdom, and Germany) were diverted to a specialized training program in a Pankisi Gorge camp. A couple of them, led by the FIS (Front Islamique du Salut, or Islamic Salvation Front)operative Khaled Ouldali, were arrested by the Georgian security services in August at the request of the French authorities. But the vast majority of the European terrorists reached their destinations safely, and by the early fall they had completed their training and orientation and began returning clandestinely to Europe. The volume of satellite and computer-encrypted communication between the jihadist command center in the Pankisi Gorge and centers in Western Europe intensified.

On November 9, acting on intelligence warnings that chemical attacks using operatives and weapons of Chechen origin were being planned for London, British security authorities arrested Rabah Kadre, the chief of the French-Algerian network in London. Kadre's initial interrogation provided enough clues for the French security forces to launch a massive dragnet for the "Chechen network" in their midst. In December, the French rounded up several French-Algerians in various suburbs of Paris. In one of the first homes searched, the French discovered vials of chemicals, including cyanide, two gas cylinders, and protective clothing for handling chemicals; later searches turned up operational triggers and fuses. Particularly alarming, in the words of the French interior ministry, were "electronic systems, [which] were in working order and permitted explosive devices to be set off by remote control using portable phones." (The same kind of remote-control triggers and fuses would be used in the Madrid commuter train bombings of March 2004.) It didn't take long for the French security authorities to identify and apprehend Menad Benchellali. At his family home in Vénissieux, the French recovered "a home laboratory for the production of ricin," as well as equipment and materials for "the manufacture of explosives and toxic gases of the cyanide type." In their interrogation, two members of Benchellali's network acknowledged that they were preparing for a chemical attack on the Russian embassy in Paris using ricin and botulin.

Thus, by the end of 2002, the French security services had dismantled two Islamist-Jihadist networks trained and organized in the new Chechnya-Georgia terrorist hub. The primary objectives of these networks were to launch strikes against the Eiffel Tower and other Parisian landmarks, as well as department stores, police stations, and the rush-hour Metro. The attacks were to be carried out with "chemical weapons," to cause as many casualties as possible among tourists and French civilians. The network was also preparing to strike the Russian embassy with chemical and/or toxic weapons. The original rationale for the strike was to be revenge for Khattab's assassination, but shortly before the strike the jihadist leadership in Pankisi Gorge asked the cell to hold off: Soon, they anticipated, there would be martyrs at Dubrovka Theater to avenge as well.

In early January 2003, the interrogation of the French-Algerian terrorists helped the British security services uncover a comparable network in Wood Green, East London—a cell whose core was composed largely of French-Algerians trained in the Pankisi Gorge camp, but which included recruits from the Muslim community in the United Kingdom. Members of the network were caught manufacturing toxic chemicals and ricin poison using the same techniques and materials taught in the Pankisi Gorge camp and used by the Paris networks. The

London network's intended targets were similar to those of the Paris networks: They, too, had the local Russian embassy in their sights, as revenge for the war in Chechnya, as well as a host of crowded public places and transportation hubs in the heart of London. Specifically, the terrorists targeted the Tube at rush hour and London's Jewish center. The captured terrorists acknowledged that they intended to use toxic chemicals.

On January 14, 2003, British security forces raided a terrorist safe house in Manchester, recovering a quantity of ricin and other equipment and materials used to produce explosives. One of the three Algerians caught in the apartment was a senior lieutenant of Rabah Kadre's named Kamel Bourgass. Once identified by the security authorities, Bourgass was asked to undergo a test to verify whether he had handled toxic substances. He lashed out, grabbing a kitchen knife and stabbing a Scotland Yard detective to death. Among Bourgass's possessions was a large envelope containing detailed instructions for producing explosives and ricin. Another London terrorist later acknowledged to investigators that he had been sent to London "by Abu-Atiya" specifically in order to "carry out chemical attacks."

Chapter 22

THE BLACK WIDOWS

THE FRENCH AND BRITISH RAIDS IN THE WINTER OF 2002 TO 2003 dampened the Islamists' hopes of making Western Europe their next locus of terrorist insurgency. Yet that winter also saw the Chechen jihad bounce back, with a series of new attacks in Russia, in the Caucasus, and against pro-Moscow officials in Chechnya itself. Over the next two years, the terrorists managed to multiply the frequency and effectiveness of their strikes and episodes of random violence, using terrorism as their primary method of confronting Russia while attempting to draw popular sympathy away from the Russians back to the jihadist cause. Yet, in so doing, the Islamist-Jihadists trampled on the exhausted population's hopes for security or economic recovery. Indeed, with their popular support dwindling, the jihadists set about trying to alienate the people from the security forces—a desperate measure designed to prevent the population from turning on the mujahedin.

This new wave of terrorism would increasingly involve the use of martyr-terrorists—particularly young female suicide bombers, commonly known as Black Widows, who became a prime instrument and symbol of the Chechen terrorist jihad. The Black Widows were a piquant symbol of the disconnect between the Islamist jihad in the Caucasus and the Chechen population in whose name the jihad was ostensibly waged. Although the Chechen Islamist-Jihadists used a wide array of terrorist tactics during their campaigns of 2002–2004, the Black Widows were a true—if atrocious—innovation of the period, and an increasing challenge for the Russians. The significance of the

Black Widow phenomenon transcended the mere fact that women were entering the ranks of would-be martyrs. It was a final, and desperate, symbol of the Islamist influence on segments of Chechen society.

Before 2002, there were two isolated cases of female terrorists in the Chechen conflict. In July 2000, Hawa (Khava) Barayeva—the niece of Arbi Barayev—drove a truck full of explosives into a military post. In November 2001, Ayiza Gasoueva killed the Russian military commander of Urus-Martan, who had tortured and killed her husband in her presence. But the phenomenon of Black Widow strikes followed the dramatic involvement of female terrorists in the seizure of the Dubrovka Theater in October 2002, exploited by the jihadist leaders for maximum public impact. The Islamist-Jihadist media highlighted revenge and solidarity with their brother mujahedin as factors in the Black Widows' motivation. "These women, particularly the wives of the mujahedin who are martyred, are being threatened in their homes. Their honor and everything are being threatened," Abu-al-Walid told Al-Jazeerah TV. "They do not accept being humiliated and living under occupation. They say that they want to serve the cause of almighty God and avenge the death of their husbands and persecuted people."

Conventional wisdom holds that this sudden and unprecedented upsurge in female terrorists was the result of the horrific conditions created by the war in Chechnya. Having lost all, or virtually all, of their male relatives to the Russians, the logic went, these women had nothing left to live for; taking revenge against their tormentors was the only salve for their despair. Sympathetic European sociologists and psychologists studying the phenomenon concluded that the Widows were driven by despair born of extreme violence toward either the women or their families, and that their relative isolation left them feeling deprived of any perspective or alternative. According to this argument, the Islamists weren't responsible for motivating these young women into suicide—only for providing them with an "opportunity for sacrifice" and revenge. "There is a line of [young women] hoping to be chosen as candidates for being suicide bombers," journalist and activist Anna Politkovskaya told CNN. "They say they want to force Russians to feel the same pain they have felt."

Yet this popular explanation is contradicted by the evidence in several ways. For one thing, the plight of civilians during the Chechen nationalist rebellion of 1994–1996 was far greater than during the Second Chechen War, which began in 1999, yet the first war produced no suicide bombers, not to mention female suicide bombers. The second war, on the other hand, was primarily Islamist, not nationalist, in character, and the female suicide bombers it produced must be understood in the context of the war's overall Islamist ideology and value system. As in the Arab Middle East, the Islamist-Jihadist recruiters and terrorist masters exploited religious devotion but also shame and despair to drive women to attempt martyrdom. In Chechnya, the oppressive circumstances of the war increased women's desperation—but it was the alien influence of enforced Islamicization that introduced and legitimized the idea of self-cleansing-through-revenge through martyrdom bombing.

Yet, as the stories of the individual women who became Black Widows suggest, the phenomenon was born not of a single cultural trend, but out of an infinite variety of influences. Some Black Widows were indeed widows, or sisters, of Chechen mujahedin, yet others came from families not profoundly affected by the war. Some lived in abject poverty, some in relative affluence. Some claimed to have been raped, tortured, or otherwise humiliated at the hands of the Russian military, yet others claimed to have suffered similar treatment by the mujahedin themselves. Some Black Widows hailed from families that were clearly aligned with the revolt—in both its nationalist and its Islamist-Jihadist strains—others were social outcasts, motivated by personal depression or despair. The intensity of religious devotion varied from one Black Widow to another. The key thing—perhaps the only thing—the Black Widows had in common was that they all ended up in the hands of the Islamist terrorist cells, submitting themselves to training for suicide missions.

The new wave of Chechen terrorism, known as Operation Whirlwind, was launched by Shamil Basayev and Abu-al-Walid toward the end of 2002. On December 27, a Chechen man and his two teenage children blew up the Russian-dominated Chechen administration building in

Grozny. With his fifteen-year-old daughter in the cabin, Gelani Tumriyev drove a KamAZ military truck loaded with a four-metric-ton bomb. Tumriyev's seventeen-year-old son drove a GAZ military jeep loaded with a three-hundred-pound bomb behind the truck. Both Tumriyev and his son were wearing military uniforms and waved military IDs and papers at the guards at the gates as they sped by the checkpoints; with a young girl in the cabin, the truck looked relatively innocuous. The two vehicles crashed into the administration building, and the resulting explosions completely destroyed the four-story building. There were at least 83 fatalities, and more than 150 occupants suffered injuries, although none of the key Chechen leaders who had been expected to be in the complex at the time was injured.

Although the operation was hailed as the work of an entire family driven to martyrdom by the anguish of the Russian oppression, Basayev himself later cast doubt on whether this was a martyrdom operation after all. He claimed to have been present in Grozny to supervise the operation, and to have activated the bombs himself by remote control. "I not only took part in the blast, I pushed the button on the remote control explosives devices that were in the vehicles. I was watching from a distance, and when the vehicles disappeared from eyesight and entered the premises of the puppet administration, I pushed the button," Basayev recalled. Apologists for the Chechen jihad later clarified that the Tumriyev family had volunteered for martyrdom, but, given the extreme importance of the strike, Basayev had taken matters into his own hands to ensure that they didn't falter at the last minute. If true, Basayev's claim to have activated the bombs raises questions about whether the Tumriyev family got cold feet about martyrdom, if they ever intended to be martyrs at all.

These weren't the only attempted strikes in late December. At the open market near the Yugo Zapadnaya metro station in Moscow, police acting on a telephone tip arrested two young Chechens in their twenties carrying antipersonnel bombs they intended to detonate among the holiday shoppers. The Chechen leaders weren't deterred by such setbacks: For the coming year of 2003, Udugov promised a new wave of "kamikaze operations" to unnerve and punish the Russians.

The fall 2002 operations, however, had momentarily exhausted the jihadists' pool of would-be martyrs. Moreover, a close examination of operations like the Tumriyev family's bombing plot suggested that the resolve of such amateur martyrs wasn't always irreversible, and the leaders were concerned about missions being aborted because of last-minute doubts. To shore up their stable of committed, reliable bombers, Basayev and Abu-al-Walid put a recently arrived corps of Iraq-trained Arab expert terrorists to work on an intense training program near Urus-Martan in southeastern Chechnya, for would-be martyr-bombers, bomb makers, and indoctrinators for would-be martyrs. The Arab experts brought knowledge of more sophisticated clandestine techniques, better bomb-making techniques, reliable fuse and detonator technologies, and, most important, the sophisticated art of sustaining and reinforcing the psychological resolve of aspiring martyrs while underground in a hostile environment, such as a Moscow safe house.

The first graduates were ready for action in April or May 2003, and soon a new wave of operations was launched within Chechnya. After testing a number of improved bombs in April, the terrorists planted a a concealed roadside bomb in Grozny, destroying a minibus and killing every passenger inside. A bomb concealed in a manhole cover destroyed another vehicle as it was passing above, killing eight. In Gudermes, a sophisticated small bomb hidden inside a couch was detonated by remote control, killing Dzhabrail Yamadayev, the commandant of the local pro-Moscow special military group. In mid-April, a minibus used to carry personnel to and from the Khankala air base was split in two and its fifteen passengers killed by the explosion of a powerful remote-controlled roadside bomb. A second antipersonnel bomb was detonated as rescue workers approached the scene, killing two and wounding dozens. The last operation in this cycle took place on May 9, in Grozny's Dinamo soccer stadium. A powerful antipersonnel bomb was installed so as to inflict a large number of casualties in the annual World War II Victory Day parade, which was about to begin. The bomb was discovered by the security forces, but detonated—probably by remote control—as sappers with the Russian special police (OMON) were attempting to defuse it. One sapper was killed and two were wounded. A later forensic examination of these bombs demonstrated a marked improvement in the technical skills and sophistication of the Chechen bomb makers.

The next phase in the escalation, from May to June 2003, saw the gradual introduction of the new class of would-be martyrs recently trained by the Arab experts. These operations took place in Chechnya and the northern Caucasus—a "friendly" environment where the more experienced mujahedin could put peer pressure on the martyr recruits, and where defection to the Russians would be more difficult, if only for fear of retribution against the martyr's family. It was during this period that the Black Widows were gradually introduced into combat, first in tandem with male counterparts and in groups for self-reinforcement, and later alone—an approach that suggested caution on the part of the terrorists' masters.

On May 12, one female and two male suicide bombers drove a KamAZ truck bomb into the Nadterechny district government building in Znamenskoye, killing sixty and wounding roughly one hundred occupants—including a number of survivors of the December bombing in Grozny, who had been moved to the supposed safety of Znamenskoye after the blast. The next day, two or three female suicide bombers launched a joint attempt to assassinate Chechnya's pro-Moscow president, Akhmad Kadyrov, during a religious ceremony in Iliskhan-Yurt. Twenty-six people were killed and roughly one hundred wounded in the series of explosions, but Kadyrov escaped unharmed. On June 4, a young Black Widow attempted to force her way onto a military bus carrying helicopter pilots near the Russian air force base in Mozdok, North Ossetia. When the suspicious driver refused to let her board the bus, she detonated her powerful bomb at the door, killing eighteen.

On June 16, another pair of suicide bombers, one female and one male, attempted to drive a KamAZ truck bomb loaded with two metric tons of high explosives into the MVD building in Grozny. A suspicious guard stopped the truck. Cornered behind preexisting barriers, and under gunfire, the martyrs blew themselves up, killing six and wounding thirty-six guards. The last operation in this cycle took place on June 24. A twenty-two-year-old Black Widow wearing a bomb vest was escorted by two male fighters in Grozny. Their plan called for the mujahedin to attack the guards of a government building, creating a diversion that would allow the Black Widow to rush inside the building and blow herself up among the security personnel gathered to respond to the firefight outside. Again, however, the three were challenged by suspicious security personnel, and a firefight ensued. The Black Widow attempted to blow herself up, but the bomb malfunctioned, exploding only partially and wounding only the bomber herself. Before she died, an interrogation by security officers on the scene gave the Russians its first clues about the Black Widow phenomenon.

Starting in the summer of 2003, the jihadists assigned Black Widows to play an increasingly prominent role in their plans for new suicide bombings in Moscow and the northern Caucasus. A new cell of trained terrorists was organized in Moscow in early to mid-May, with the intention of launching a new round of bombings starting in early July. As a rule, the Black Widows-in-training arrived in Moscow from the Caucasus a few days before the operation. They were kept in a safe apartment, where they were given additional briefing and indoctrination before being sent out on their mission. The network was run by three operatives. The matron Lida or Lyuba (the testimony of surviving Black Widows varied), also known as "Black Fatima," was responsible for the psychological tempering of the young women (often drugging them with spiked orange juice and/or tranquilizer pills), and prepared them for their operations. Black Fatima handed the women their bomb vests (or handbags) and then escorted them to their designated target sites, dropping them off nearby. The network's ideological and theological training was in the hands of a male member known as Ruslan or Igor, who wrote and videotaped the last wills of the Black Widows on the eve of their missions. He also provided escort to prevent security lapses. The final supervisor, Arbi Khabrailov, known to the women only as "Andrei," was the network's security manager, explosives expert, and bomb maker. For security reasons, the would-be Black Widows had little contact with "Andrei."

By early July, the Moscow network was ready to launch operations.

On July 5, Lida or Lyuba personally escorted the first suicide bombers to their destination. With their hair fashionably dyed, and wearing tight and revealing clothes, two Black Widows—one nineteen years old and the other twenty-six—mingled with other young people on their way to the Wings rock concert at the Tushino airfield near Moscow. As they waited in line, they realized that they would soon be faced with security checks, and they blew themselves up in quick succession, killing fourteen people and wounding sixty. Witnesses later reported seeing Lida/Lyuba escape from the scene. On the night of July 9, another twenty-three-year-old Black Widow was unleashed—this time in an area of crowded restaurants and coffee shops on Moscow's main street. Stylishly dressed and carrying a handbag bomb, she walked around for a long time—apparently enjoying the new experience of an evening in Moscow and wrestling with second thoughts about blowing herself up. When she finally decided to activate the bomb, the fuse malfunctioned. In desperation, she ran to another restaurant to try again, but security forces cornered her and she surrendered peacefully. (A sapper attempting to disarm her bomb was killed when it detonated during the procedure.) This young woman cooperated fully with the Russian authorities and added to their understanding of the phenomenon.

The capture of this latest Black Widow caused Basayev to withdraw the remaining members of the Moscow network to the safety of Chechnya. Instead, a new network was sent to infiltrate Moscow, in hopes of reconstituting the terrorist campaign in time to meet the original timetable for Operation Whirlwind. But the hasty change of plans resulted in the capture of the key operatives while the network was still in its preparation phase. On August 8, Moscow police noted a blue Zhiguli driving around in a suspicious manner and stopped the car. In the car, police found three Chechens—two men and a woman—with fully loaded but unarmed bomb belts and a map marked with potential targets. Under interrogation, the Chechens acknowledged that they were organizing a new network and had been reconnoitering potential targets when they were stopped. The network had been planning to send a number of Black Widows to crash a stolen fire truck turned massive car bomb into one of several iconic Moscow buildings. The speedy collapse of this network, and the prisoners' cooperation with their interrogators, caused Basayev and Abu-al-Walid to suspend operations in Moscow until a more secure and professional network could be formed there.

For the next four months, as preparations for Operation Whirlwind continued in Moscow, they would turn their short-term focus back to Chechnya and the northern Caucasus. The terrorists managed to launch a series of new strikes throughout the region, but their diversity and erratic nature reflected indecision within the Islamist-Jihadist leadership over tactics and methods. At the same time, this diversity also reflected the vast array of options that were available to the Islamist-Jihadists virtually at will, at least on their home turf. Even though they had lost the bulk of their grassroots support, the jihadist leaders were able to use expertise recently acquired from Iraq-trained experts to adapt to the prevailing conditions, relying increasingly on small clandestine networks and cells to launch their strikes with reasonable reliability.

It was no accident that a Black Widow spearheaded this new cycle of terrorism in Chechnya. On July 27, a twenty-six-year-old Black Widow blew herself up outside the garrison of Ramzan Kadyrov's Kadyrovsky Spetznaz in Totsin-Yurt, killing herself and an innocent bystander. The woman in question had reportedly been in training for a Moscow operation when the network collapsed; too far gone to be psychologically "defused," she had to be committed to a local operation.

Yet the Chechen jihadist leaders soon returned to more proven methods. On August 1, a male would-be martyr crashed a truck bomb into the Russian military hospital in Mozdok, North Ossetia, killing fifty occupants and wounding more than eighty. On August 25, a car bomb was used in Dagestan to assassinate Magomedsalikh Gusayev, a Dagestani government minister involved in the suppression of the local jihadists. On September 3, the jihadists planted a powerful bomb on the railroad tracks outside Kislovodsk, Russia. Activated by a passing locomotive, the bomb blew up several passenger cars—but the train was moving slowly, and only six passengers were killed and fifty-four wounded. On September 15, a married couple from Nogai attempted to drive a truck bomb into the new FSB building in Magas, Ingushetiya;

prevented by security from reaching the building, they blew themselves up in an empty parking lot, killing three bystanders and wounding twenty-eight.

The last strike in this cycle took place on December 5, and it involved a group of recently trained and indoctrinated Chechen martyr-bombers—a man and three women. Mingling with the passengers, mostly students, on a crowded early morning commuter train out of Yessentuki (in the Stavropol region of southern Russia), they blew themselves up in quick succession, killing forty-two and wounding more than one hundred. The above phase was conducted in the Northern Caucasus—Chechnya and neighboring regions—by locally based networks. Concurrently, the jihadists consolidated a new network in Moscow, a far more professional team. The moment the Moscow network was ready to strike out, the Chechen leaders wound down their regional operation and launched the new cycle in Moscow.

The next phase of the cycle would take place in Moscow, and its first strike came at around 11 A.M. on December 9, when a still-unidentified lone Black Widow blew herself up outside the National Hotel near the Kremlin and the Russian State Duma in Moscow, killing six and wounding fifteen. She was in her mid-thirties—older and more mature than the first round of Black Widows—but also dressed fashionably and carrying her bomb in a shoulder bag. The organizers of the operation were taking no chances: Several eyewitnesses described seeing the Black Widow with an older woman resembling Black Fatima shortly before the attack, asking travelers at several Metro stops for directions to the Duma. Clearly disoriented, after leaving Fatima, the Black Widow herself stood alone in front of the National, asking passersby for directions to the Duma—which was actually just across the street. Exasperated, she blew herself up where she stood.

In a later communiqué, Shamil Basayev would take credit for both the Yessentuki and National Hotel bombings in the name of his Riyad-us-Salikhin Battalion, calling them "pre-planned combat operations . . . carried out by our Brigade's fighters." The operations were milestones for Basayev, confirming that the new crop of would-be martyrs—again, predominantly Black Widows—was prepared to strike in Moscow and other challenging locations. This time, however, the Moscow-based networks would be more diverse and their operations sustainable, despite improvement in the work of the Russian security services.

For security reasons, Basayev decided on a short and intense new cycle of strikes, from a month to six weeks in duration, to ensure that all missions were completed before the Russians could close in on the terrorist network. The first strike took place on February 6, 2004. A martyr-bomber couple—a man and a woman—carried a sophisticated bomb in a suitcase to the green-line Metro train near Moscow's Paveletskaya station. It was a highly professional bomb made of at least five to seven pounds of plastique high explosive mixed with trinitrotoluene (a highly unstable but powerful additive) and metal pieces for shrapnel. The explosion destroyed the car, killing at least forty passengers and wounding more than a hundred. For the next strike, on February 18, the terrorists changed methods, blowing up two natural gas lines in Ramenskoye, south of Moscow, using conventional bombs with timers. On March 5, another Black Widow was dispatched for a martyrdom mission, but her performance was one more reminder of the risks of young suicide bombers: A twenty-one-year-old who had just arrived in Moscow from a remote Chechen village, she was sent into the center of Moscow with a bomb bag but was mesmerized by the city center, and she was arrested for suspicious behavior before she could detonate her bomb.

Rather than risk sending any more young, green terrorists for new strikes in Moscow, Basayev decided to finish this round of strikes using a number of operatives already in Moscow. On March 11, the Moscow network launched a complex attack designed to assassinate MVD deputy minister Sultan Satuyev. A few terrorists with assault rifles opened fire on Satuyev's car and others in his motorcade to slow them down and divert their attention. Next, a suicide bomber driving a car bomb was supposed to exploit the confusion and ram the minister's car—but the would-be martyr got excited and detonated his bomb too early to kill anyone but himself.

On March 15, the terrorists used a number of timer-fused bombs to blow up three electrical transmission towers in the Leninsky district, just outside Moscow. For the first time, they left a calling card—a

Chechen flag. With Russian security services launching a dragnet to rein in the network, the mujahedin folded operations just in time, safely withdrawing key operatives to the northern Caucasus before the Russians caught up with them. Soon thereafter, Basayev claimed responsibility for the wave of bombings.

Although it would later influence the broader Islamist movement, by the spring of 2004 the spate of Black Widow attacks had begun to ebb. In their quest to understand the phenomenon, Russian security services were left to ponder the testimony of a handful of female suicide terrorists they had managed to interrogate and the reflections of friends and family members who were willing to talk.

Luiza Asmayeba, the twenty-two-year-old wounded on June 24, 2003, in a shootout with police in Grozny, was clearly driven by personal shame. Before her death, she told security authorities that she had been repeatedly raped and impregnated by Chechen mujahedin and saw in martyrdom the only way to cleanse herself from that shame. After her death, an autopsy confirmed that she was pregnant. In contrast, Zulikhan Yelikhadzhiyeva, who killed herself on July 5 in the Wings concert attack, had been hardly touched by the war. She lived with a relatively affluent family, studied at the village's medical vocational school, and worked for a local clinic; none of her family members had been killed in the conflict. One of her childhood friends told a Russian security service interviewer that it was "inconceivable" that Zulikhan had adopted any form of extremism. "She studied," the friend said. "She was a cultured girl, a modern girl. She could not have had anything like this in her mind."

Indeed, the true story of Yelikhadzhiyeva's transformation from young Chechen woman to suicide terrorist was far more personal than ideological. Zulikhan Yelikhadzhiyeva, it turns out, was kidnapped on order of one of her half-brothers—a commander in one of Basayev's groups—for a surprising reason: family honor. As letters found on her body revealed, Zulikhan had fallen in love with her stepbrother, a mujahed named Zhaga. They escaped together and had sex, but soon she was consumed with a fear of "falling into the hands of my relatives." If her transgression were discovered, Zulikhan wrote, she could no longer "live in this dirty world and [would] go to hell" for her sins. The mujahedin convinced her that the only way to ensure reaching heaven was to "become a shakhida on the path of Allah." In one letter, she implored Zhaga to stop fighting the Russians and "just pick up a [martyr's] belt and become a shakhid on the path of Allah and we will be together [in heaven]. . . . I will anxiously wait for you. I love you, love you, and will love you there in the sky." On the basis of this letter, Zhaga was captured alive by the Russian security forces.

Another personal story was that of twenty-three-year-old Zarema Muzhikhoyeva, who surrendered after attempting to blow herself up in central Moscow on the night of July 9, 2003. Like Yelikhadzhiyeva, Muzhikhoyeva was clearly driven to despair by anguish over her family. Brought up by her grandparents after her separated parents abandoned her, Zarema had married young and became pregnant almost immediately. Two months later, her husband was shot dead in a business dispute. Soon after her birth, her daughter was taken away by her late husband's childless brother, and his family informed Muzhikhoyeva that as a single mother she was unlikely to find work or a place to live in their village. In desperation, she tried to escape with her daughter to Moscow, only to be caught at the airport by her husband's family. They dragged her back home, returned her daughter to her brother-in-law, and subjected her to disgrace and threats. "They beat me," Muzhikhoyeva told interrogators. "They kept repeating, 'It would be better if you die.' So I thought, 'Why not?'" Contacting a woman she knew to be close to the jihadists, Muzhikhoyeva offered herself as a suicide bomber and was taken to the mountains to meet with Basayev. When the terrorist leader promised to give her family one thousand dollars to help raise her child after she blew herself up, Muzhikhoyeva was sold.

There were some Black Widows who insisted that they were motivated by Islamist-Jihadist ideology. Yet even among these women the knowledge of the Koran and the tenets of Islam was shallow at best. For example, twenty-one-year-old Zara Murtazaliyeva, who was captured in Moscow on March 5, 2004, had been given extensive indoctrination about the tenets of jihad, yet she could not elucidate even the most basic concepts of Islam and jihad to her interrogators. Equally illuminating was an early September 2003 interview with "Kowa"—a young Chechen war widow and mother who resolved to commit martyrdom. "I have only one dream now, only one mission—to blow myself up somewhere in Russia, ideally in Moscow," Kowa told British interviewers. "To take as many Russian lives as possible—this is the only way to stop the Russians from killing my people. Maybe this way they will get the message and leave us alone, once and for all." Kowa's Islamist mujahed husband had introduced her to Islam and made her practice, and after his death she had elected not to kill her husband's killer but take revenge against Russia. She sought and received religious training to help her ready herself for martyrdom. "I am preparing for my mission by reading texts and contemplating what I am about to do," Kowa concluded. "I am just waiting for the order from my commander."

Mark Franchetti of the London *Sunday Times*, who later conducted an extensive interview with Kowa, noted the pervasiveness of her indoctrination. "The most chilling and in many ways depressing thing about meeting her was the complete lack of emotion with which she spoke about what she was about to do," Franchetti told the BBC. "Especially considering she was so young, and she has a small child, she was a brick wall—no emotions whatsoever. She spoke in a pre-programmed way, almost as if she was dehumanized." Kowa had clearly volunteered for martyrdom, but Franchetti saw definite signs of coercion in her account. "I have no doubt that she made the decision herself, but I think to a certain extent she has been also brainwashed from a religious point of view," Franchetti explained. "[Her religion] is quite superficial—when I tried to press her about her religious faith she was not really able to explain it very well."

Whatever their relationship with Islamist ideology, the Black Widows had made a lasting mark on the Chechen jihad—and would spearhead at least one more dramatic episode in the ongoing conflict with Russia.

Chapter 23

THE ROAD TO BESLAN

IN THE SPRING OF 2004, SHAMIL BASAYEV AND THE ISLAMIST-JIHADIST leaders resolved to suspend further bombings in Russia and shift their attention back to Chechnya itself—where they were desperately losing ground. By that fall they had instigated a new and dramatic campaign of attacks—one that succeeded in its immediate goal of stunning observers around the world, but which, over the long term, would see the movement's grassroots support in Chechnya ebbing away.

Chechen society had been slowly regaining equilibrium as everyday Chechens turned their efforts from rebellion to reconstruction, and pro-Russian authorities in the region were gaining support. The war continued, but Russian special operations and raids had the rebels on the defensive, limiting their effective areas of operation to the mountainous south. The rebel presence in the inhabited plateau proved merely irritating to the pro-Moscow authorities, and even a series of assassinations of local officials and small-scale acts of sabotage throughout the northern Caucasus by jihadist cells only alienated the population further from the Islamists' cause.

The leaders of the new pro-Moscow Chechnya—starting with Chechen president Akhmad Kadyrov, who had been installed by the Russians as an interim leader in 2000, but confirmed in a popular election in 2003—were former rebels who had joined Moscow in order to end the carnage and start rebuilding their country. These leaders were increasingly trusted by the population, and a growing number of former Chechen nationalist rebels followed suit, aligning themselves with Kadyrov and his pro-Moscow administration.

To counter this shift, the Islamist-Jihadist leaders began planning a campaign of assassinations of pro-Moscow leaders throughout the northern Caucasus, designed to decapitate the up-and-coming regional leadership. "We will throw Kadyrov's head at Maskhadov's feet this summer," Basayev declared in April.

Yet the jihadists had assassination problems of their own. In mid-April 2004, according to Arab Islamist sources, Abu-al-Walid—the commander of the foreign mujahedin since Khattab's 2002 assassination and one of Basayev's closest Islamist-Jihadist allies—was killed during a routine Russian aerial bombing of a mountain camp in southern Chechnya. He and his companions were praying outdoors when a bomb exploded among them, killing several Arab mujahedin. All indications are that this was a lucky strike for the Russians: Walid had decided to stay at the remote camp at the last minute, so the Russians probably had no advance warning of his whereabouts. Walid's death was confirmed a few days later by his family in Saudi Arabia and the relatives of his Chechen wife.

The killing of Abu-al-Walid was a severe blow to the Arab Islamist-Jihadist group in the Caucasus. None of the key Arab commanders had such close relations with the supreme international Islamist leaders. The leadership hastened to nominate Abu-Hafs al-Urduni as Abu-al-Walid's successor, but not until mid-September would he be given the formal title of Commander of the Eastern Province and the Mujahedin and Al-Ansar in Chechnya. During this time, the role of another prominent jihadist, Abu-Omar al-Seif, grew as the newly appointed Abu-Hafs referred to him regularly for advice on ideological and theological issues.

The Islamist-Jihadists' own assassination campaign got off to a shaky start. On April 6, a suicide bomber tried to ram a car bomb into the armored limousine of Ingush president Murat Zyazikov en route from his Nazran home to his office in Magas. The car bomb crashed into the passenger side of the president's limo but failed to destroy it, wounding six but failing to injure Zyazikov. Then, on the night of

May 3, terrorists made two attempts on the life of Suliman Yamadayev—a nationalist commander of the Chechen National Guards and a longtime Maskhadov supporter and opponent of Islamicization who had since aligned himself with the Russians. Two powerful remote-controlled roadside bombs were detonated as Yamadayev was driven home to Gudermes. Both missed their target.

On the morning of May 9, 2004, the jihadists' assassination campaign peaked with a great success—the virtual decapitation of the pro-Moscow Chechen government. Back in April, the Islamist-Jihadists had infiltrated the renovation of Grozny's Dinamo sports stadium, embedding a powerful remote-controlled bomb encased in a 152mm artillery shell within the concrete structure of the stadium's VIP section. The renovations were finished on May 7 without the bomb being discovered—two days before the entire pro-Moscow Chechen leadership would be at the stadium for the annual Victory Day celebrations. All the terrorists had to do was wait patiently for the leaders to gather in the VIP section, and then push a button.

The explosion killed Chechen president Akhmad Kadyrov, State Council head Khusein Isayev, State Council security chief Adam Baisultanov, finance minister Eli Isayev, and three others. Fifty-four were wounded, including General Valery Baranov (the commander of the Russian forces in the northern Caucasus), Grigoriy Fomenko (the commander of Russian forces in Chechnya), and several other members of the Chechen government, including economics minister Abdul Magomadov, presidential press secretary Abdulbek Vakhayev, and local MVD chief Alum Alkhanov. The gory spectacle was broadcast live throughout Russia, adding to the shock factor. "We apologize to the president of the Chechen Republic of Ichkeria Aslan Maskhadov for being unable to literally throw Kadyrov's head at his feet as we had promised a month ago," Basayev gloated.

But it was too late. As dramatic as it was, Basayev's assassination campaign was insufficient to reverse the Caucasian population's disillusionment with the Islamist-Jihadist cause and their appetite for normalization, even in the context of the Russian Federation. Despite incessant urging by the Islamist-Jihadist media, Kadyrov's assassination never ignited a significant rebellion among the Chechen people, and the pro-Russian government's quick recovery proved that the days of grassroots Chechen revolt were over.

For Basayev and the Islamist-Jihadist leaders, it was finally becoming clear that there was nothing the Chechens could do within their own borders to shatter Russian resolve. And so they returned to their original conviction: that only spectacular terrorist acts within Russia—especially Moscow—might compel the Russian population to pressure the Kremlin to abandon Chechnya.

On June 13, Basayev and his Arab assistant, "Amir Khasan," chaired a meeting of the commanders of the Riyad us-Salikhin brigade, studying the lessons of Kadyrov's assassination and reviewing future plans. In a statement e-mailed to Kavkaz-Tsentr, Basayev noted that the assembled commanders had developed plans for "a series of special operations against the occupying forces which will be a severe blow to the enemy both in the military and political sense." He promised that "our blows will be very painful for the Putin regime and will take it by surprise."

The result of their plans was a ten-day period, in the fall of 2004, in which Russia was subjected to a pair of intense and spectacular terrorist acts—perhaps the most dramatic of the war. The downing of two passenger jets leaving Moscow, perpetrated by two female suicide terrorists, marked a kind of parting shot for the Black Widow phenomenon, while the bloody siege of the school in Beslan, Northern Ossetia, seized global attention more fully than any previous attack. These twin attacks, just days apart, hinted at the shape of things to come for Islamist-Jihadist terrorism—not just in Russia, but around the world.

The attacks shared two important characteristics. First, the strikes were perpetrated by Chechen-led networks comprising mainly native Caucasians, with only a few Arabs involved, although some of the attackers used weapons and means supplied by Osama bin Laden's international Islamist-Jihadist networks and developed in other jihadist theaters. Second, the stated justification for the strikes included not just the Chechen Islamists' decade-old jihadist ideology, but also a new global offensive recently declared by bin Laden. In both ideological and practical terms, these new attacks marked the latest step in integrating the global jihad with localized Islamist-Jihadist causes.

The new wave of attacks was neither accidental nor arbitrarily timed. Rather, it was the first fruit of a vastly upgraded Islamist-Jihadist regional command center recently completed in Chechnya—a facility that enshrined a major reversal in priorities for the Islamist-Jihadist high command.

In the spring of 2003, with conflict mushrooming in Iraq, the top Islamist leaders had called for virtually all available jihadist resources—mainly expert terrorists—to be deployed to Iraq and Saudi Arabia. With the initial success of the Iraqi Intifadah, however, bin Laden and the Islamist-Jihadist high command changed course again, accelerating and expanding their global efforts. With Russia gaining ground in the antiterrorist struggle in Central Asia, the top Islamist leaders felt compelled to compensate the Chechens for their important contribution to the Palestinian and Iraqi Intifadahs, the war in Afghanistan-Pakistan, and the Islamist-Jihadist buildup in the Balkans. The jihad against Russia was once again a top priority.

The structure of the new command cell was defined by Ayman al-Zawahiri and key commanders during a stopover in the Caucasus in mid-August 2003. After a trip to Tehran to coordinate the anti-American resistance efforts in Iraq with Iranian intelligence, on August 17 or 18—the eve of the first spectacular terrorist strikes in Baghdad—Iranian intelligence chief Ali Younessi personally instructed a team of Iranian senior intelligence operatives to organize a clandestine route back to Pakistan, designed to ensure that Zawahiri's stay in Tehran could not be traced. The Iranians provided Zawahiri with the disguise and documents of an elderly Iranian Shiite cleric, and Iranian intelligence helped him across the border to eastern Turkey, where another cell of Iranian intelligence sheltered him in a safe house. There, Zawahiri was provided with yet another new identity, this time of an elderly

Sunni cleric. Capitalizing on the flow of Islamist volunteer mujahedin to Chechnya, the Iranians escorted Zawahiri all the way to al Qaeda's bases in the Pankisi Gorge in northern Georgia on the Russian border. From there, Zawahiri made his way to the Fergana Valley and, via northern Afghanistan, onward to Pakistan.

Along the way, Zawahiri took the occasion to inspect and approve the functioning of the new jihadist command cells in the Caucasus. On the eve of his departure, Zawahiri organized a team of commanders, drawn from the jihadists' operation in Iran, whom he would bring to the Caucasus to run the new upgraded jihadist headquarters there. Among these commanders was the Saudi Muhammad Abu-Omar al-Seif, a veteran emissary of the international Muslim Brotherhood in Chechnya and a functionary of Saudi Islamist foundations long involved with Chechen affairs; Zawahiri nominated him as "al Qaeda's emissary" in Chechnya and the northern Caucasus. Zawahiri also endorsed the command position Abu-Hafs al-Urduni was holding in the Pankisi Gorge, and that of Abu-Hajr, a Saudi who was another key commander in Chechnya. Abu-Hafs deployed to the Caucasus in summer 2003.

By the winter of 2003–2004, especially after visible failures such as Basayev's National Hotel attack of December 2003, the Chechen Islamist-Jihadist leadership was anxious to improve its ranks of commanders and terrorists by bringing in more Arab experts from both the Middle East and Western Europe (the latter arriving through Bosnia-Herzegovina). By the spring of 2004, with Abu-al-Walid dead and the assassination of Kadyrov failing to deliver any real shift in the Russo-Chechen dynamic, the Chechen leaders were desperate for a breakthrough. The ascent of Abu-Hafs al-Urduni suggested a newly enhanced role for the Arab volunteers, who would introduce new sabotage techniques and professional handlers, while once more inflecting the Chechen struggle with their own Islamist-Jihadist aspirations. The international Islamists still saw Chechnya as another potential front for al Qaeda, and in particular as a gateway to Europe. The Chechens, for their part, were desperate to reinvigorate their jihad before the civilian population's growing acceptance of Russia became permanent.

Together, these factors would inform the spectacular strikes of the fall of 2004, and the new statements of jihad that followed.

A close examination of these near-concurrent operations points up the growing international cooperation within the various branches of the Islamist-Jihadist movement in executing such terrorist operations—even those launched by local groups, ostensibly in the name of local causes. In practical terms, the pair of strikes in Russia in the fall of 2004 were accomplished with a mix of local and imported operatives, weapons, and expertise, orchestrated by Osama bin Laden's Islamist-Jihadist leadership.

On the evening of August 24, 2004, two young Chechen women boarded separate Russian passenger aircraft flying out of the same Moscow airport. At roughly 10:55 P.M., each woman detonated an explosive device, bringing the two planes down a couple of minutes apart. Satsita Dzhebirkhanova, age thirty-seven, blew up a Tu-154 jet, Siberia Airlines Flight 1047 en route from Moscow to the Black Sea resort of Sochi. Amanta Nagayeva, age thirty, blew up a Tu-134, Volga-Aviaexpress Flight 1303 from Moscow to Volgograd. At least ninety people were killed in the attack; there were no survivors. Shamil Basayev later identified the suicide bombers as members of the "regional Shakhid unit of Moscow" of "the battalion of Shakhid, Riyad us-Salikhin," and implied that he himself was the battalion's commander and thus responsible for the twin attacks.

Nagayeva and Dzhebirkhanova fit the profile of the Black Widow. They were young women ensnared in miserable lives without a prospect of a positive breakout. They were single, past the age of likely marriage, in a society with a surplus of women but that still shunned single women. Both worked in the same market in Grozny, selling trinkets for a meager income, traveling periodically to Baku to buy clothes and other items to sell. They also shared an apartment with two coworkers—Amanta's sister Rosa Nagayeva and another woman named Mariyam Taburova. These women had spent their lives together, mutually reinforcing their shared frustrations and hopelessness. All of these factors conspired to make them susceptible to the Islamists' promise of transcendence through martyrdom.

Around August 20, the four women had left their Grozny apartment. They were last seen together in the bus station in nearby Khasavyurt, Dagestan, and it is assumed that they took the bus from there to Moscow. None of the women has been seen since. No relatives inquired about either Nagayeva or Dzhebirkhanova after the crash, and no one claimed their remains.

On August 24, Nagayeva and Dzhebirkhanova resolved to embrace martyrdom simultaneously. Nagayeva was the last person to buy a ticket for Flight 1303; claiming a personal emergency, she convinced the airline to sell her a ticket only an hour before takeoff, without proper inspection of her passport. After Nagayeva bought her ticket, Dzhebirkhanova rushed to change her ticket to one for Flight 1047, so that she and Nagayeva would be in the air at the same time.

The bombs that brought down the planes were composed of about four hundred grams of plastic explosives, primarily hexogen. The main charge was the size of a bar of soap, easily concealed in a woman's bag; the simple fuse, which resembled a pen, could have been inserted into the charge at the last minute. Both women sat in the tail end of their respective aircrafts, where presumably they could have slipped into a restroom to assemble their bombs. When they detonated the devices, their bodies were torn apart—suggesting that they held the bombs against their bodies. Each explosion was strong enough to tear the skin of the fuselage, starting a structural collapse that doomed the aircraft. Those who planned the operation would have known that both the Tu-154 and the Tu-134 had their engines at the rear, putting extra stress on the structure of the tail area and making it more vulnerable to small explosions. In both cases, the tail section ended up far ahead of the nose—the result of the catastrophic structural collapse that followed the blast. In both planes, residues of hexogen were found on the wreckage of the tail section and rear fuselage section.

Forensic evidence proves that the bombs used on the two Russian aircraft were similar to the bomb used by Abderraouf Jdey, a Canadian citizen also known as Farouk the Tunisian, who brought down an American Airlines aircraft soon after taking off from JFK to the Dominican Republic on November 12, 2001, and to the bomb in the shoes of Richard Reid, who was captured trying to bring down an American Airlines flight from Paris to Miami a month later, on December 22. The fact that such materials were available to the Chechen suicide bombers was direct evidence of the operational cooperation between Chechen, Palestinian, and other Islamist-Jihadist forces; the only other source known to have built a comparable bomb is HAMAS, in the Gaza Strip.

Reid's shoe bombs were based on Iranian technology, using a new type of high explosive. An acknowledged bin Laden operative, Reid had received his bombs in the Palestinian Jebalya camp in the Gaza Strip. At the time he was a trainee-guest of Nabil Aqal, a senior commander of HAMAS's Izz al-Deen al-Qassim Brigades and a protégé of HAMAS leader Sheikh Akhmad Yassin. The bombs' key components had been smuggled to Gaza from Lebanon via Egypt by Arafat confidant Jamal Sema Dana. HAMAS's support for Reid was a major deviation from Yassin's long-standing insistence that the organization limit its operations to the Palestinian theater, contributing to international operations only under the guise of a global jihadist entity. Following the second Intifadah in fall 2000, HAMAS was anxious for foreign support from the global Islamist-Jihadist movement, and Yassin was willing to "pay" by increasing its direct involvement in the international jihad.

The roughly 150 Chechen and Arab "Chechen" mujahedin who were dispatched by Basayev and Khattab to Gaza and the West Bank after the outbreak of the Intifadah included expert bomb makers, snipers, and fortification builders. Ultimately, these Chechens had a profound impact on the overall character of the Intifadah. Significantly, a few of the Arab "Chechens" had even belonged to the HAMAS Izz al-Din al-Qassim Brigades before leaving Gaza and Nablus for advanced training in Afghanistan and Pakistan back in the late 1990s. By early 2004, HAMAS was trumpeting its solidarity with the Islamist-Jihadist forces in Chechnya, issuing recruitment posters and videos featuring both Basayev and Khattab as role models. The martyrdom in Chechnya of several Palestinians and Jordanian-Circassians—natives of "the land of Jihad"—were celebrated. By September 2004, at least three

HAMAS posters found in Gaza featured portraits of bin Laden, Yassin, and Khattab as the key leaders of the Islamist-Jihadist movement. HAMAS videos also contained a fatwa by the Chechen mufti Muhammad bin-Abdallah al-Seif, permitting suicide terrorism and authorizing the recruitment of women as bombers.

The other shoe dropped on September 1, 2004, when a group of Islamist-Jihadist terrorists seized a school in Beslan, North Ossetia, in the name of the Chechen rebellion. For three agonizing days, they held thousands of hostages, mostly children and their teachers; the standoff ended in carnage when a series of shots erupted, bombs began exploding, and a school building collapsed on the trapped hostages. Hundreds of children were killed and injured, in a uniquely horrific spectacle that was followed around the world.

The idea of creating international shock-waves by seizing a large group of schoolchildren and forcing authorities to negotiate over them in the presence of anxious and anguished parents—or, as an alternative, killing them in front of their parents and society at large—did not originate with the terrorists in Beslan. The tactic was first studied and practiced in Bosnia-Herzegovina in the late 1990s by a team of elite mujahedin—both Arab volunteers and Bosnian-Muslims—under Abu-Maali, then the commander of the mujahedin in Bosnia, a friend to Khattab and patron to Abu-al-Walid. The Bosnian mujahedin even completed exercises in a school, taping an instruction video on how to seize a school, take hostages, and then fight security forces attempting to secure the hostages' release. They shared this knowledge with their brethren in Chechnya, furthering a cooperative relationship between the Islamist-Jihadists in the Balkans and Chechnya that dated back to the mid-1990s, and was manifested in the large quantities of jihadist literature and videos to be found in the main Islamist mosques and bookstores in Bosnia-Herzegovina.

The principle of taking schoolchildren as hostages may have grown out of the Chechens' dialogue with the Kvadrat organization, founded in Sarajevo in 1995 by senior representatives of Islamist-Jihadist charities to inculcate children orphaned during the Bosnia-Herzegovina civil war in jihadist ideology and terrorist tactics. A couple of years later, Sheikh Abu-Omar adopted the principles of the Kvadrat in Chechnya, drawing a growing number of graduates from Bosnia to join the al Qaeda-supported Islamist-Jihadist forces in Chechnya. By 2004, the Kvadrat was sustaining and maintaining a comprehensive Islamist-Jihadist pipeline shuttling money, technology, and mujahedin between Chechnya and Bosnia-Herzegovina—the jihadist gateway to Europe.

The Beslan operation was organized not in Chechnya but in Nalchik, the capital of Kabardino-Balkariya. The terrorists arrived in Beslan via Nazaran, Ingushetiya, where they picked up additional personnel. By August 20, they had assembled near the village of Psedakh, Ingushetiya, awaiting final orders from Basayev. The hostage takers made the last leg of their trip to Beslan in a single GAZ-66 truck.

The thirty-two terrorists who participated in the seige in Beslan were a multinational team—hardly a band of Chechen nationalists. Nalchik was selected intentionally, as the location of an Islamist-Jihadist support network capable of organizing and launching such a challenging operation. (The local support networks in Kabardino-Balkariya are composed mainly of Circassian families that are related to the Circassian community of Jordan; it was this Circassian connection that had allowed the Arab Islamist leaders to monitor the Nalchik-based networks since 2002.) In the spring of 2004, as Basayev and Abu-al-Walid began contemplating a new wave of attacks, it was only natural that they would appeal to the Islamist-Jihadist supreme leaders for assistance and resources that weren't likely to have been penetrated by Russian intelligence—and the Nalchik networks were a perfect solution.

Although the perpetrators of the September 1 attack in Beslan claimed to the media that they were operating in the name of Chechen independence, other statements they made flatly contradicted the claim. The sole terrorist captured in Beslan, Nur-Pasha Kulayev, told his interrogators that "the attack was ordered" by Aslan Maskhadov and Shamil Basayev in order to destabilize the entire northern Caucasus. In the last briefing before the strike, Kulayev reported, one of the leaders of the strike—a terrorist he knew only as "The Colonel"—made this explicit. "They gathered us in the woods and 'The Colonel' said that we should take over a school in Beslan. That was our order," Kulayev recounted. "When we asked why we were doing this, what our goal was, 'The Colonel' answered us, 'Because we need to start a war across the Caucasus.' "

In other words, the goal of the Beslan strike was never to further the Chechen rebellion, but to further the Islamist jihad throughout the Caucasus. In that spirit, only a minority of the terrorists involved were actually Chechens. Even today, lingering questions remain about the exact number of the terrorists—most likely between thirty-two and thirty-five—as well as their real identities and backgrounds. Basayev claimed to have handpicked the Chechen Kulayev brothers as the operation's commanders. However, the Russian negotiators, including some Chechens, ascertained that the operation included no Chechens in prominent positions, but at least one Arab in a command position. Although he had chosen and trained the perpetrators himself, Basayev credited "the second [Shakhid] battalion under the command of Colonel Orst-khoyev" with responsibility for the Beslan operation, and claimed that the unit included "twelve Chechens, two Chechen women, nine Ingush, three Russians, two Arabs, two Ossetians, a Tatar, a Kabardin, and a Guran. . . . The Gurans are the people who live in the Baykal area and, in fact, they are virtually Russified." Russian intelligence claimed that the strike force comprised about thirty-five terrorists—including "Chechens, Ingushes, Kazakhs, [ten] Arabs, Uzbeks and Slavs." Arab Islamist sources tend to agree with the original Russian claim that a large number of Arabs were among the hostage takers. Ali Hammad identified one of his former subordinates—a Turk named Abu-Abdallah al-Turkiy who served in Bijelo Bučje, near Travnik, in 1992–1993 and moved to Chechnya in early 1996—among the Beslan terrorists.

Similarly, there were no Chechens among the operation's four on-site commanders, although reports from Russian officials and private individuals who had contact with the hostage takers, from the hostages themselves, and from the official Russian investigation provide a confused and sometimes contradictory portrait of the command elements of the strike. The one thing these four commanders' backgrounds do illustrate is the transformation of the leadership of the Chechen jihad.

Although three other terrorists from the northern Caucasus—Polkovnik (Russian for "colonel"), Abdullah, and Magas—presented themselves at various times as the "leader" or "commander" of the Beslan operation, it is clear that the actual commander was an anonymous Saudi known as Abu-Zeit (or Abu-Dzeit). According to Arab Islamist sources, Abu-Zeit was a close associate of the two leading Islamist-Jihadist commanders in the Caucasus, Abu-Hafs and Abu-Hajr, and was also trusted by Abu-Omar al-Seif, who helped to fund the strike. The Arab Islamist sources stress that these Saudi senior commanders personally selected Abu-Zeit to lead the strike. Basayev, for whom Abu-Zeit had once worked as a bodyguard, also gave his consent. Although Abu-Zeit had no contact with the negotiators, his voice was heard by monitors intoning Koranic verses in Saudi-inflected Arabic, as well as barking instructions in fairly good Russian. His combination of leadership and complete anonymity earned him the nickname "Fantomas" from Russian intelligence experts monitoring the crisis.

The most visible and vocal of the "commanders" was Ruslan Khuchbarov, the "Colonel" whom Kulayev had described. The media portrayed the Colonel as the de facto leader of the Chechens participating in the strike, and the commander of events inside the school. Born in 1972, Khuchbarov was an ethnic Ingush from Galashki, Ingushetiya. In the mid-1990s, he moved to the city of Orel, where he lived in the gray zone between society and criminality. Although he held no job, he frequented expensive restaurants, was a notorious womanizer, indulged in heavy drinking, and occasionally used drugs. In 1998, he killed two Armenians in a dispute over a woman. With law enforcement closing in on him, Khuchbarov escaped to Chechnya, where he was picked up by the mujahedin and converted to Islamism. Within a couple of years, Khuchbarov became a devout jihadist. In 2002–2004, he participated in a few terrorist operations in his native Ingushetiya.

According to former Beslan hostages, "Colonel led the operation inside the school." They described him as highly competent but sadistic, demanding uncompromising obedience from both the hostages and the terrorists, and willing to go to extremes to impose his orders. According to Kulayev, Polkovnik shot and killed another male terrorist and later used a remote control to detonate the belt bombs of two Black Widows when they objected to taking children as hostages and refused to kill children. In contacts with mediators during the siege, the Colonel demanded to be addressed by would-be negotiators as "Shakhid Polkovnik" ("Martyr Colonel") and "Sheikhu Polkovnik" ("Sheikh Colonel"). Some of the terrorists addressed him as Ali. Yet Khuchbarov wasn't everything he insisted he was: Russian security services and intelligence correctly concluded that Polkovnik was a kind of "right-hand man" to Magas, the third commander. Although official records claim that Polkovnik was killed during the assault on the school on September 3, there are persistent rumors that he shaved his beard and managed to escape at the height of the chaotic assault.

In their telephone conversations, which were monitored by Russian intelligence, the terrorists repeatedly referred to a senior commander they called Magas (after the capital of Ingushetiya), and Arab Islamist sources identify Magas as the most senior terrorist commander answering to Abu-Zeit. Magas is the nom de guerre of Ali Musaevich Taziyev, a former Ingush police sergeant and member of an elite police unit that provided bodyguards for senior officials of Ingushetiya. On October 10, 1998, Taziyev, his partner, and the wife of an adviser to the Ingush president were kidnapped while she was shopping. A Chechen organization sent a ransom note. Soon thereafter the woman was released to her husband and the body of the other bodyguard was found. Taziyev was widely assumed to have been killed, but in reality he had been won over to the cause of jihad. After proving his loyalty by participating in several operations against Russian forces in Chechnya, he ended up in Abu-al-Walid's camp. A few years later, Magas was given command of his own unit of roughly thirty mujahedin from Ingushetiya, Chechnya, Kabardino-Balkariya, and Dagestan, as well as two Arab supervisors and instructors. They raided targets mainly in Ingushetiya. A large number of the mujahedin at Beslan were apparently members of Magas's unit.

Although Russian authorities initially insisted that Magas was killed during the assault, this was not the case. Some sources now suggest that he was not at the Beslan school at all, but directed key elements of the operation by phone. By early 2005, a growing volume of intelligence reports suggested that "Amir Magas" (as he was dubbed by Shamil Basayev) was responsible for numerous terrorist strikes in Ingushetiya. In June of that year, Magas apparently played a major role in the assault on several police stations and government buildings in Nazran, Ingushetiya, in which almost a hundred people were killed. Magas resurfaced in mid-May 2006 as one of two terrorists who organized the assassination by suicide car bombing of Ingush deputy interior minister Dzhabrail Kostoyev in Nazran, a blast in which six others were also killed. Magas was killed by Russian special forces on the night of July 9/10, 2006, in the ambush that also killed Basayev.

The most controversial of the four Beslan commanders was Abdullah, who was ostensibly in charge of logistics and transportation for the operation. Abdullah, whose real name was Vladimir Anatolievich Khodov, was born in 1977 in the North Ossetian village of Elkhodovo, less than fifty miles from Beslan. A Ukranian by birth, and the son of a Soviet army officer, Khodov slid into petty criminality with his brother as a teenager. In 1995, Khodov's brother was jailed for stabbing a man to death; in prison, he shared a cell with a group of Muslims and converted to Islam. Khodov, who visited his brother frequently, also converted. Khodov was indoctrinated, perhaps even recruited, while studying Islam in Dagestan in the late 1990s, but his newfound faith was not enough to change his criminal lifestyle. With the police looking for him under suspicion of rape and other crimes, Khodov went into hiding in Chechnya. Stressing his Islamic orientation, Khodov was sheltered by Basayev's mujahedin and gradually rose in the ranks. In 2003–2004, using the name Abdullah, he was involved in petty terrorist activities in northern Ossetia. Former hostages at Beslan characterized Abdullah as "one of the most ruthless of the gang."

Vladimir Khodov's body was positively identified after the assault, but the Khodov saga was far from over. In a late August 2005 interview, Basayev claimed that Khodov was actually a Russian agent sent to infiltrate his group, but that he had turned against the Russians. According to Basayev, Khodov had been arrested at his brother's funeral, and then approached by security services offering to make him an informant instead of sending him to jail. According to Basayev's story, Khodov pretended to have escaped to Chechnya, gradually endearing himself to the mujahedin, and eventually reached Basayev himself. "To earn my trust," Basayev claimed, "[Khodov] organized attacks in Vladikavkaz. Then, following orders from the secret services, he set a trap for us. He proposed that we take hostages in Parliament and at the Ossetian government central offices in Vladikavkaz. Russian Special Forces would have been waiting for us outside the city and would have killed all of us. But finally, after spending a month with the Chechen mujahedin, Khodov decided to confess and tell us that he was working for [the Russian secret services]."

Instead of killing him, Basayev decided to turn Khodov into a double agent. "Once we had won him over to the cause of Islam, I suggested that he tell his bosses that no one was interested in his idea to take hostages in Vladikavkaz. So, to make him seem more trustworthy in my eyes, they helped him organize an attack against soldiers in Vladikavkaz and against a passenger train. Afterward, he told them that he had 'earned my trust,' and so, through him, in Spring 2004 we began organizing—together with the FSB—the hostage-taking in Parliament and the government offices," Basayev boasted. Basayev claimed that Khodov was able to get free-passage documents for the Beslan terrorists, but he insisted that the strike he and Khodov were preparing for was politically motivated, scheduled for the September 6 anniversary of Chechen independence.

Basayev never explained how or why these "plans" ended up as the September 1 attack on a school full of children. Indeed, all evidence suggests that he was lying about the original objective and target date of the strike. The Beslan school had been the objective of the strike from the start, and actual preparations for the school's seizure and the hostage operation were already under way by late June or early July. The terrorists' support network used Ingush channels—construction contractors renovating the school over the summer—to hide weapons and explosives inside the school gym, where the hostages would ultimately be held. A later examination of the terrorists' assault rifles showed that some of them had been stolen from an Ingush MVD armory during a raid by Ingush and Chechen terrorists on the night of June 21, 2004. Hostages debriefed by the Russian security services reported that soon after entering the gym, several of the terrorists removed floorboards and retrieved "pre-planted extra weapons and explosives" they must have known were there. Moreover, an analysis of the series of explosions at the end of the siege concluded that the two very powerful bombs that exploded in the roof section of the gym building were so heavy, and so deeply buried within the roof's structure, that they must have been planted there during the summer renovations.

The attack on School No. 1 in Beslan was launched on September 1, 2004, the first day of the new school year. In Russia, this is a festive event attended not only by children and teachers, but also by parents and local notables, who gather in the school's yard or gym to celebrate. The attackers had clearly planned for this: Dressed in fatigues and wearing black ski masks, the terrorists crossed the railroad tracks behind the school to the sound of the music signaling the beginning of the ceremony. Firing in the air, they herded the crowd into the gym, where other terrorists were already waiting. When teachers and parents tried to block them, the terrorists shot them. When local policemen tried to help a few children trying to escape, terrorists appeared at second-floor windows and fired at them. Inside the gym, the terrorists immediately started separating men from women and warned the hostages repeatedly that they would all die there together. "We have nothing to lose. We came here to die," they declared.

By now, nearly twelve hundred hostages—most of them children—were packed into the gym's main hall. After terrorizing the hostages into passivity, the terrorists began retrieving their hidden weapons and explosives. Using the male hostages as human shields and porters, they started fortifying and booby-trapping the gym in preparation for a long siege. All together, some fifteen antipersonnel bombs were installed in the school. Among them were two big bombs hung from the basketball hoops with tape, a few smaller bombs taped to the beams, and several small bombs that were placed around the hostages. All the bombs were connected with a series of electric wires that ended into two spring-loaded detonator pedals, which were held down constantly by the feet of two terrorists; if any of these terrorists were incapacitated and stopped applying pressure to the pedals, the hostages were warned, the entire room would explode. The hostages were kept without food and water in the stifling heat of the gym. Gradually they all stripped down to their underwear; some children fainted from thirst, and others drank their own sweat and urine, but the terrorists refused to provide water and threatened to shoot those who asked. In time, male terrorists started dragging teenage girls into rooms adjoining the main gym hall, where they were held and raped repeatedly, while the rest of the hostages were forced to listen to their screams.

Russian security and antiterrorism experts rushed to the scene, followed by Russia's Alfa antiterrorism unit. After studying available intelligence about the school layout, the number of hostages, the booby traps, and the terrorists' behavior, the experts grew apprehensive that any rescue attempt would result in heavy casualties among the children. An Alfa officer on the scene commented that the Dubrovka Theater siege had been "a kindergarten compared to this." The terrorists only aggravated the tension when they started randomly firing assault rifles and lobbing RPG rockets from second-floor windows in all directions—making it difficult for the Russian security services to collect real-time intelligence.

Determined not to provoke the terrorists, the Russians did not return fire or otherwise react to these provocations. Instead, Moscow tried to appeal to the hostage takers. From the evening of September 1 through midday on September 2, several local leaders and prominent citizens were summoned to Beslan and tried in vain to elicit a list of demands from the hostage takers. Still, Moscow remained hesitant about a military option, fearing heavy casualties among the weakened and dehydrated children.

On the afternoon of September 2, the terrorists decided to try to break the stalemate. Gathering about twenty of the male hostages, they took them to the second floor of the main school building, where they killed the hostages—in sight of the security forces and negotiators—and threw their bodies into the school yard.

The following morning, mediators managed to convince the terrorists to permit half a dozen emergency services personnel to remove the men's bodies from the school yard. As they approached the building at 1 P.M., a major explosion occurred. Terrorists opened fire on the emergency services people, killing two. The first explosion was apparently triggered when the tape holding one of the bombs to a basketball hoop failed and the bomb fell to the floor. (Another theory is that the emergency services people hit a trip wire nobody knew about.) The blast of the explosion knocked one of the terrorists, whose foot was pressuring the pedal—thus starting a chain reaction of explosions. Screaming and hysterical children started running away from the bombs, only to be shot at by the terrorists. It was during this chaotic moment that the two Black Widows refused to kill children; Polkovnik detonated their belt bombs by remote control, killing them and several hostages nearby.

Outside, pandemonium broke out. Ever since the school was seized, hundreds of parents and local security officials—many of them also related to the hostages—had armed themselves with assault rifles and shotguns and surrounded the school. Now, when the first half-naked children, some burned or wounded, started streaming out of the school under terrorist fire, these armed civilians rushed the school, hoping to save their loved ones and take revenge on the terrorists. In so doing, they not only blocked the Alfa team's efforts to reach the gym before the terrorists could detonate all the bombs, but some of them even shot at the security personnel. Indeed, most of the Alfa units' casualties—both fatalities and wounded—were the result of civilians shooting them in the back.

By now, armed civilians were running around the school complex—some of them trying to find and evacuate surviving hostages, others shooting in all directions in an attempt to gun down terrorists, but causing several casualties among themselves and the security forces. Unsurprisingly, a few terrorists may have exploited the chaos to escape. Cornered by the onslaught, some of the terrorists resorted to extreme violence. One eighteen-month-old baby was repeatedly stabbed by a male terrorist after he ran out of ammunition. Finally, the surviving terrorists withdrew to the rear of the gym building and triggered the two big bombs in the roof section, bringing the roof down on many of the hostages who were still inside the building. The majority of the casualties among the hostages were caused by this collapse.

By the time it was all over, some 330 people—half of them children—had been killed, and more than seven hundred—most of them children—were wounded. Between twenty-six and thirty-one of the terrorists were killed, and one captured. The security forces were frustrated by their inability to reach the terrorists quickly enough to prevent the two roof bombings. The Alfa assault team was stalled because "armed locals got in the way," explained a senior security official. "There would have been a lot fewer victims if the local civilians had not got involved," lamented an Alfa unit officer. It was hard to blame the people of Beslan for trying to "do something" to alleviate the horror they were facing, but in such circumstances the cure may only have complicated the disease.

Significantly, the first claims of responsibility for the airliner and Beslan attacks came not from Chechen sources, but from authoritative Islamist-Jihadist channels associated with bin Laden's elite forces. These claims, written in Arabic, justified the strikes in terms of the global anti-Western jihad, characterizing the anti-Russia jihad as part of the greater confrontation.

On August 26, Kataeb al-Islambuli (the al-Islambuli Brigades) issued a communiqué claiming responsibility for the downing of the two Tupolev aircraft. Associated with the Iran-based Muhammad Shawqi al-Islambuli, the al-Islambuli Brigades had previously claimed responsibility for numerous terrorist attacks in Pakistan against both Western and Pakistani objectives. The title of the communiqué—"That Was a Message to Putin, the Other Messages Will Follow"—articulated the essence of the new jihadist campaign, defining the objectives of the strike in all-Islamist terms, portraying Russia as a prominent foe, and mentioning Chechnya as one of many Jihadist fronts. "The Russian slaughtering of Muslims still continues and will not stop until the war whose purpose is to shed blood begins. Our mujahedin were able . . . to deliver the first strike, which will be followed by a series of other operations in the wave to provide support and cooperation to our brothers in Chechnya and other Muslim areas suffering from the infidelity of the Russians," the message declared.

On September 5, the Beslan operation was claimed by the Ansar al-Zawahiri (The Supporters/Devotees of al-Zawahiri). Once again, the message's title—"The Ossetia Operation Is but the Beginning"—portrayed the strike as an early event in a protracted jihad aimed not only at Russia but at Europe as a whole. "The Ossetia operation is but the beginning, and we threaten Europe with a fierce street war," the communiqué asserted. "The Crusader and barbaric Russia has been severely punished so as to be an example for any world state that kills innocent Muslims." Although the vast majority of victims were Russia's children, the Ansar al-Zawahiri stressed that the primary objective of the carnage was to make Russia an example to the governments of Europe. The coming "street war and gang war," the message announced, would be "launched by the heroic mujahedin against the security forces and the European army in the streets of Caucasus and Europe. The European leaders should learn the lesson so that the destiny of their states will not be similar to the destiny of Russia, which is currently burning from north to south and from east to west due to extreme panic, terror, and fear." The atrocities committed against the Russian citizens were characterized as revenge for crimes committed by the enemies of Islam against Muslims all over the world—global dynamics in which Russia was a minor actor. "The killing of the Russian murderous crusaders is a right for the Muslim martyrs and the Muslim women who were raped by the Russian mercenaries in Chechnya, the Zionist Army in Palestine, and the U.S. army in Iraq," the message declared. "The Russian people have to bear the consequences of their actions."

The last warning in Arabic, issued on September 18 by the Chechnya al-Khattab Group, articulated the ideological and operational priorities of the anti-Russia jihad and its role within the global jihad. The warning was a statement from Abu-Hafs, the Commander of the Eastern Province and the Mujahedin and al-Ansar in Chechnya. In this account, the latest strikes were part of a long-term global jihad aimed as much against the United States as against Russia. "In consideration of the Russian and U.S. parties exceeding the limits of their aggression against the honor and dignity of Islam and Muslims in Chechnya, Palestine, Iraq, Indonesia, Afghanistan, and other Muslim countries, within sight of the international community and without awakening the conscience of any Arab leader . . . the commanders of the Chechen mujahedin and al-Ansar in Chechnya declare the beginning of simultaneous attacks against Russian and U.S. interests." Abu-Hafs emphasized that the coming escalation would be designed "to make Russia and the U.S. experience the taste of death and to make the month of Ramadan the month of conquest with our martyrdom operations everywhere on earth." The quest for an independent or Islamic Chechnya went unmentioned.

Only after the Islamist-Jihadist leadership had issued these communiqués did Shamil Basayev acknowledge his responsibility for the strikes. In his first message, posted on the Kavkaz-Tsentr website on September 19 but dated September 17, Basayev expressed lofty objectives and a desire for peace with Russia. From an ideological point of view, however, the strikes were classic Basayev operations. Basayev signed this statement with his Arabicized name and title: "Abdallah Shamil Basayev, the Amir of the Riyad us-Salikhin [Gardens of the Righteous] Martyrs' Battalion."

In this first message, Basayev identified the perpetrators of the recent strikes, and confirmed that they were accomplished at his direction. He explained that "the battalion of Shakhid, Riyad us-Salikhin has carried out several successful operations on the territory of Rusnya [a derogatory name for Russia]. The regional Shakhid unit of Moscow is responsible for the blasts on Kashirskoye Shosse [highway] and the Rizhskaya underground railway station in Moscow. The downing of [the Tu-134 and Tu-154] airliners was carried out by the special operations unit. And the second battalion, under the command of Col Orstkhoyev, was responsible for the Nord-Vest and Beslan operations." Basayev stressed that he had personally recruited the operation's commanders and checked their bombs. At the same time, he insisted that the Beslan strike had not been intended to harm the hostages, but rather to force Putin's hand into reaching a political settlement favorable to the Chechens.

Not only were the Russians responsible for sparking the final conflagration that ended with hundreds of children dead, Basayev claimed, but Moscow had missed the message of the operation altogether. "Putin screamed like a stuck pig and said that Rusnya was under attack and that war was declared on Russia," Basayev sneered. Basayev, too, described this newest cycle of terrorist strikes in Russia as an integral component of a worldwide terrorism offensive, though in his account it was aimed to avenge Russian hostility and global indifference. "They are fighting us without any rules with the direct connivance of the whole world, we are not bound by any commitments, and we will fight as we know how and in accordance with our rules. If the world community indeed wants the tragedy not to be repeated, all you need to do is to demand that your very own international norms be observed by Putin," Basayev asserted. "We have no big choice, they offered war to us, and we will fight it till victory despite what they say about us and how they label us."

Basayev hardened his message and vowed more terrorist strikes in an interview conducted by the Kavkaz-Tsentr website and published on November 1, 2004, on the Daymohk website. This time, Basayev conceded only "a degree of responsibility" for the strikes and stressed that he "[did] not think that I am guilty of that outcome." Yet he hastened to add that the carnage provided a kind of ultimate justice that the rest of the world refused to acknowledge. "The Russians have been holding the entire Chechen people hostage for five years and nothing happened. But we held one thousand people hostage only for three days to stop the genocide of the Chechen people, and 'the whole world is shocked.' What is this if not hypocrisy?"

Under the prevailing conditions, Basayev said, he saw no alternative but to further escalate the war. He urged Putin to abide by "international law" when waging war against Chechens, because only "then automatically we will do so," never "unilaterally." Clarifying this point, Basayev issued a veiled warning about the possible use of WMD in future strikes. "I would like to note that the Russians had many times used chemical and bacteriological weapons against us as well as various poisons, and we feel free to retaliate," he explained. Basayev also hinted at an expansion of the Chechen jihad beyond Russia into other parts of the West. "We also feel free in our actions against those who support the Russian occupying formations in their policy against us by offering their property, themselves, advice, etc. In particular, leaders of other countries must know that by pleasing Putin and declaring the war on us without sorting things out, they expose their own citizens." Basayev stressed that "there is no difference between fighting hundreds of millions or billions; on the contrary, the latter creates more opportunities." The Chechen jihad might be better served by surging out of its current confines, in other words, to take on targets worldwide.

The spectacular terrorist strikes of fall 2004 were classic Shamil Basayev operations—based on the lessons and legacy of Basayev's earlier terrorist strikes, and justified not as part of a nationalist or even an Islamist Chechen rebellion, but as the kind of geopolitical tactic recommended by Basayev's ideological mentor, Magomed [Muhammad] Tagayev, to force a dramatic change in the strategic posture in the Caucasus. Starting with the Budennovsk strike in the summer of 1995 (designed to prevent a Chechen defeat by forcing political negotiations while opening the door to Islamicization), Basayev had returned to this strong-arm strategy again and again: in the bombing of the Moscow apartment houses in fall 1999 (to provoke Moscow into regional confrontation, despite the Islamists' failure in Dagestan), in the seizure of the Dubrovka Theater in the fall of 2002 (to reassure the Islamist-Jihadist leadership that Chechnya and the Caucasus remained a key jihadist front, avenging Moscow's contribution to the American war on terrorism), and down to the strikes of fall 2004, which marked the further integration of the Chechen anti-Russia jihad into the global Islamist jihad against the West.

The Chechen people, in short, had become pawns in the worldwide designs of the Islamist-Jihadists. The process of Chechenization, in other words, was complete.

Chapter 24

SELF-DEVOURING

BY EARLY 2005, THE CHECHEN LEADERS FACED A PROFOUND CRISIS. Popular support for the nationalist rebellion in Chechnya had been steadily eroding for nearly a year. Chechnya was undergoing a profound change, one that transcended the impact of the elimination of a few key old-hand commanders by Russian forces or the growing numbers of defecting leading and junior commanders. After more than a decade of devastation, Chechnya had entered a period of accelerating reconstruction, thanks to the Russians, and stability and security finally seemed within reach—even if under Moscow's rule. The fact that the new pro-Russian government in Grozny was being run by former nationalist rebels reassured average citizens that their basic interests would not be sacrificed. The people of Chechnya felt increasingly alienated from the rebels, and recruitment declined in most of Chechnya.

Aslan Maskhadov, the old-guard nationalist who had maintained a tenuous hold on power in Chechnya for more than a decade, was finally losing his authority, in part because of the growing stream of defections to the Russians—a trend that intensified throughout 2004, even after Akhmad Kadyrov's assassination. The number of clashes between the Chechens and the Russian security forces dwindled, and what few violent incidents there were usually involved isolated, small-scale firefights in secluded areas that had no impact on the Chechen population, or terrorist bombings that only alienated the civilian population in Chechnya, Dagestan, and Ingushetiya, who saw them as painful incidents of fratricidal attrition. The Chechen people were exhausted; the increasingly futile war had lasted too long.

Sensing this trend, Maskhadov saw an opening to reestablish his own relevance on the Chechen political scene. On February 2, 2005, he offered Moscow a cease-fire intended to reverse the backlash against the fall 2004 terrorism offensive. He even ordered a unilateral cease-fire to last until February 22—the eve of the anniversary of the Soviets' 1944 mass exile of Chechens to Central Asia. "This is a demonstration of good will and an invitation to the Russian side to end this war at the negotiating table," Maskhadov's statement read. "By this step the president has shown the world that the Chechen Republic's armed forces are not dispersed groups, as Moscow wants to depict us, but following the orders of a Supreme Commander. . . . Only a strong force could take such a step." Of course, Maskhadov's cease-fire applied only to offensive operations in Chechnya itself, not to the various Islamist-Jihadist forces in neighboring republics—as was evident from several follow-up statements from Dagestan and Ingushetiya.

Maskhadov's offer was supported by an order from Basayev to the Chechen forces to "end all offensive military actions" until February 22. Basayev's order stressed that this was neither an unconditional or complete cease-fire, nor a first step toward cessation of hostilities and possible reconciliation. The order stipulated that for reasons of expediency, for the next three weeks, Chechen forces should refrain from carrying out "diversionary [terrorist] attacks in Chechnya or in the rest of Russia, attacks on Russian bases, Russian convoys or vehicles, or on 'traitors or unbelievers.' " Basayev's order specified that the Chechen forces should "continue mining approaches to their bases and to continue operations to destroy people or machinery of the enemy forces who are spying or attacking mountainous forested areas."

In practical terms, the declared cease-fire was a nonissue, since fighting had ground to its usual halt in the harsh winter. Still, the cease-fire was greeted with skepticism. As security officials in the northern Caucasus noted, "Maskhadov is just one of the leaders of the Chechen resistance, and not even the strongest." They described Maskhadov as a "fading leader," and dismissed the cease-fire as "just an attempt to show he's still relevant." Indeed, only a few days passed before the rebel ambushes, bombings, and assassination attempts resumed.

Basayev's cease-fire was a pro forma attempt to portray the Chechen leadership as a unified body. Within a couple of days, however, Basayev was not only openly challenging Maskhadov and the cease-fire but promising to escalate his terrorist activities in Russia and abroad. In an interview with Britain's Channel 4, Basayev stressed that Beslan-type terrorist strikes, designed to inflict heavy civilian casualties, would remain a major instrument of the Chechens' war against Russia. Although he blamed the Russian security forces for the carnage and civilian casualties in Beslan, Basayev reminded Britons of the "Shield Fatwa," which bin Laden had used to justify heavy civilian casualties in past and future terrorist strikes. Russian civilians, he argued, were "accomplices in this war," and thus constituted a legitimate target for the Chechen jihad. "You must understand us correctly," he said. "We are at war. Russians approve of Putin's policies. They pay their taxes for this war, send their soldiers to this war, their priests sprinkle holy water on the soldiers, and bless their 'holy duty,' calling them heroic defenders of the fatherland. But we are just 'terrorists.' How can they be innocent?"

Under these conditions, Basayev stressed, the widespread shock value of spectacular strikes was necessary to compel the Russian people to force the Kremlin to reconsider the war in Chechnya. Leveling charges of kidnapping and "genocide" against the Russians, he promised, "Cynical though it may seem, we are planning these operations, and we will conduct them, if only to show the world again and again the true face of the Russian regime, the true face of Putin with his Satanic horns."

In the Channel 4 interview, Basayev also attempted to reconcile his belligerent language with the Maskhadov cease-fire initiative that he had endorsed—by adding a caveat he knew Moscow would never accept. "We are ready to stop the war, and, as Maskhadov says, to start negotiations without preconditions," he announced. "But there is one condition—that is, the non-negotiable and full withdrawal of the occupying Russian troops from our territory." Basayev also dismissed negotiation initiatives such as Maskhadov's as impractical, given the character of the Putin regime and Moscow's determination to occupy Chechnya indefinitely.

On February 7, Maskhadov responded through an interview with the Moscow newspaper *Kommersant.* He reiterated his call for the unilateral cease-fire as a first step toward political negotiations, in which he expected the Kremlin to make the main concessions. "I hope for a reasonable reaction," Maskhadov explained. "If our Kremlin opponents demonstrate a sober approach, we will finish the war at the negotiating table," he said. "If not, the blood will flow for a long time." Trying to portray himself as an equal to Putin, Maskhadov claimed that his new initiative was aimed to help both of them. "This [call for a cease-fire] is an attempt to prove our allegiance to peace and to call the Russian leadership to reason. I believe that what is going on in the Caucasus today leads to disaster. At the same time I am not sure that President Vladimir Putin is truly aware of the depth of the abyss Russia and the entire Caucasus are being drawn into. That is why my call is addressed, first and foremost, namely to the president of Russia," Maskhadov explained. He also promised to rein in the Islamist-Jihadist forces. "Now that Basayev has bowed to my cease-fire order, I can consider that I succeeded in averting many terrorist acts unacceptable to us," Maskhadov stated. Yet this effort would only succeed once Moscow demonstrated its support for him and his political initiatives. Time, he warned, was running out.

It wasn't to be. On the very next day, Basayev and the Islamist-Jihadist leaders began challenging Maskhadov with a stream of statements and communiqués from the northern Caucasus, all of them reiterating that the jihad was regionwide, and thus beyond Maskhadov's control. The first was a statement issued by the Sharia Unit of Dagestani Resistance Forces (Mujahedin), denying any intention to obey the cease-fire. Other groups followed with similar communiqués. By mid-February, Russian military officials considered the cease-fire to be an empty, and failed, political gesture. The fighting resumed as the weather improved, and military officials reported that "the order of Aslan Maskhadov, ex-leader of the 'Ichkeria regime,' on the termination of combat operations" was being "ignored by the militants" all over Chechnya. The daily average number of shootings and acts of sabotage remained steady, Russian military officials reported. "Fifteen to twenty cases of shooting are registered every day."

The public rift between Basayev and Maskhadov convinced the Islamist-Jihadists that Maskhadov was a damaging liability. It wasn't that they feared that Maskhadov might strike a deal with the Russians; he was too deeply implicated in previous terrorist campaigns to pull that off. But he had used his prominence to focus local and foreign attention on the "Chechen cause," clouding the Islamist leaders' ongoing campaign to portray the Chechen struggle as entirely jihadist in character.

The leaders had immediate, practical concerns on their mind: Over the winter, they had built up a strong new force of expert terrorists in Georgia, especially in the Pankisi Gorge. By early 2005, some forty foreign volunteers and more than a hundred Chechen/Caucasian mujahedin were ready to cross into Chechnya as a first step toward launching a new cycle of terrorist strikes. In one series of raids in mid-February, for example, Russian security forces killed and/or detained two teams of Arab mujahedin and nine other Caucasian mujahedin near the Chechen-Georgian border. By late February, the Russians were becoming increasingly apprehensive, as special forces raids and ambushes were netting a steady stream of expert terrorists recently crossed over from Georgia. It was clear that the Islamist-Jihadist leaders were bringing in reinforcements for a major escalation later in the spring.

But the challenge faced by Moscow was far greater than an increased flow of individual terrorists. The Islamist-Jihadist insurrection in the northern Caucasus was metastasizing. For years, Basayev and the other leaders had been promising to launch a new jihad throughout the northern Caucasus, to establish a unified Islamic state between the Black and Caspian seas—but for a decade the "Chechen" forces responsible for the terrorist activity in the region had been a loose aggregation of natives and foreign volunteers, transient and sporadically organized. Now, the various insurgencies and separatist struggles not only were spreading fast but were increasingly unified and interdependent. Basayev had used the winter of 2004–2005 to consolidate a streamlined regional command structure, one that would give him more power to sustain a jihad as large as his ambition.

Especially important was the jihadists' expanding network of localized jamaats ("war councils" or "jihad councils"). Each local jamaat operated atop a pyramid of Islamic religious, educational, and social services organizations, which interacted in turn with the population at large. Lavish funds from Islamist charities, mainly from the Arabian peninsula, fueled the machine. This dependence on foreign funds—most of which were dispensed via Muhammad Abu-Omar al-Seif—gave the Basayev-led Islamist-Jihadist leadership a potential foundation for a centralized alternative power structure that spanned the entire northern Caucasus. To the Islamist-Jihadist leaders, the winter of 2004–2005 marked a turning point in their consolidation of regional influence. In mid-February, an emissary of the Islamist-Jihadist leadership assured Arab supporters and donors that "the North Caucasus is being less and less controlled by Moscow" as a result of these activities.

The network of jamaats extended deep into Dagestan, Chechnya, and Ingushetiya, reaching out to the populations of areas firmly controlled by the Russians. The jamaats were also spreading into republics with mixed populations such as Kabardino-Balkariya and Karachayevo-Cherkessia, and even into the predominantly Christian North Ossetia. In Kabardino-Balkariya, for example, the rising power was thirty-nine-year-old Musa Mukoshev, a graduate of Koranic studies in Saudi Arabia and Jordan. Mukoshev returned in 2002 to organize the Yarmuk War Council, and began preaching—first on religious and social issues, but gradually introducing jihadist themes. In early 2005, the declared objective of the Yarmuk jamaat was "to establish an Islamic state in the Caucasus." In February, a confidant of Mukoshev described the work of the Yarmuk jamaat as "only the beginning. Soon the youth will become real Mujahedin." Jamaats affiliated with Basayev's Islamist-Jihadist leadership were also established in Russia. The most important of these was the Nogai War Council, which traced its origin to the clandestine Nogai Battalion of the First Chechen War, which had supported Basayev's terrorist raids into southern Russia. During the winter of 2004–2005, the Nogai Jamaat was transformed into the headquarters of a clandestine network spreading beyond the Neftekumsk District, which borders Chechnya, throughout much of Stavropol Province, Southern Russia.

All of which explains Shamil Basayev's defiance of Maskhadov's cease-fire proposal.

Just when the Islamist-Jihadist leaders' aspirations finally seemed within their grasp, Maskhadov had suddenly shifted public attention back onto the narrow issue of Chechnya and the specter of a Western-style political process. In a late February written interview with Radio Liberty, Maskhadov sounded another conciliatory note, asserting that a "thirty-minute, fair, face-to-face dialogue" with Putin would result in an agreement ending the war. "In my opinion, for this dialogue to begin, it would be enough to reach agreement on the following issues: guaranteeing the security of the Chechen people and protecting Russia's regional and defense interests in the North Caucasus," Maskhadov wrote. "If we are able to open the eyes of our opponents, the Russian leaders, peace can be established." Maskhadov then appealed to the West to help "deliver" Moscow. The interview was a clear attempt to go behind the back of Maskhadov's Islamist-Jihadist "partners" to curry U.S. and Western political support as he grappled with Putin and the Islamist-Jihadist leaders. For Basayev, another cycle of debates with Maskhadov was the last thing he needed.

And so it was decided that Maskhadov had to go. Of course, the Islamist-Jihadists could not afford to have Maskhadov eliminated in an internal power struggle; he would be of far greater use to their cause as a martyr—and the most expedient way to effect that was to betray him to the Russian security forces. To the Russians, Maskhadov's elimination would be merely a symbolic success; though he was well meaning and committed to the destiny of his people, it was clear by now that he no longer had the power to act on behalf of his people. Ever since 1996, when he scuttled Chechnya's chance to acquire de facto independence after the Khasavyurt Agreement, he had been losing ground to the Islamist-Jihadists. The jihadists viewed him as an opportunist who elected to make deals with the devil to shore up his claim on the limelight. Even if the Kremlin had accepted Maskhadov's last-minute call for negotiations, and even if half an hour with Putin would have been successful, there was no way that Maskhadov could have delivered the Chechen side, let alone impose anything on the Islamist-Jihadist leadership.

When the Islamist-Jihadists tipped the Russians off about Maskhadov's whereabouts, then, Moscow saw no reason not to act. On

March 7, 2005, Russian special forces raided a house in the village of Tolstoy-Yurt in northern Chechnya. Adjacent to the house was a deep bunker that Maskhadov used as his hiding place; his few bodyguards put up stiff resistance, and the Russians had to blow up the bunker before they could break in. When they did, they found Aslan Maskhadov dead within. There is still debate over whether Maskhadov was killed by the explosion or accidentally by his bodyguards during the firefight; the only apparent injuries to his body were a few bullet holes. According to Mona Abdel-Malek Khalil of Egypt's *Al-Ahram Weekly*, however, Chechen sources alleged that either Maskhadov had "ordered his followers to shoot him dead, in order not to be taken alive, or his supporters decided to kill him so that their commander avoided a disgraceful arrest." Significantly, Tolstoy-Yurt is located in a part of Chechnya that has been under the tight control of Russian forces since 1999. Maskhadov had chosen to hide here, away from his "struggling people," simply because he knew that had he stayed in the areas controlled by the Islamist-Jihadist mujahedin he would have been assassinated long ago.

The Islamist-Jihadist leaders leaped to proclaim Maskhadov's killing as a turning point for the jihad in the Caucasus. From their very first public statements, both Basayev and Movladi Udugov used their eulogies for the "martyr Maskhadov" to announce a new escalation in the jihad. On March 8, in the first official announcement of Maskhadov's death, Udugov was clear about its implications. "Moscow has not been making a secret of [the] extreme importance of what happened," Udugov noted approvingly, because "for quite a long time the figure of [the] Chechen president has been determining in any affairs in the Caucasus." (Note that he refers to "the Caucasus," not just Chechnya.) Yet Udugov ridiculed the fallen martyr's attempts to reach terms with Russia. "Aslan Maskhadov was the only person who believed that there was still something to talk about with Moscow. Now there is no such person in Chechnya." Maskhadov's death, he said, "killed the last illusion in those Chechens who . . . still believed in . . . civilized forms of communication with today's regime in Moscow." The only remaining recourse was total war. "A new period in the history of Russian-Chechen military confrontation has started," he announced. "This period does not imply [either] any negotiation, [or any] cessation of the war. The war . . . will be stopped only when the threat from the North is eliminated once and for all."

Shamil Basayev did not react to Maskhadov's death until March 11, with a brief and formal statement that declared Maskhadov a martyr on the path of Islam and then turned immediately to issues of succession. Predictably, Basayev seized the opportunity to announce that Maskhadov's death marked a formal and irreversible shift from Maskhadov's Chechen liberation struggle to the broader jihad aimed at transforming the Caucasus. "Mujahedin (Warriors) come and go, and those who have been fighting for the sake of Maskhadov can now rest," Basayev noted, "but those who have been fighting for the sake of Allah—for them Jihad is continuing." Basayev also announced that the new president of Chechnya would lead the regional jihad, and that "all Muslims of Chechen Republic of Ichkeria, the Caucasus and the entire Russia" would be required to swear a loyalty oath to the new president. Other interested observers concurred. Murad Batal al-Shishani, a Jordanian-Chechen writer, predicted that Maskhadov's death would expedite Basayev's regionwide jihad. With Maskhadov out of the way, al-Shishani wrote, "the extremist wing will get the chance to rearrange itself and direct more attacks at Russia"; jihadist attacks beyond Chechnya would "probably increase, especially now that these extremists are free from any obligations to Chechnya's political leader."

One important by-product of Maskhadov's removal was that it cleared the way for the further Islamicization of the popular power structure in Chechnya. This trend was manifested by the ascent of a young and little-known Islamist named Abdul-Khalim Sadulayev as Maskhadov's successor, and by the ensuing rise of the Chechen jamaat leaders. Although the jamaat leaders emerged from localized *teips* (tribal organizations) and key clans, from whom they drew their power, during the 1990s they gravitated to the Islamist-Jihadist cause; with the eruption of the Second Chechen War, they became the most important grassroots support system for Basayev, Khattab, and the Islamist-Jihadist leaders.

An Islamist judge, Abdul-Khalim Sadulayev was the leader of the only jamaat in the city of Argun—an early hotbed of Islamist fervor within the Chechen revolt. Among his closest allies and fellow jamaat leaders were the radical Akhmadov brothers, who headed the jamaat in Urus-Martan; Arbi Barayev, who headed the jamaat in Alkhan-Kala; and Islam Khalimov, who headed the jamaat in the village of Khatuni. The jamaats' traditional responsibilities included religious and social welfare activities, but these groups also took command over militant terrorist units; the choice of Sadulayev to succeed Maskhadov was a potent symbol of the transformation of the separatist conflict into a regional jihad.

Abdul-Halim Sadulayev issued his first communiqué as president on March 14. In what was fundamentally a religious text, he opened with an obligatory eulogy to Aslan Maskhadov's martyrdom, and attempted to reconcile Maskhadov's legacy with his own call for jihad: "Our brother Aslan showed to the enemy how negligible are his vain efforts to defeat the one who—with the name of the Supreme on his lips—went on the way of Holy Jihad, protecting the true Faith, the sacred Native Land and the Honor of the Nation," he noted. Sadulayev expected that Maskhadov's martyrdom would "inspire our mujahedin, who are fighting for the liberation of the Fatherland, to acts of heroism." He also promised that the coming phase of the jihad would involve an uncompromising total war. "The Kremlin [has] made it understood that it still does not want peace on the Chechen soil, that it will continue to aim at the complete destruction of our people as a unique ethnos. The Chechen nation has the capability to break the pride of its enemy in the form of the Russian empire, and force it to peace." He predicted future terrorist strikes within Russia, saying that such actions were justified by Russian attacks on Chechens. "We have the right to act against our enemy by methods which are acceptable to God, in order to protect the right of the Chechen nation to independence. I swear by Allah, that we will protect our people from the Russian genocide, whatever price we will have to pay for it!"

With Maskhadov gone and Sadulayev on the rise, whatever moral and traditional policies had been holding the radicalization of the Chechen struggle at bay were gone for good.

Chapter 25

GOING INTERNATIONAL

STARTING IN THE FALL OF 2004, THE ISLAMIST-JIHADIST MOVEMENT underwent its most profound transformation since 1998–1999, when Osama bin Laden first called for a jihad against the West, aimed at establishing an Islamist caliphate. The attacks of September 11, 2001, were but the most dramatic implementations of this first phase in the global Islamist jihad. It was the Muslim world's reaction to the U.S. invasion of Iraq that emboldened the Islamist-Jihadist supreme leaders and their allies in the key sponsoring states to begin pursuing the next phase of their jihad, launching an uncompromising surge against all foes of Islam, internal and external.

During 2004–2005, the Islamist-Jihadists and their sponsor states launched the total war of which they had long warned—a war in which they saw no possibility for compromise, or even coexistence, with their enemies. The northern Caucasus jihadist movement, led by the Chechen rebels, was now reborn as an important and fully integrated element of the global jihad—no longer a nationalist campaign, but a source of manpower, energy, and fighting strength for a wider campaign.

In late August 2003, Osama bin Laden decided to shift the emphasis of the Islamist movement to the jihad in Iraq, declaring a worldwide mobilization of the Islamist movement to defeat the United States in Iraq, confront Israel, and spark a historic upsurge of militant Islam. Bin Laden's emissaries ordered that this overall effort, and especially the jihad in Iraq, should take precedence over other jihads. To accomplish this, bin Laden now dispatched to Iraq a senior mujahedin commander and former chief of the training complexes in the Khowst area known as Abdul-Hadi al-Iraqi. Until his defection in the early/mid-1980s, Abdul-Hadi was a veteran member of the Iraqi Baath Party's security and intelligence apparatus. After becoming attracted to radical Islam, however, he escaped from Iraq to join the Afghan jihad, leaving behind a sizeable number of closet supporters among his Baathist colleagues. Bin Laden sent Abdul-Hadi not just to escalate the Islamist jihad, but also to reach out to his former Baathist colleagues, now fighting the Americans. He also ordered the activation of an experienced cadre of veterans from Afghanistan, Chechnya, Bosnia-Herzegovina, and other jihad theaters, ordering them to travel—along with new recruits—to "pre-determined assembly points" from which they would be transported to "the decisive Jihad front"—Iraq.

One of the first commanders to react to bin Laden's call to arms was his senior commander in Chechnya, Abu-al-Walid. Starting in mid-August, he dispatched several Chechen bomb-making and sniper experts to Iraq via Syria, to help local cells improve their capabilities without protracted expert training. The Chechen experts he sent were extremely knowledgeable about the Soviet-origin weapons and munitions that made up the bulk of the arsenal available to the mujahedin in Iraq. Several hundred experienced Saudi mujahedin were also diverted from Chechnya and Central Asia to Iraq—again, via Syria—to bolster Islamist ranks. By early September, some of the Islamist detachments operating in the Mosul, al-Khaditah, Ramadi, and Fallujah areas, included more Arab than Iraqi mujahedin. Ayman al-Zawahiri's brief stopover in the Chechen camps in northern Georgia must have contributed to Walid's decision to commit so many assets and resources to the jihad in Iraq.

The Chechens' presence on the battlefield was noticed almost immediately. The growing effectiveness of improvised explosive devices (IEDs) against heavy American tanks, which saw increased use starting in the fall of 2003, was the direct result of experience gained in Chechnya, experience that was passed along to Arab mujahedin from the

Middle East trained in the Chechen camps. On the afternoon of October 28, 2003, an M-1A1 tank was destroyed by a roadside bomb near Balad, forty miles north of Baghdad; two crew members were killed and a third wounded in the attack. After numerous failures over the previous six months, it was the first time the jihadists had managed to destroy an M-1A1. Attesting to Abdul-Hadi al-Iraqi's involvement, the attack took place in an area dominated by Baathist guerrilla forces. The highly specialized anti-tank bomb was constructed by Chechen expert terrorists; Russian raids on Chechen training camps had yielded detailed instructions on how to construct such antitank bombs, and similar bombs had been used successfully by Islamist terrorists against Russian tanks in Chechnya and against Israeli Merkava tanks in Gaza, after Chechen experts arrived there. After the successful Balad strike, roadside bomb strikes using these techniques (and similar techniques used in Chechnya) multiplied throughout Iraq. In the months that followed, a steady stream of Jordanian-Circassians and Saudis with extensive combat experience in Chechnya arrived in Iraq, bringing their proven expertise in urban warfare tactics and sophisticated IED construction.

The proven expertise of the Chechen fighters, as well as the strategic value of the Caucasus jihadist front, also led the Palestinian HAMAS to start throwing its support to the Chechen home front. Starting in the spring of 2004, HAMAS started recruiting Palestinian volunteers to help the Chechen jihad, even from areas like the Gaza Strip, where they were immersed in their own pressing jihad. HAMAS distributed posters and videos highlighting the Islamist jihad in Chechnya as a struggle equal in importance to those in Palestine and Afghanistan, stressing that Russian fighters were anti-Islam and "terrorist in nature," and portraying Basayev and Khattab as peers of HAMAS founder Sheikh Yassin and Osama bin Laden. By mid-2004, HAMAS had smuggled several Palestinian recruits to Jordan, where they joined the flow of Jordanian-Circassians and Saudis traveling to Chechnya. The goal of this influx was to help season the foreign mujahedin on the Chechen front, offering them advanced training and combat experience before they returned to help the anti-Israeli jihad. HAMAS leaders even rallied to the Chechen cause: Starting in mid-2004, HAMAS political-military leader Khaled Mashaal made several trips to Qatar and other Gulf states on fund-raising missions for the Chechens—collecting tens of millions of dollars in the process.

Perhaps the most significant sign of the Chechen influence on the Middle East jihad was HAMAS's decision in mid-2004 to begin recruiting women martyrs, a move inspired directly by the example of the Chechen Black Widows. The immediate impetus was Reem al-Riyashi's successful January 2004 suicide bombing at the highly secured Erez crossing in Gaza. After several attempts by male mujahedin to reach the crossing had failed, Riyashi—the mother of two young children—had "volunteered" to commit martyrdom rather than face an inevitable honor killing for adultery. Her success captured the attention of HAMAS, leading the group to establish its own Black Widows unit in the summer of 2005.

Before committing to the use of Black Widows, however, HAMAS leaders engaged in an intense debate over the theological implications of female martyrdom operations. The dominant theological document used in this phase was a fatwa issued by Muhammad bin-Abdallah al-Seif, whom HAMAS identified as "the mufti of the Chechen mujahedin." Al-Seif's elaborate and scholarly fatwa addressed all aspects of suicide terrorism, relying on several theological sources and arguments captured in notes by Abu-Musab al-Zarqawi, as well as in fatwas written by the jihadist leaders in Iraq, to justify the growth of suicide terrorism there. Al-Seif's fatwa, which HAMAS considered as authoritative as those issued by senior Saudi Arabian and Kuwaiti clerics, endorsed Chechen suicide bombings, and specifically sanctioned female suicide attacks. Al-Seif based his religious instructions on the case of Hawa (Khava) Barayeva, the first Chechen female suicide bomber, stressing that the Chechen female suicide bombers' attacks were theologically justified because their strikes were an integral part of the jihad against the Russians. Reviewing comparable jihad fronts, al-Seif mentioned the Palestinian jihad against Israel as another one that warranted the use of suicide bombers—a comparison the HAMAS leaders seized upon as a theological approval of their operations. Al-Seif's fatwa leaves no doubt that Islamist-Jihadist leaders in Chechnya and Iraq were engaging in extensive discussions on both theological and tactical subjects—marking the growing significance of the Caucasus theater in theological terms.

In mid-2005, HAMAS disclosed the establishment of a women's suicide unit within the Martyr Izz al-Din al-Qassam Brigades. The unnamed commander of the unit, a middle-aged woman who had lost sons in suicide operations and had a son in an Israeli jail, stressed that jihadist martyrdom operations should be considered acts of self-fulfillment rather than revenge for abuses or losses. Jihad and motherhood, she explained, are "all one path. We raise our children and perform our domestic duties, the duty of encouraging devotion to religion, as well as the other everyday duties, and the epitome of them is Jihad for the sake of Allah. Jihad is a duty that every Muslim is required to fulfill if [one] can. Our joining the military organization is one of the essential everyday tasks." The commander stressed the inspirational role of previous suicide attacks for the members of her unit. "The martyr Reem al-Riyashi is like a crown on our heads and a pioneer of the resistance. Nobody can fathom the magnitude of her sacrifice. . . . By the name of Allah, we hope to become like her at once."

The current phase of this global jihad was launched in the late fall of 2004. The Islamist-Jihadist leaders were emboldened by their strategic triumph in Iraq and by Osama bin Laden's call for the use of WMD (especially nuclear weapons) against the United States. Theological writings portrayed the Iran-sponsored Islamist-Jihadist forces in the Middle East as the key to a historic offensive aimed at reviving the caliphate and reinstating its five traditional divisions: the Land of Greater Syria, the Land of Greater Egypt, the Land of the Twin Rivers, the Land of the Twin Mosques, and the Land of the Berbers. Evicting the United States from the region, destroying Israel, and even toppling non-Islamist Arab regimes would be necessary steps toward these higher objectives. The long-term worldview of bin Laden's shura were clearly articulated in the "Working Strategy Lasting Until 2020," a blueprint developed in early 2005 under the guidance of a former senior Egyptian officer who goes by the name of Muhammad Makkawi. The

Working Strategy largely endorses consolidating the jihad in the Hub of Islam, a move long favored by Ayman al-Zawahiri.

By the spring of 2005, there were clear indications that the Islamist-Jihadist camp had adopted the goal of achieving victory over the West and establishing a caliphate by the year 2020. Based on tenets articulated by al-Zawahiri, the Working Strategy called for the Islamist-Jihadist caliphate to replace the United States as the world's sole hyperpower. In military terms, it called for the defeat of the "West" (as led by the United States and including Russia as a key power), through two types of war: (1) a gradual attrition of U.S. and allied forces through a myriad of debilitating quagmires—particularly in Iraq, the Caucasus, and Israel; and (2) a series of spectacular strikes in the United States, Russia, and other Western targets, leaving the West stunned and demoralized.

The Islamist-Jihadist leaders expected the initial stage of this campaign to culminate soon in a cataclysmic confrontation in the Middle East, with Iran playing the dominant role in the jihadist camp. For the mullahs of Tehran, the Islamist-Jihadist triumph in Iraq was already serving as a war-by-proxy, delivering a historic victory that would ultimately consolidate Iran's status as a regional empire controlling the shores of both the Persian Gulf and the Levant, some ninety percent of the region's energy reserves, and the overall thrust of Shiite Islam. The importance of the mullahs in Islamist-Jihadist strategy in today's Middle East cannot be overstated.

Meanwhile, the Islamist-Jihadists were pouring reinforcements into their strategic springboards in the Balkans (against Western Europe), in Chechnya (against Russia and Eastern Europe), and in the tri-border region of Argentina, Brazil, and Paraguay in Latin America (against the United States). As the movement moved people and materiel among the various Islamist-Jihadist centers, Chechnya emerged as the strategic meeting point between the strategic arch eastward to Afghanistan-Pakistan (code-named Livo) and the strategic arch westward to the Balkans and Western Europe (code-named Kvadrat). Chechnya thus became a key center for the spread of Islamist-Jihadist doctrine and practices worldwide.

The ongoing importance of the Working Strategy was demonstrated by a May 30, 2005, letter from "Sheikh al-Zarqawi" to "Sheikh bin Laden" titled "A Message from a Soldier to His Commander." In the letter, Zarqawi noted that the fighting in Iraq was being conducted according to a master plan that bin Laden had approved, and reiterated the importance of the movement's twin-track strategy—combining ongoing action throughout the Hub of Islam with terrorist strikes within enemy territory. This dual approach, Zarqawi reminds bin Laden, was based on historic precedent—namely, a decision in 633 CE by the first caliph, Abu-Bakr al-Siddiq, to keep driving Arab-Muslim forces forward, despite adverse conditions, while ordering Osama Bin-Zayd to push northward in an audacious military expedition, fighting the Byzantines on their turf in order to force the enemy to withdraw despite the overall superiority of their forces. The surge had succeeded in putting the Byzantines on the defensive and facilitated the subsequent Arab occupation of the entire Middle East and North Africa. "The Ummah is now facing similar hardship, and as Allah made it possible, back then, for [the] Osama Army to be sent to fight the Romans [i.e., the Byzantines] on their turf and for Abu-Bakr to fight the apostates elsewhere, He is surely capable of making it possible now," Zarqawi wrote. "O God, make the expedition of Osama [bin Laden] proceed toward its goal. . . . We await your orders as to the next stage of the plan."

In a private letter dated July 9, Ayman al-Zawahiri responded to Zarqawi's "Message." Zawahiri seconded Zarqawi's focus on establishing the caliphate in the Middle East as a foundation for attacks against the West. "It has always been my belief that the victory of Islam will never take place until a Muslim state is established in the manner of the Prophet in the heart of the Islamic world, specifically in Greater Syria, Egypt, and the neighboring states of the [Arabian] Peninsula and Iraq; however, the center would be in Greater Syria and Egypt," Zawahiri wrote. He portrayed the caliphate as "a bird whose wings are Egypt and Syria, and whose heart is Palestine," but emphasized that efforts to establish the caliphate should not come at the expense of the worldwide jihadist fronts, with their crucial role as springboards into the West. "As for the battles that are going on in the far-flung regions of the Islamic world, such as Chechnya, Afghanistan, Kashmir, and Bosnia, they are just the groundwork and the vanguard for the major battles which have begun in the heart of the Islamic world," Zawahiri told Zarqawi.

Zawahiri, the driving force behind the Working Strategy, also issued a public reply to Zarqawi in a televised message on August 8. Zawahiri heralded the escalation of the war in the West, especially Europe, in the context of a global jihad also directed at the Arab Muslim regimes who were cooperating or allied with the West. The objective of the jihad, Zawahiri stated, was to cleanse the entire Muslim world of all vestiges of Western presence, and terrorism in the West would be a key instrument in achieving this goal. Zawahiri put special emphasis on the growing threats to the United States. "What you have you seen, O Americans, in New York and Washington and the losses you are having in Afghanistan and Iraq, in spite of all the media blackout, are only the losses of the initial clashes. If you continue the same policy of aggression against Muslims, God willing, you will see the horror that will make you forget what you had seen in Vietnam," he warned.

Once the Western powers were banished from the Muslim world, Zawahiri promised, the Islamist-Jihadist forces would be able to destroy the existing regimes and replace them with the Islamist caliphate. Toward this end, he urged the entire Arab Muslim world to mobilize to rid itself of regimes cooperating with the West against what he termed the true interests of the Muslim world. "I call upon the Nation's opinion leaders—the scientists, writers, intellectuals, craftsmen, merchants, and officers—to come together and discuss means of disposing of these corrupt and corruptive regimes that weigh down on our Nation's chest. I call on them to start preparing for change no matter how much time and effort is needed. I call on them to spread calls for change among all sects of the Nation, and to know that victory always comes at a price. A dark fate awaits us if we do not resist them [the regimes]," Zawahiri warned. Zawahiri's message was an unmistakable call to arms—and, like previous messages from major Islamist-Jihadist leaders, it was not uttered before the relevant Islamist-Jihadist forces were prepared to back up its promises with action.

* * *

Even as Abu-Musab al-Zarqawi was defining the grand strategy of the jihad, the Islamist-Jihadist leaders in Chechnya were attending to a grand strategy of their own, for the next phase of the Jihad in the Caucasus and beyond. In a major document issued on August 6, 2005, Movladi Udugov retraced the history of the wars in Chechnya, in terms designed to demonstrate that the Chechens had always been part and parcel of a jihadist civilization. The title of the document—"Security in Exchange for Independence"—said it all: Unless Russia and the West acquiesced to—and facilitated—the establishment of a Sharia state in Chechnya, they would have no security at home.

In the statement, Udugov argued that Chechnya has always been a Muslim state, and that its quest and struggle for freedom were essentially jihadist in character. "No National Liberation movement of the Chechen Nation has ever existed outside of Jihad," he asserts, building on this assertion to claim that Islamophobia had motivated Russian aggression of recent years, particularly during the Second Chechen War. "The 'hidden motive' of the new Russian aggression against the Chechen State and its people is the fact that **independent Ichkeria started moving towards revival of the Shariah** as the only acceptable system of social and legal relationship for the Chechen society: **the system capable of stabilizing the internal situation and to resist foreign aggression**" [boldface appears in the original]. This Russian Islamophobia makes it impossible for the Kremlin to acquiesce to the very existence of an independent Chechen state under any conditions. "The organizers of the first and the second wars against the Chechen state perfectly realized that by the very logic of its historical development independent Ichkeria would sooner or later revive the Shariah **due to objective requirements of national survival within the national and state system.**" Udugov argued that the war had only accelerated the Islamicization of Chechen society. **"The Chechen nation got consolidated and turned to its historical experience, where Islam was the main value as the religion and where the Shariah was its practical embodiment in the relations within the society."** And since an independent Chechen state not ruled by Sharia law was inconceivable to the Chechens, according to Udugov, there could be no negotiated settlement or compromise between the Russian government and the Chechen Islamist-Jihadist leadership.

Another major factor aggravating the crisis in the Caucasus and contributing directly to Russia's boldness in pursuing its war in the Caucasus, Udugov argued, was the position of the United States and Western Europe in the conflict. The West had aligned itself against the Chechen rebels in pursuit of its own self-interests, he noted, but their support had enabled Russia to continue the war without major sacrifice. "If the higher political circles of the Western states were adhering to the cynical position that 'Chechnya is Russia's internal affair,' . . . Russia would not have been able to be waging a long-term war in the Caucasus with the obvious scarcity of its economic and political resources and with total absence of moral resources." According to Udugov, Western governments were "not even making a secret of the fact that [the] Russian military clique is continuing its genocidal war in the Caucasus exclusively with financial, economic, political and informational support from the Western states." The West was convinced that its economic and strategic interests in the Caucasus could be guaranteed only by Russia, and was therefore willing to support Russia's enduring presence in the Caucasus.

Ultimately, however, Udugov stressed that it was the West's inherent Islamophobia that drove and reinforced Western support for Russia. "With all of their internal disagreements," Udugov insisted, "members of the Western alliance keep some basic strategic settings unchanged, one of them being: **not to allow an independent Islamic state with the Shariah rule appear on the map of the world.** This is the general setting in the policymaking of the Western leaders, dictated by imposed false fear before Islam. And this is the source of the inconsistency, inertia, discrepancies and outright immorality [of] their attitude toward the genocidal war of Russia against the Chechen State." Still, Udugov reiterated, the importance of the West's economic interests in the Caucasus and Russia should not be discounted. "The Western alliance is believed to have its own 'set of interests' in both Russia and the Caucasus. In Russia, first of all it is energy resources and control over the nuclear weapons. In the Caucasus it is oil and safe access to the Caspian oilfields." The West was increasing its active support for Moscow, Udugov argued, both because of the inherent Islamophobia of the West and because of its economic interests in the Caucasus. And these interests were responsible for "the [West's] silent approval of genocide and almost outright financial and political help to Moscow in extermination of the Chechen people." This account portrays the United States and the West as an enemy in the jihad for Chechnya and the Caucasus, legitimizing the jihadists' right—even obligation—to launch new strikes on the West.

The only key to ending hostilities, Udugov argued, could be found in the principle of "Security in Exchange for Independence." The key to this model lay not in reconciliation or peacemaking, but in a mutual pursuit of diverse self-interests. The two mutual interests of the warring parties, Udugov argued, were "**stopping the bloodshed, and SECURITY.** . . . It is SECURITY that all interested parties are trying to achieve. Russia and the Western alliance are demanding security for their state interests, and Ichkeria is demanding security of its national sovereignty." Udugov argued that any proposed agreement should cover not just Chechnya but the entire Caucasus, because that is where both Western and Islamist-Jihadist interests lay. "The Russian-Chechen war . . . can fully and realistically be called the New Caucasus War. Therefore efforts on a much larger scale must be put in the foundation of the settlement of Russian-Chechen military conflict, rather than the ones that have been offered to this day."

The core of Udugov's proposal was the creation of what he called "All-Caucasus Security Structures," which he claimed would preserve the state interests of Russia and the Western alliance in exchange for Chechen independence. The All-Caucasus security structures would "serve as an instrument in the settlement of other conflicts in the Caucasus region," he noted. Lest any of this sound like a throwback to the earlier generation of Chechen nationalism, Udugov reiterated that the "Islamic ideology" was "a natural and necessary form of national self-defense and ethnic survival in the national and state system" of Chechnya, and that an "independent national state based on the Shariah Law is the goal." Udugov chided the Western powers for "supporting the genocide of the Chechen people," describing them as "the Kremlin's accomplice in these crimes." Udugov warns that the West will suffer if it refuses the Chechen offer, while the very nature of the jihad would protect the Chechens against lasting damage from Western retaliation.

Udugov's plan was dressed up in an intriguing proposal, but beneath its rhetoric was an all-too-evident threat to both Russia and the West. Until Chechnya was granted independence and allowed to establish an Islamist state, his logic went, neither Russia nor the West would enjoy security. And the only way Chechnya and the entire Islamist-Jihadist movement could deprive Russia and the West of security was through the conduct of international terrorism.

Starting in late August 2005, the Islamist-Jihadist movement was increasingly consumed by the prospect of a dramatic breakout—a Great Ramadan Offensive. Several Islamist-Jihadist leaders and commanders were convinced that this new offensive would overshadow even the dramatic impact of September 11. The Great Ramadan Offensive would include a "fateful confrontation" with U.S. and Israeli forces in the Middle East, as well as a series of unprecedented strikes in Western Europe, Russia, and perhaps even the United States. Throughout the Muslim world, the anticipation was reinforced by a series of virtually simultaneous communiqués starting around August 24, 2005. It was the first time that so many significant doctrinal-theological statements had been issued at once—confirming the sense that a well-coordinated and devastating global onslaught was on the horizon.

The first shot across the bow was a new recorded message by Abu-Musab al-Zarqawi, apparently made on August 21 and announced around August 24. "The mujahedin in Iraq have, praise be to God, moved the battle from the ground [in Iraq] to the land of the cross," Zarqawi announced. "Jihadist units have been founded in all of Western Europe, to defend the powerless within the nation. For the crimes the Crusaders have committed against the Muslims, they will reap in their own homes [countries], God willing."

A few days later, a statement on Abu-Musab al-Zarqawi's doctrine as leader of the al Qaeda Organization in the Land of the Two Rivers laid the groundwork for the coming strikes. Al Qaeda's immediate goal in Iraq was to ensure that the newly established Iraqi state could never function properly, by isolating the government and its supporters—local and foreign—from the bulk of the Iraqi people, and by eliminating militias and "traitors" among the Shia and "clearing the stage of all possible rivals before the withdrawal of the U.S. military." Yet Zarqawi's ultimate objective in pursuing such activities was not to seize power in Iraq, but rather to turn it into a springboard for the establishment of the caliphate throughout the Middle East, and for launching new strikes in the West. The real objective of the al Qaeda Organization in the Land of the Two Rivers (also known as "Al Qaeda in Iraq") was to establish "another base that will export jihad to all parts of the world the same way the mother al Qaeda in Afghanistan was."

The next signal was the announcement, dated August 23 but not published until August 25, that the Islamist-Jihadist rebels were establishing an "emergency government" in Chechnya. In the decree, "President" Abdul-Kalim Sadulayev announced that Shamil Basayev had "reassumed the role of deputy prime minister," and radical ideologue Movladi Udugov had been named press minister. The decree also announced the establishment within the government of a "Power Block" in charge of "the National Security Service, Anti-terrorist Center, Ministry of Internal Affairs, etc.," to be administered by Basayev, and an "Information Block" in charge of "the Ministry of Information and Press, Ministry of Media and Information," to be run by Udugov. This was clearly a war leadership council, called to meet the challenges ahead—including a likely new cycle of terrorist strikes against Russia.

The explanation for the new Chechen war cabinet appeared in an August 27 communiqué titled "The War Can Be Stopped Only by War" over the signature of a pro-Chechen radical from Russia named Boris Stomakhin. The communiqué stressed that the war in Chechnya was entering a new era, in which audacious offensive strikes would take the place of diplomatic efforts. "The center of gravity of the war," Stomakhin announced, was returning "from Strasbourg to Johar [Grozny] and the mountainous forests of Ichkeria. . . . Henceforth the war will be opposed with war." The communiqué criticized the European Union's efforts to mediate a negotiated solution to the war and held the Europeans responsible for the continued bloodshed in the Caucasus. The communiqué contrasted the hypocrisy of the late Aslan Maskhadov and his supporters with the "righteous men—Chechen mojaheds, asserting their ground and freedom with the weapon in hands" who had fought under Shamil Basayev. The Islamist jihad was the true representation of the real interests of the Chechen people, the communiqué asserted, and under Basayev's leadership the Chechens would now "revenge the aggressor, the invader, the chastiser not only here, on the native ground, but also at his home, in his den—everywhere where the hand of the avenger may reach his relatives and close people! eye for eye, tooth for tooth!—and no mercy to enemies!" The Stomakhin communiqué echoed the Shield Fatwa used by Osama bin Laden and his inner circle to justify terrorist strikes against civilians, rejecting the peace lobby's argument "that in the destruction of a quarter of [the] Chechen people only Putin with his pack is guilty, but not the Russian people who have selected this Putin." The communiqué concluded by vowing that a jihadist war would "start over again and again—until either Chechens do not remain on earth, or bloody and criminal Russia doesn't fall to pieces forever [sic]!"

The third signal was an August 26 message from Muhammad al-Dief (Abu-Khalid), commander-in-chief of the Martyr Izz al-Din al-Qassam Brigades, the military wing of HAMAS, "on the occasion of the Zionist retreat from the Gaza Strip and the northern West Bank." Significantly, this message not only committed HAMAS to continue the jihad until Israel was eliminated, but tied the jihad against Israel to the anti-U.S. jihad in Iraq and other Islamist-Jihadist causes worldwide. Dief explained that "the liberation of our beloved Gaza Strip" was only the first step in a jihad that would culminate in "liberating Jerusalem, the West Bank, Acre [Akko], Haifa, Jaffa, Safad [Zefat], Nazareth, Ashkelon, and all parts of Palestine" through the destruction of Israel. The West Bank, however, would be the theater of the next violent confrontation with Israel. "The sacrifices of our people in Jenin, Nablus, Hebron, Ramallah, Jerusalem, Bethlehem, Tulkarm, Qalqiliyah, and in all our cities, villages, and refugee camps in the West Bank portend that the next stage will, with the help of God, be to drive out the occupation from your areas." Dief promised the Israeli Arabs that their liberation was imminent: "We will not give up and we will not resign until the liberation of all our sacred soil in our beloved land which is munificent in its giving, and so are we in our giving as we continue along the road, for the march will not stop." And he saluted "our beloved people in patient and enduring Iraq," hailing the insurgency there as an integral part of HAMAS's regional struggle in the Middle East. Dief stressed that the jihad in Palestine was equally relevant to jihadists in Iraq. "The liberation of the Gaza Strip is for us and for you and for all free people in the world, a lesson to learn. Do not give up your weapons. Resist your enemies until you rub their noses in the dust," he advised the Iraqi mujahedin. "May God help you, give you strength, unify your ranks, strengthen your determination, and make your views and aim correct."

Then came a number of statements from the Islamist-Jihadist supreme leadership, which sought to put this flurry of messages in context. On September 1, 2005, Al-Jazeera TV broadcast a videotaped message, in English, from suicide bomber Muhammad Siddiq—the leader of the London attacks of July 7, 2005—along with several excerpts from a videotaped speech in Arabic by Ayman al-Zawahiri. Significantly, this was the first videotape to be issued by al Qaeda with English subtitles—a clear indication that its target audience was in the West, not just the Muslim world.

Siddiq's message had been recorded for broadcast after his martyrdom. Addressing both Westernized Muslims like himself and the general population in the West, he defended future strikes on the West on the basis of the Shield Fatwa, insisting that there were no innocent civilians among the people of the West—even among women, children, and the elderly. "Your democratically elected governments continuously perpetuate atrocities against my people all over the world, and your support of them makes you directly responsible, just as I am directly responsible for protecting and avenging my Muslim brothers and sisters. Until we feel security, you will be our targets, and until you stop the bombing, gassing, imprisonment, and torture of my people, we will not stop this fight," Siddiq declared. He identified the London strike as part of the larger mission of the jihad, tying the attack to the situation in Iraq by citing Zarqawi as a source of inspiration for the strike. In deciding to sacrifice his life, Siddiq said, he drew inspiration from "today's heroes, like our beloved Sheikh Osama bin Laden, Dr. Ayman Al-Zawahiri, and Abu Musab Al-Zarqawi, and all the other brothers and sisters who are fighting in Allah's cause."

The "new" message from Zawahiri was clearly edited together from brief portions of a longer speech—the same one used in the August 8, 2005, Zawahiri message broadcast on Al-Jazeera. The message contained no new points, but it refined and better articulated key themes the Islamist-Jihadist movement has long been attempting to convey to the West. He stressed anew that the center of the jihad had moved into the West and that the ongoing campaign of strikes was retaliation for Western transgressions against Islam. Zawahiri addressed his British audience as representatives of the entire West. "I speak to you today about the blessed London raid, which came as a slap in the face of the arrogant Crusader British hegemony, and which made it take a sip from the cup which it has long made the Muslims drink from. Like previous great raids in New York, Washington, and Madrid, this raid has moved the battle to the land of the enemy after it kept moving the battle for many ages to our land, and after its armies and forces occupied our land in Chechnya, Afghanistan, Iraq, and Palestine, and after it continued to occupy our land for many ages while it remained safe in its land," Zawahiri warned.

Zawahiri made special mention of the role that Iraq had played in inspiring the London strikes. "O people of the Crusader coalition, not only does Blair think nothing of the blood of Muslims in Iraq, Palestine, Chechnya, and Afghanistan, but he also thinks nothing of your blood, because he sends you off to the crematorium in Iraq, and exposes you to death in your own home, because of his Crusader war against Islam. Bush, Blair, and all those who march behind their Crusader-Zionist banner should know that the mighty mujahedin of Islam have pledged before their God to fight them until either victory or martyrdom." Zawahiri noted that the West, too, had harmed civilians during its various wars against Muslim countries, justifying the jihadists' avenging strikes against Western civilians. "Treatment in kind is just. Who have been killed in our countries? Are they not the women and children in the Al-Amiriyah Shelter, and in the mosques of Khost, Gaza, and Al-Fallujah?"

Zawahiri returned to a theme repeated in all the recent communiqués—namely, that there were no innocent civilians in the West. "These civilians are the ones who pay taxes to Bush and Blair, so they can equip their armies and give aid to Israel, and they are the ones who serve in their armies and security services. They are the ones who elected them, and even those who did not vote for them consider them legitimate rulers who have the right to give them orders and must be obeyed, and who also have the right to order strikes against us, killing our sons and daughters, and to wage war in their name, and to kill Muslims on their behalf. . . . The lands and interests of the countries that participated in the aggression against Palestine, Iraq, and Afghanistan are targets for us, and whoever wants to be safe [should] stay away from them. The one who warns bears no guilt," Zawahiri concluded.

Then, in early September 2005, an unexpected event confirmed to the entire Muslim world that Allah approved of the jihadists' intentions. When Hurricane Katrina hit the Gulf Coast of the United States—causing immense damage and suffering—the Islamist-Jihadist camp greeted it quite literally as an act of God. Given the economic and military disparities between the Muslim and Western worlds, the Islamists had long hoped for a divine intervention of sufficient strength to overwhelm their nemesis. "Soon, God willing, Allah and his angels are going to begin the destruction of evil America," the Islamist Imams predicted. To expedite the process, they argued, "all that the mujahedin have to do is to fight and kill infidels wherever they can, and . . . Allah, [when] pleased with them, will do the rest. He will destroy their cities, one after the other." To the Imams, the destruction of New Orleans looked like a divine answer to their prayers. As soon as the images of Katrina's devastation were beamed around the world, the Islamist-Jihadist Imams couldn't resist issuing an "I told you so" message, asserting that "Allah has punished America with winds and water." The Muslim world soon dubbed the storm "Mujahed Katrina," giving the Islamist-Jihadist movement one of its most powerful and lasting images.

By mid-August 2005, the intensity and volume of Islamist-Jihadist electronic communication—what Western observers called "chatter"—exceeded that of summer 2001. An unprecedented flurry of messengers scurried around the world, presumably carrying last-minute instructions and orders for the redeployment of forces. There were surges in traffic of mujahedin from Chechnya to Iraq, and from Iraq to Western Europe via Bosnia-Herzegovina. The West European intelligence services picked up on the increased activity among the Islamist-Jihadist networks, led by veterans of Iraq and Chechnya, but there were few arrests; the magnitude of the threat was undiminished.

Chapter 26

IN THE THEATER OF GLOBAL JIHAD

BY MID-2005, AS THE RHETORIC WAS REACHING ITS PEAK, THE INTEGRATION of the Chechen jihad into the global Islamist-Jihadist movement was nearly complete. With the intensity of the fighting in the Caucasus once again subsiding, growing numbers of both Chechen and Chechen Arab mujahedin assumed a major role in the key jihadist fronts of Afghanistan/Pakistan, Iraq, and Western Europe. There was a growing sense that the Islamists-Jihadists were once again on the offensive, as the U.S.-led West was hopelessly ensnared in military quagmires, incapable of uprooting the Islamist-Jihadists in their midst. The series of theological and strategic declarations in 2005 set the tone for the current of worldwide Islamist-Jihadist terrorism—in which Chechen and Chechen-trained mujahedin played an ever-growing role that which continues today.

The dramatic change of posture was most evident in Afghanistan/Pakistan, the theater where American-led forces first clashed with, and ostensibly defeated, the international force of Islamist-Jihadist mujahedin. As early as the first months of 2004, as the indigenous Pashtun tribal forces resumed their insurrection against both Kabul and Islamabad, the Islamist-Jihadist movement went on the offensive. Foreign mujahedin—including both veterans of the 2001–2002 invasion and recent international arrivals—were instrumental to this surge. As before, Chechen mujahedin quickly distinguished themselves as especially indomitable fighters. According to Pakistani intelligence sources, by the spring of 2004 there were roughly six hundred foreign mujahedin in the South Waziristan area—two hundred Uzbeks and one hundred Uighurs, with Chechens, Arabs, and others making up the remainder. The presence of Chechen-trained mujahedin from Jordan along with ethnic Chechens made such estimates unreliable, but the number of Chechen mujahedin was put variously at anywhere from a few dozen to roughly one hundred.

On March 25, 2004, Ayman al-Zawahiri highlighted the importance of Pakistan in a lengthy message broadcast on Al-Jazeerah. Refuting the rumor that he was on the run, Zawahiri addressed the ongoing fighting in Pakistan, stressing its connection to the jihad for Central Asia and the Caucasus. The unfolding struggle for an Islamic Pakistan was pivotal for the overall fate of the Islamist-Jihadist cause, he advised, noting that "the frenzied Crusader-Zionist campaign in Afghanistan, Iraq, Palestine, and Chechnya basically targets Pakistan . . . because the United States does not want Pakistan to be a distinguished power in Central Asia." Zawahiri warned against U.S. attempts to stifle the Islamist presence in Afghanistan, reminding listeners that an Islamist Afghanistan/Pakistan would serve as a springboard into Central Asia and the Caucasus.

Washington was fully aware of this reality, Zawahiri argued, and General Pervez Musharraf was trying "to stab the mujahid Islamic resistance in Afghanistan in the back." But the Bush-Musharraf conspiracy was being met by grassroots resistance: "The Muslim Pakistani people have extended a helping hand to their brethren the Afghan mujahedin and their supporters, including Arabs, Uzbek, Turks, Chechens, and Muslims from other nationalities." Having failed to defeat the jihad, Zawahiri noted, "America [had] commissioned Musharraf with the task of taking revenge on the border tribes, especially the valiant and lofty Pashtun tribes, in order to contain this popular support for Jihad against its Crusader campaign." Yet mujahedin worldwide, including Chechens, had rallied to the cause of the Pashtun tribes, taking an active role in their anti-Pakistani jihad.

In the fall of 2004, the Chechens' contribution to the jihad in Pakistan and Afghanistan occasioned the establishment of a distinct

Chechen militant group within the ranks of Osama bin Laden's elite force, the International Islamic Front for Confronting the Crusaders and the Jews. The Chechen group, known as the Majlis-al-Shura Al-Askari-e-Mujahedin Al-Shishan [Military Advisory Council of the Mujahedin of Chechnya], immediately established a distinct presence among the Islamist-Jihadist forces in both Pakistan and Afghanistan. Pakistani intelligence officials noted the decisive role that this mixed force of Chechens, Uzbeks, and Uighurs would play during battles in Waziristan in October and November, citing them as the most professional and hardened fighters facing Pakistani security forces. A sprinkle of Uzbek, Chechen, and Uighur mujahedin also appeared in the ranks of the Taliban and Gulbaddin Heckmatyar's Hizb-i-Islami Afghanistan, offering their expertise and manpower to the various Afghan forces in eastern and southern Afghanistan, where most of the fighting against U.S. and coalition forces was taking place.

Chechen mujahedin were now dispersed throughout Pakistan's badlands. In late November 2004, a few Chechens attempted to rob a money changer in Quetta, whom they accused of stealing money from the jihad. One of them was arrested and subjected to a few days of torture by the Pakistani police, finally identifying their safe house. As the police were closing in, most of the Chechen mujahedin escaped. But four of them remained, and they managed to hold the vastly larger force of police at bay for more than a day, inflicting heavy casualties before they were overwhelmed and killed by the police. The Waziristan police chief announced that his forces had recovered "grenades, explosives, and bomb-making material" from the house. In March 2005, Pakistani intelligence agents uncovered a plot by Chechen mujahedin in Waziristan to kidnap and execute a senior Russian diplomat and his family in Islamabad. Chechen mujahedin were later observed casing key Russian diplomatic facilities throughout Pakistan. A police raid on a safe house in Islamabad turned up communiqués in the name of the Majlis-al-Shura Al-Askari-e-Mujahedin Al-Shishan taking responsibility for the operation—but the Chechens involved, as well as a Saudi accomplice, managed to evade the Pakistani dragnet and find refuge in Waziristan.

The growing presence of Chechen mujahedin was noted by

American and Pakistani officials as the fighting escalated in the spring of 2005. In one early May engagement in eastern Afghanistan, a small group of Chechens and Pakistanis held their ground and gave cover to retreating Afghan and Arab mujahedin as superior U.S.-led forces closed in on them. According to the governor of Zabol Province, two Chechens and one Pakistani were killed in the standoff (and about forty more Afghans were killed by U.S. aerial bombing of the withdrawing force). "These were well-trained, well-armed people . . . not just a rogue group," Colonel James Yonts told the Associated Press. "They didn't flee, they stood and fought."

By early summer, every substantial clash between Pashtun tribal forces and U.S. and Pakistani security forces saw Chechens and Uzbeks play a leading, and often decisive, role. The Chechens always led from the front and elected to hold the line while covering the withdrawal of their fellow mujahedin. Their success led to an increase in the flow of foreign mujahedin into camps in Waziristan, from where they launched operations in both Pakistan and Afghanistan. One senior Western security official noted that there had been "an increase in foreign fighters: Chechens, Arabs, Middle Easterners" in the region, explaining that Western intelligence "[could] see this from the dead bodies" and also "from the radio traffic we pick up in different languages."

One milestone in the escalating war in Afghanistan was a major special operation against elite U.S. forces in the eastern province of Kunar, on June 28, 2005—a series of clashes that signaled, more clearly than the ongoing resistance in other parts of Afghanistan, a new phase in the war. The decision to launch the strategic offensive was based on the jihadists' new confidence in their grassroots support in Afghanistan. In mid-June, Mullah Akhtar Usmani assured the Saudi daily *Al-Sharq al-Awsat* that the Taliban was enjoying greater support from the Afghan people because of American "brutality against Muslims and their bias against Muslim countries. . . . The Taliban are everywhere. In some places they are very dominant and in others they are not. They are dominant in the eastern, southern and southwestern provinces." This widespread grassroots presence, the jihadists felt, would help to sustain and support their offensive in Afghanistan.

The June 28 attacks, a series of ambushes that destroyed a U.S. Navy SEAL patrol and an MH-47 rescue helicopter, were a complex operation requiring excellent intelligence and organizational skills, and an audacity that went far beyond that behind most Afghan operations. Indeed, the Kunar operation was the result of a strategic decision by the jihadist high command to introduce new forces into the theater, including a group of expert terrorists, among whom were Chechens and West Europeans, who were dispatched into an area just across the border in Pakistan. This group was command by Abd-al-Hadi al-Afghani, the amir of Arab mujahedin in Afghanistan. Asadullah Wafa, the governor of Kunar Province, noted that the attacks were "carried out mostly by foreign terrorist groups—Arabs, Chechens and Pakistanis. They come from the other side of the border, launch their attacks and cross back into Pakistan within thirty minutes."

The Chechens were especially useful in Afghanistan because their Russian-language proficiency allowed them to communicate directly with older chieftains and commanders who had learned some Russian during the 1980s, and because they were comfortable in the mountainous conditions of the region. British, French, and other Western European volunteers were also sought after, for their skills in handling electronic and other high-tech instruments. Significantly, these expert mujahedin cooperated with locally recruited and commanded forces, rather than elements of the national "parties." In Kunar, they usually cooperated with the mujahedin force named after Bira'a bin Malik (an early companion of Prophet Muhammad), which was commanded by Mullah Muhammad Ismail, a former Taliban commander turned local warlord. Such localized resistance forces posed a great challenge to the United States, competing for the same population that U.S. and Afghan forces were trying to engage and recruit. In turn, this opened the door for jihadist spies to penetrate U.S.-Afghan forces.

These foreign elements were crucial to the June 28 clash in Kunar Province. Islamist sources insisted from the beginning that they had advance knowledge that Allied forces were planning to insert a deep reconnaissance team—including four U.S. Navy SEAL troops with a few Afghan guides and translators—and that at least one of these Afghans was a mujahedin spy. The early warning allowed jihadist commanders to coordinate operations with local forces—including the non-Afghan experts from across the border. Abd-al-Hadi al-Afghani and a small team of senior expert terrorists dictated the roles to be played by the non-Afghan mujahedin "through specialized groups from several countries."

On June 28, a composite force of Afghan mujahedin and non-Afghan expert terrorists waited in scattered positions for the SEAL team to be inserted by helicopter into the rugged and isolated Shorek Darra in Kunar's Manogay district. Once on the ground, deep in enemy territory, the SEAL team was surrounded and subjected to intense fire from several directions. According to jihadist reports, two of the SEALs were killed in the firefight, while the other two managed to evade the ambushing forces. Mullah Muhammad Ismail, one of Kunar's military commanders cooperating with the jihadist forces, later reported that he had known that "the American soldiers were on a spying mission as they were taking pictures and carried different instruments to perform their job."

One of the ambushing force's goals was to capture the American soldiers and their equipment. (On August 5, Al-Arabiyah TV broadcast a series of interviews with the British and French mujahedin in this group, in which they were shown analyzing recently captured military documents, plans, and maps for the U.S. command in Afghanistan—items reportedly retrieved from a computer carried by a SEAL.) Although the Americans were killed before they could be captured, the fighters had succeeded in spiriting away their binoculars and other equipment. The Afghan mujahedin were able to track down the third SEAL (who had been wounded in the initial clash), cornering him and ultimately killing him after a long firefight. (The jihadists had expected to capture this SEAL alive, prompting a series of premature false claims by Taliban commanders that one of the SEALs had been captured and eventually beheaded.) The fourth SEAL evaded the mujahedin for more than a day, until he was offered shelter in an Afghan village elder's home in accordance with the tenets of *Pushtunwali* (the tribal code of conduct). The elder sent emissaries to notify Kunar provincial governor Asadullah Wafa, although it wasn't until July 2 that the SEAL was finally picked up by U.S. forces.

After the initial ambushes, the events of June 28 continued to unfold as the jihadist commanders had planned. As expected, the SEALs radioed back to their base, calling for an emergency rescue and reinforcements. According to jihadist reports, the Afghan spy—with the help of an expert terrorist—used the captured equipment "to activate an emergency signal for extraction by a helicopter," and the jihadists were able to activate a distress beacon for the rescue helicopter to home in on as they tried to locate the SEALs. The MH-47 Chinook, with eight SEALs and eight Army crewmen onboard, flew straight into a second jihadist ambush near Asadabad, with devastating results. "Taliban Mujahedin shot down the aircraft while it was flying close to the ground," Commander Mullah Rauf explained, "Using only small arms and simple [RPG-7] rockets." Mullah Muhammad Ismail reported that "the helicopter caught fire after being hit by rockets fired by the Taliban. None on board survived." (Afghan jihadist sources acknowledged privately that Chechen expert mujahedin had fired the lethal RPG barrage and that the operation was attributed to the Taliban to boost morale among Afghans.) Mullah Ismail reported that the jihadist forces managed to vacate the area before a pair of AH-64 Apache attack helicopters and a pair of A-10 Warthog attack jets swarmed in after the downing of the Chinook. "The U.S. jet fighters and helicopters were still scrambling over the area and some bombing had already taken place" by the time his forces safely reached their hideouts, Mullah Ismail asserted. The non-Afghan team—including a few Chechens, two British, and one French mujahedin—was back in its Pakistani safe haven by the end of the day.

The downing of the Chinook was a strategic first for the mujahedin—a distinct shift from the defensive attrition and terrorizing of the American forces and Afghan population to new offensive operations against foreign forces, designed to force their withdrawal and, eventually, to replace the regime of Hamid Karzai in Kabul with an Islamist regime. Indeed, the Taliban and jihadist leaders were quick to claim responsibility for the Kunar operation and to stress its ramifications. "As Kunar was the stronghold of jihad versus the Russian communists," explained local mujahedin commander Mullah Salar Haqyar, "thank God it is the same against the Americans right now." Leading

Taliban commander Mullah Dadullah told the Pakistani daily *Ausaf* that the Kunar incident was the opening shot in a decisive offensive, declaring that "the current spree of attacks against the United States will continue till September"—when the next Afghan parliamentary elections would be held—and thereafter. The dramatic success in Kunar had "paved the ground for victory of the Taliban and now people are willing to cooperate with them. . . . Now attacks will be launched against the enemy's army from all sides," Mullah Dadullah predicted. He also claimed that "there is disorder in the official Afghan Army and its high-level officers are in contact with the Taliban."

Mullah Dadullah, described as the "military coordinator for the Taliban," elaborated on the significance of the Kunar clash in a July 18 interview with Al-Jazeera TV. The unfolding jihadist offensive was characterized by "significant changes in tactics, types of weapons, financial support, and support of the Afghan people," he said, emphasizing the growing grassroots support among Afghans and their ability to supply a regular flow of accomplished fighters for the cause. "All Afghan people are Muslims; they all have weapons and know how to use their weapons. The majority of them are supporter[s] of [the] Taliban and their Jihad against the enemies of Islam. Few have been drawn to the dollars [i.e., attracted by cash incentives to join the American side]. There are also Arab mujahedin in Afghanistan." Mullah Dadullah announced that the jihadist forces "currently have advanced weapon systems and [they] are getting even more advanced weapons, [and] logistic and hi-tech support systems in the next few months." The jihadist command structure was also adapting to meet the challenges of the new offensive. On July 25, spokesman Mofti Latifollah Hakimi explained that the Taliban had "established two 14-member military councils from Kunar to Ghazni and from Ghazni to Balkh," which would "be holding consultations with the Mujahedin Supreme Council in all affairs."

By the summer of 2005, the Islamist-Jihadist leaders were convinced that these new forces and weaponry would enable them to defeat the U.S.-led forces through a new series of swift and audacious strikes. "Taking cities is not part of our present tactics," Mullah Dadullah explained on Al-Jazeera. "Our tactics now are hit and run; we attack certain locations, kill the enemies of Allah there, and retreat to safe bases in the mountains to preserve our mujahedin. This tactic disrupts and weakens the enemies of Allah and at the same time, allows us to be on the offensive. We decide the time and place for our attacks; in this way the enemy is always guessing." The Kunar operation was an early stage of this offensive.

Although Mullah Dadullah and the other Afghan senior jihadist leaders never said so explicitly, the Kunar operation had been made possible by the availability of Arab, Chechen, and Pakistani expert terrorists operating freely out of sanctuaries in Pakistan. Indeed, according to Pakistani security officials, by late July of that year Abd-al-Hadi al-Afghani had several hundred Uzbek, Chechen, West European, and Arab mujahedin in the Shikai area of Waziristan, Pakistan, alone, running regular cross-border operations while training recruits, planning operations in the West, and operating a sophisticated computerized propaganda shop. The Islamist-Jihadist leaders in the area increasingly stressed their close cooperation with the global jihadist movement—in the form of al Qaeda—as a reciprocal relationship, the Afghan contribution to the worldwide jihad against the United States. "Cooperation between us and Al Qaeda is very strong. Many of our Arab mujahedin brothers are fighting alongside of us to establish the religion of Allah. We will accompany Al Qaeda anywhere to fight the enemies of Allah," Mullah Dadullah told Al-Jazeera in mid-July. Even the soldiers of the mujahedin recognized that their effort was intended less to reverse whatever little nation-building effort had taken place in Afghanistan, neutralizing the effects of the September 2005 parliamentary elections, but had more to do with reviving the Afghans' active participation in global jihadist causes. Speaking on Al-Arabiyah TV in early August, one West European mujahed who had participated in the June 28 operation stressed that he was part of "this blessed Jihad with Mullah Omar and Sheikh Osama bin Laden."

The close cooperation between Chechen and West European mujahedin in Afghanistan/Pakistan was apparent in Western Europe as well. In early August 2005, the Pakistanis arrested a senior operative called Osama bin Yussaf in Faisalabad. Bin Yussaf was running a call center that coordinated electronic and telephone communication with

Islamist-Jihadist cells in Western Europe, including veterans of the Chechen wars. The Pakistanis discovered detailed maps of Italian, German, and British cities saved on bin Yussaf's hard drive. Meanwhile, a group of Western European Islamist-Jihadists in Chechnya established a center for the creation and distribution of instructional material for mujahedin in Western Europe. By mid-2005, this center had become the primary source of instructional material for the conduct of clandestine operations in the West. The detailed material was distributed through the Internet, as well as on CDs and DVDs. Among the products seized by the Russian security forces were detailed instructions on the use of everyday items (such as easily available kitchen tools and appliances) as weapons and of readily available materials (found in food stores, repair shops, and the like) to create explosives and sophisticated bombs.

In summer 2005, German intelligence raised an alarm about the mounting threat of terrorism in Europe, and the growing significance of the Chechen connection. "Terrorism is coming home," warned a German senior intelligence official. "And it's coming home to those countries whose governments may have believed they were immune from terror because for years they have provided safe haven to notorious Islamic extremists." Berlin foresaw the greatest challenge coming from young second- and third-generation immigrants living in Europe who had become captivated by the idea of global jihad, including many who had acquired combat experience in Iraq. German intelligence stressed that many of these young European jihadists "have excellent contacts to Islamic extremists in Chechnya." These terrorists, German intelligence described, "are highly mobile, networked across the entire continent, supported by sympathizers and powerful financiers, but also able to operate independently. This new generation of holy warriors has already established sufficiently deep roots in Europe to be able to move about freely and without attracting attention. Many have German, Spanish, British or French passports. They often speak several languages, are employed and develop their attack plans in their free time."

A concurrent investigation conducted in France under Judge Jean-Louis Bruguiere into "the Chechen networks"—that is, networks of French Islamist-Jihadists who had trained and fought in Chechnya—

stressed their skills in conducting spectacular operations. A report prepared by the Directorate of Territorial Security (DST), France's internal security and intelligence forces, in late October 2005 noted their training in "planning chemical or bacteriological attacks using toxic substances, such as cyanide, botulin toxin, or ricin," and observed that the jihadists "also have criminal plans targeting French airports." The European mujahedin had also been trained to use a wide array of surface-to-air missiles—no idle threat, as French intelligence confirmed that by 2002 the French jihadists had acquired two SA-18-type shoulder-fired antiaircraft missiles, using the Chechen Mafiya to smuggle them into France. By the fall of 2005, the report conceded, it was no longer known "whether [these missiles] have been destroyed, taken to other European countries, or simply stored somewhere in France with a view to a future attack."

The extent of the threat posed by Chechen-trained networks was demonstrated in December 2005, when Swiss and French security authorities prevented a plot by Western European jihadists—all of North African origin—to shoot down an El-Al plane in Geneva. The terrorists planned on using RPG-7s smuggled from Chechnya, following tactics perfected by Chechen mujahedin in Afghanistan. Earlier that fall, an undercover agent had overheard three North African migrants affiliated with Salafist organizations boasting of their plans to smuggle weapons from Chechnya and use them to shoot down an Israeli airliner. The ensuing investigation led to the arrest of seven terrorists in Switzerland, along with similar cells in France and Spain with whom they were in contact. At the time of their arrest, three of the terrorists were actively casing the Geneva airport in search of a suitable point from which to launch RPG-7 rockets at the airliner as it was parked or taxiing. According to a French security source, two of the arrested—a twenty-year-old Algerian and a thirty-five-year-old Libyan—led investigators to a safe house in Basel where the RPG and other weapons smuggled from Chechnya were already hidden. Swiss security authorities deny that any weapons were seized and insist that the network was rounded up before it could bring in weapons from Chechnya. Whatever the case, El-Al stopped flying to Geneva for several months in the winter of 2005–2006.

The Chechen-sponsored networks were in Europe to stay.

Chapter 27

PACIFICATION IN CHECHNYA, ERUPTION IN THE CAUCASUS

THE YEAR 2005 SAW THE EMERGENCE OF TWO COUNTERVAILING trends in the northern Caucasus, reflecting the schizophrenic nature of the latter-day Chechen revolt. Chechnya itself was a place of growing peace and social stability, the result of an effective pacification and reconstruction program. Though the atmosphere has been punctured by sporadic acts of violence—primarily terrorist bombings—the pro-Russian government in Grozny enjoyed the growing acceptance of both the Chechen population and the local governments of the northern Caucasus. The key to legitimacy was the government's ability to impose peace in larger segments of Chechnya, while helping to rebuild and develop local economies. Mindful of this trend, the jihadist leaders have repeatedly attempted to use violence to disrupt, or at least slow, the normalization process. By most accounts, their violent strikes have only further alienated the population from its self-declared saviors.

Devoid of grassroots support, the Chechen Islamist-Jihadist leaders began emphasizing the northern Caucasus as a whole, not just Chechnya, as the focus of their regional jihad. They also relied increasingly on isolated Islamist pockets—the jamaats—as safe havens for launching terrorist strikes against both local authorities and the public at large. They appeared to abandon their efforts to win over the population, converting only a select few to the jihadist cause. By contrast, the Russian and local pro-Russian security forces, which had long since recaptured popular support, went on the offensive throughout the North

Caucasus. The pro-Russian Chechen security forces showed particular improvement, working with Russian special forces and security services to maintain the upper hand in the region despite sporadic outbursts of terrorism. The Chechen mujahedin and their foreign allies were pushed back to the remote mountains of southern Chechnya, to safe havens in neighboring northern republics, and across the Georgian border. By 2005, the indigenous Chechen revolt had all but ceased to exist—and the Islamist-Jihadist networks throughout the North Caucasus, including Chechnya, were virtually dependent on an umbilical cord of foreign aid from across the Georgian border.

The spring of 2005 found both sides trying to settle on their strategy for Chechnya after Maskhadov. The Islamist-Jihadist commanders, led by Basayev, vice president Dokka Umarov, and President Abdul-Khalim Sadulayev, scrambled to prevent the near collapse of rebel forces in Chechnya, as the majority of Chechens were now rejecting the continuation of the war. The new posture was articulated by the Chechen writer Zalpa Bersanova, who spent fifteen years studying public opinion and social change in her native country. In mid-April, Bersanova noted that the recruitment of mujahedin was "shrouded by shame and secrecy," as families tried to hide any sign that a relative had joined the jihad. The continuing attacks had compromised the quality of their daily life, Bersanova noted. "The overwhelming majority of Chechens reject these acts as shameful and angrily deny that they are being committed in their name."

And yet the violent attacks on Russian forces escalated again in early May 2005, increasingly committed by non-Chechen mujahedin. On May 2, a patrol of Russian special forces in the Nozhay-Yurt district was attacked by two mujahedin mobile squads; both sides suffered casualties in the firefight that followed, and the Russian forces had to abandon their mission. Only one of the squads was Chechen, commanded by "Amir Akhmad"—Akhmad Avdorkhanov, the commander of Chechnya's President Guards. The second squad was Dagestani, under the command of "Amir Rabbani"—Rappani Halilov, the commander of the Dagestan mujahedin. The Chechen population continued to supply the Russians with tips on impending attacks, including terrorist strikes. On May 5, Chechen citizens near Grozny led Russian security officers to an expertly prepared truck bomb with about 2,640 pounds of high explosives fitted into the truck's frame and chassis. "The only thing left to do was to put a suicide bomber behind the wheel and turn on the electric detonator," Major General Ilya Shabalkin told the Interfax press agency. The Chechen informers also identified a few mujahedin, including experts connected to al Qaida.

The military clashes and skirmishes continued in several regions of Chechnya, most notably in Nozhay-Yurt, Shatoy, and Achkhoi-Martanovskiy. But when Shamil Basayev promised a "fiery summer," he was careful to hint at strikes in the northern Caucasus and Russia rather than Chechnya. Similarly, Dokka Umarov announced that the mujahedin had resolved "to carry out military activities on the territory of the opponent," stressing that Maskhadov's policy of operating exclusively within Chechnya was no longer valid, and that large-scale operations would soon be launched on the territory of Russia. In mid-May, Abdul-Khalim Sadulayev signed an order creating the united "Caucasus Front" of "the Ichkerian Military Force," unifying the Ossetian, Ingush, Kabardino-Balkariyan, Stavropolian, Karachayevo-Cherkessian, Adygean, and Krasnodar "sectors" of "the western front of the Ichkerian Military Force" with the "commanding fronts" within Chechnya itself. This order reflected the Islamist-Jihadist leaders' growing difficulty in sustaining operations in Chechnya without support from the rest of the northern Caucasus. Not that the leadership was more successful in running operations at the heart of Russia: On May 27, having failed to deliver the "fiery summer" he had repeatedly promised, Basayev claimed responsibility for a major power outage in Moscow, even though it was caused by failure of Soviet-era equipment. "Our sabotage units delivered a major blow to one of the most important life-support systems of the Russian empire," Basayev claimed in an e-mail to the Kavkaz-Tsentr website.

By early June, there were growing signs of desperation among the remaining rank and file of the Chechen revolt. Frustrated by public rejection and its own failure to strike effective blows in Chechnya, the movement began to devour itself. Many mujahedin became preoccupied with dreams of kidnapping, torturing, or assassinating former mujahedin who had chosen to stop fighting and return to normal life. At first, many of these former mujahedin had merely accepted the existence of the pro-Moscow authorities as inescapable. As fratricidal violence—including harsh countermeasures by the security forces—mounted, however, more and more ex-mujahedin were driven to turn against their former brothers in arms. "The only way to stay safe is to join Chechnya's [pro-Russian] security services," several former rebels explained.

With all their bluster about escalating the war against Russia, the Chechen commanders were painfully aware of the widening gap between them and the population. These commanders kept insisting that the public was still committed to the jihad, and attributed the loss of popular support to Russian repression. In a mid-July interview with Radio Free Europe/Radio Liberty, Chechen vice president Dokka Umarov articulated these arguments. Despite the mounting challenges and growing losses, he could see "no alternative" to continuing the armed struggle. Yet Umarov no longer envisioned the Chechens delivering a decisive strike or clear-cut defeat to the Russians. Instead, he expected to carry on a prolonged campaign of low-level violence, designed to grind the Russians down. Ultimately, he acknowledged, only the Kremlin's own decision to withdraw, not a Chechen military victory, could bring about a Russian withdrawal, and he saw no reason that the Kremlim would make such a decision anytime soon. He blamed Moscow for creating the impression that "there can be no hope the war will end," but insisted that "the situation isn't hopeless." Despite the grim reality, Umarov pointed to the continued flow of younger volunteers to fill the ranks of the fallen mujahedin as a sign of the health of the jihad. At the same time, he acknowledged that even many people who had actively supported the First Chechen War were now "succumbing to the fear and hopelessness that pervades daily life" and leaving the jihad. "Today, they are in some way broken by the fear produced by the FSB and other government structures. Everything [the Russians] do in Chechnya they do to break people's spirit and break their sense of human worth. And they've achieved real success because they do terrible things to people," Umarov concluded.

In its struggle to sustain the jihad in the northern Caucasus, the Islamist-Jihadist leadership increasingly focused on neighboring Dagestan as its main anti-Russian front. With a militant Islamist insurgency simmering since the late 1990s, Dagestan was a promising theater: Not only did the nation suffer from all the ills that had afflicted Chechnya until recently—a police force torn by internal strife, corrupt local governments, a poor and desperate population, and violence-prone Islamists—but Dagestani society was also consumed by fratricidal clashes and feuds among its dozens of ethnic groups. By late June, there were as many terrorist strikes and acts of sabotage in Dagestan as in Chechnya. Statistics prepared by Russian security services anticipated that by the end of 2005, between seventy and ninety incidents would have occurred in each republic.

The Russians were clearly caught unaware by that summer's escalation in Dagestan, which included a series of bombings and train derailments. By early August, the country had seen more than seventy terrorist attacks—more than double the number in the same period in 2004, and more than in any region of Russia, including Chechnya. Significantly, in Dagestan the jihadist violence had begun attracting support from the population, and radical Islamist entities—most notably the Shariah Jamaat—enjoyed expanding populist appeal.

In Moscow, Russian senior security officials warned that if Dagestan's deterioration into chaos continued, "Russia will have lost civilian control over half the North Caucasus." Left uncontained, they predicted that the spreading jihadist violence in the northern Caucasus "would also threaten to destabilize the oil-rich states to the south" of Russia, affecting the global energy economy. Regional experts warned of dire long-term ramifications as Islamist-Jihadist teachings spread from Chechnya throughout the northern Caucasus. Shamil Beno, a former Chechen foreign minister now working as a political analyst in Moscow, noted that three Islamic movements—"the state-controlled and pro-Moscow *muftiate*, the traditional Sufi *tariqats,* and the radical Islamist *jamaats*"—were competing for popular allegiance. Since mosques were "the only social institution that is functioning more or less normally in [Chechnya]," Beno explained in an August 10 interview with *Nezavisimaya Gazeta*, the outcome of this competition would ultimately define "Chechnya's future" and that of the entire Caucasus. "The radicalization of Islam in the Northern Caucasus will gradually reach a critical level," Beno concluded. "And that in its turn can lead to the beginning of a new war, the consequences of which will be unpredictable."

Meanwhile, the Chechen leaders were preoccupied with their new success in Dagestan and Ingushetiya. On August 7, Sadulayev met with commanders of jihadist groups from what the pro-rebel Chechen press agency defined as "different regions of the North Caucasus" to discuss their next moves. "Sadulaev heard the reports from the commanders about how to provide all squads with everything needed for effective military and sabotage operations," the agency reported. Sadulayev and his fellow senior commanders worked out a strategy for regionwide escalation, including plans to supply and otherwise replenish the jihadist forces. "The deterioration of the entire regional situation results from a coordinated strategy of both Chechen and Dagestani rebels," Russian senior security officials lamented.

On August 27, Sadulayev delivered a lecture "to the Chechen mojahedin and their foreign representatives," the text of which was later distributed to Islamist-Jihadist centers worldwide by the Information and Analysis Center. "The situation in Chechnya at the present moment is determined by the war," Sadulayev stated, "a Jihad against the occupation forces and the national traitors." He insulted the pro-Russia Chechen authorities and security forces personally, clearly apprehensive about their growing influence in Chechnya. In discussing the mujahedin's operations, he completely ignored any distinction between the war in Chechnya and the spreading Islamist-Jihadist insurgency throughout the Caucasus. "As far as the Chechen Resistance is concerned, the mojahedin have stepped up their combat operations against . . . the occupation forces and their henchmen—national traitors in Chechnya, Dagestan, Ingushetia, Kabarda-Balkariya [sic] and other regions," Sadulayev explained. He claimed that Moscow was already aware of the jihad's regionwide character. "The Russians realized that the war would not be contained within Chechnya and they would have to fight outside its borders," Sadulayev said. Growing numbers of Russian troops were being deployed to Dagestan, Ingushetiya, Kabardino-Balkariya, and Karachay-Cherkessia, to contain this newest incarnation of the Chechen jihad. This regional surge had effected one profound shift in the jihad: the suppression of the Chechen jihad by "the national traitors"—the pro-Moscow Chechen authorities—would no longer be sufficient to pacify the Caucasus. "Even if the holy war ends in Chechnya, the resistance will not end there," Sadulayev declared.

Indeed, during the summer, the Islamist-Jihadist leaders concentrated on consolidating the remote components of the Caucasus front. While the bulk of the fighting was still conducted by Chechen, Ingush, and Dagestani mujahedin, the bulk of the resources were invested in building up Islamist infrastructure in Karachay-Cherkessia and Kabardino-Balkariya, where a move was on to expand the clandestine Islamist-Jihadist cells and armed groups. The movement drew strength from the overall socioeconomic stagnation in the northern Caucasus, as disaffected youth who had given up on all other social prospects began joining the jamaats and their clandestine cells in search of both personal solace and vengeance against society. By now there were roughly fifteen hundred to two thousand youth in the jamaats of Karachay-Cherkessia, and three to four thousand in the jamaats of Kabardino-Balkariya. Perversely, the jamaats of Kabardino-Balkariya also benefited from the gradual exodus of many of the Jordanian and Saudi Circassian units back to the jihads in Iraq and Saudi Arabia—which allowed Ilyas Gorchkhanov, the local Jihadist leader, to promote local Islamists to higher and more responsible command positions, in turn shoring up the movement's grassroots popularity. By fall, the local jihadists, with some help from experienced teams from elsewhere in the northern Caucasus, had made Nalchik the leading clandestine base of the Islamist-Jihadist movement in the Caucasus.

Yet the growing activity and successful recruiting efforts in Kabardino-Balkariya led to new popular pressure to pursue the jihad. The Islamist leadership had doubts about the advisability of such operations: Despite their confidence in the leadership of the Nalchik-based underground networks, they feared that any such operation would call

Moscow's attention to a bastion of clandestine work that had gone largely unnoticed.

But the theoretical debate gave way to reality on the morning of October 13, 2005, when violence erupted in Nalchik. That morning, some two hundred mujahedin, most of them local, attacked the local airport and several government buildings—particularly police and army headquarters—throughout the city. One police official called them "meticulously planned and synchronized attacks." Street fighting erupted as small groups of mujahedin seized and barricaded themselves in several buildings. According to a number of eyewitnesses, only some of the mujahedin wore the camouflage uniforms usually associated with Chechen resistance. Several were wearing civilian clothes, and dropped their weapons and attempted to melt into the local population as the situation deteriorated.

The authorities rushed fifteen hundred Russian troops and five hundred special forces soldiers to Nalchik and quickly closed in on the mujahedin. Fearing a repeat of the Beslan tragedy, the Russian authorities concentrated on securing and evacuating all schools in Nalchik before taking on the mujahedin. Most of the fighting was over by nighttime, as Russian teams shelled and then stormed most of the buildings held by the mujahedin. But several pockets of resistance held their positions in a number of isolated buildings, where they would stay through October 14 and into the 15th. By the time the fighting was over, seventy-two mujahedin—including Gorchkhanov, twelve security personnel, and twelve civilians—were killed, and a few hundred wounded, many of them civilians caught in the crossfire. About thirty mujahedin were captured by the security forces. Between fatalities and the fighters who were captured, the mujahedin had lost roughly half their entire force—not counting the wounded who were able to flee Nalchik. Despite the loss of numerous experienced commanders and cadres, the Nalchik assault was nevertheless viewed as a success by the Islamist-Jihadist leaders, who welcomed the evidence that jihadists in the Caucasus were still capable of mounting sizeable operations.

Initially, however, the Islamist-Jihadist leadership in the Caucasus proved uncertain about how to cope with the events unfolding in Nalchik. At first, Kavkaz-Tsentr claimed the operation as part of Sadulayev's recently announced regional strategy. "Forces of the Caucasus Front—a unit of the Chechen Republic's Armed Forces—went into the town [of Nalchik], including attack brigades from the Kabardino-Balkarian Yarmuk [Islamic Brigade]," read the first statement. As the fighting intensified and more pockets of resistance succumbed to Russian assaults, however, the Islamist press agency limited its coverage to vague reports, reflecting the leaders' attempt to reserve judgment about the events in Nalchik.

Only after the success of the seige was clear did Kavkaz-Tsentr publish "a short statement from the headquarters of [the Islamist] Caucasian Front" taking credit for the attack. The Islamist leaders acknowledged that "the forces of Mujahedin of the Caucasian Front (the Kabardino-Balkaria sector and the units of other sectors of CF), in pursuance of the order of the Commander-in-Chief of Armed Forces of Mujahedin of Sheikh Abdul-Khalim [Sadulayev]," had "carried out a successful military operation in the city of Nalchik." The statement asserted that "as a result of effective actions, the units of Mujahideen have fulfilled all assigned tasks," including attacking fifteen buildings affiliated with the security services. The jihadists claimed to have inflicted heavy casualties among the "kafirs [Russian infidels] and munafiqs [apostates, or local forces]," and to have caused widespread damage.

On October 14, Kavkaz-Tsentr reported that fighting was still raging around Nalchik. The new report identified the on-site leader as Amir Sayfulla, and announced that "the basic attack force of the Mujahideen consists of subdivisions from the Kabardino-Balkarian sector of the Caucasian Front, with associated forces from the Karachai-Circassian sector of the CF and several mobile subdivisions from other sectors of the Caucasian Front." The next day, the Islamist agency published an analysis of the Nalchik operation based on conversations with Amir Assadula, "one of the commanders of the mobile Mujahedin unit" participating in the Nalchik fighting. According to Assadula, the primary objective of the strike was to seize weapons and ammunition from the armories throughout the city. He measured success by the quantities of small arms, grenade launchers, "special weapons," and ammunition "confiscated in the buildings" attacked by the mujahedin. In conclusion, Amir Assadula stressed that "the Mujahedin units successfully accomplished the assigned task."

On October 16, "Amir Abdallah Shamil Abu-Idris"—Shamil Basayev—issued his own statement describing and explaining the Nalchik operation. According to Basayev, 217 mujahedin of "the Kabardino-Balkarian sector of the Caucasian Front" carried out the Nalchik operation, which he claimed had lasted only two hours—between 9:14 and 11:15 on the morning of October 13. The units had "simultaneously stormed 15 military objects of all power structures" throughout Nalchik. Although earlier Kavkaz-Tsentr reports had mentioned the Arsenal weapon store as an objective of the raid, Basayev now insisted that "arms shops were stormed by groups unknown to us." Basayev claimed that his forces had inflicted far heaveir casualties than acknowledged by the Russian authorities, and that the mujahedin casualties amounted to only "41 Shaheeds, insha Allah. These all are our wounded Mujahedin who could not move and consequently conducted combat up to the end, by remaining in city."

Basayev noted that the Nalchik strike had almost been abandoned before it started. "Five days before the operation," he reported, "a serious information leakage" had prompted the Russians to rush heavy reinforcements to Nalchik. The mujahedin commanders held a council on October 11 and decided to go ahead with the operation as planned despite the loss of surprise and the prospect of heavier casualties. Basayev explained that he had personally "carried out the general operative management" of the mission, and that "Amir of the Kabardino-Balkarian sector of the Caucasian Front Sayfullah was the commander of this operation." He also stressed that most of the operations' senior commanders had also joined in the fighting. "During the storm of FSB building my naib [assistant] on operative work, Amir of Ingush Jamaat Gorchkhanov Ilyas, who commanded the assault groups of Mujahedin 'Center,' became Shaheed, insha Allah." The Amirs of the Ossetian and the Krasnodar sectors were also slightly wounded in combat. The "Amir of Caucasian Front did not participate in the operation, since he is busy [with] other work that I have assigned to him," Basayev added.

Alluding to Amir Assadula's explanation of the operation, Basayev noted that a "sharp shortage of ammunition" had forced the mujahedin to limit themselves to "a cleanly assault operation" in which they attacked and swiftly withdrew, rather than their customary "blocking" operation, in which forces remain for a protracted and punishing siege. According to Basayev, the operation's protracted firefights had not been anticipated, but developed in the course of the battle. The "unprecedented courage and heroism of our brothers in Nalchik spikes confidence and determination into us to continue the struggle up to a full victory," Basayev concluded.

After a long investigation, however, Russian security forces eventually managed to establish the identity of the mujahedin casualties and prisoners, and when they did it became clear that Basayev's main claim—that the Nalchik operation had involved forces from the entire Caucasus Front—was a lie. Most of the mujahedin were members of the Kabardino-Balkariyan Yarmuk Islamic Brigade, just as Kavkaz-Tsentr's original statement on the morning of October 13 had reported before it was withdrawn. According to Russian security officials, most of the detainees explained that they were committed to "creating a strict Islamic state independent of Moscow"—confirming Moscow's fears that the Islamist-Jihadist revolt was taking roots throughout the northern Caucasus.

Indeed, the strongest impact of the Nalchik attack was felt in nearby Karachayevo-Cherkessia, and particularly in its capital of Cherkessk. The mounting anger and frustration of the local youth was driving them into the folds of the jamaats and their clandestine military cells. The number of recruits increased after Nalchik until its overall membership exceeded two thousand. Moreover, the overt display of Islamist traits became a form of social protest. Efforts by security authorities to suppress the jamaats by arresting some of the more militant Imams only aggravated an already tense situation. The arrested leaders were quickly replaced by equally radical Imams from Chechnya, who were more adept at clandestine work. Their preaching invariably increased hostility toward the authorities. The jamaats also ran a growing number of "unofficial mosques"—mosques not registered with the government's spiritual directorate—which were becoming increasingly popular as a form of social protest. The Imams and the youth surrounding them were so militant that regular Imams and public officials were afraid to approach the mosques and challenge them. By late November, Russian experts and security officials were warning that Cherkessk was "likely to be the next Nalchik," that the merest spark could ignite the whole city.

Luckily, winter set in before sparks could fly in Cherkessk, freezing all regional operations. And this winter seemed to sap whatever momentum had attached to the Islamist surge in the northern Caucasus. The Islamist-Jihadist leaders' inability to sustain routine operations in Chechnya, Dagestan, and Ingushetiya, or to initiate at least one more successful operation elsewhere in the northern Caucasus, played into Russian hands. When violence resumed throughout the northern Caucasus in early spring 2006, the Russian security forces and the pro-Moscow Chechen authorities were firmly in control of the situation. Creeping exhaustion led the fighting units and the public in Chechnya, Dagestan, and Ingushetiya to curtail routine operations, and led the public to increase its assistance to the security forces. The jihadist resistance never managed to take the initiative, and the Islamist-Jihadist leaders' hopes of energizing the jihad by dispersing their efforts throughout the northern Caucasus also failed.

Even though the grassroots Islamist-Jihadist movement continued to expand, then, the Islamist-Jihadist leaders proved largely incapable of mounting major strikes—and the northen Caucasus settled into a tense but stable acceptance of Russian rule. The era of sporadic terrorism was far from over, but the chances of a successful grassroot challenge to the Russian rule had effectively disappeared.

The most important development in preserving this rocky peace began in the fall of 2005, when the predominantly Chechen local security forces—both intelligence agencies and special forces—took the lead in assertive counterinsurgency operations throughout the northern Caucasus, as well as a series of cross-border raids. While the Kremlin had made strenuous efforts to organize and run effective Chechen security forces from the outbreak of the Second Chechen War in summer 1999, only now did these forces become a meaningful instrument—capable not just of tactical achievements on the battlefield, but of affecting strategic postures throughout the region. Capitalizing on this profound transformation of the war, from a battle between Russians and Chechens into clashes of varying intensity between natives of the Caucasus (bolstered at times by small Russian units), the Russians seized the strategic high ground, apparently for good.

At the same time, a profound evolution was occurring within the resistance movements in Chechnya, Dagestan, and Ingushetiya. Fewer and fewer resistance fighters had been with the movement since the war in Chechnya: By the spring of 2006, there were roughly a thousand fighters in Chechnya, Dagestan, and Ingushetiya—of whom some 100 to 150 were foreign, mostly Arab, mujahedin. Another five hundred "Chechen" mujahedin were outside Russia—about two-thirds of them in northern Georgia, and the rest mainly in central Asia and Afghanistan/Pakistan. The fighters in Chechnya, Dagestan, and Ingushetiya were now operating in cells of twenty-five to thirty fighters each, constantly on the move to avoid Russian forces. These forces were divided into three groups:

1. *Veterans of the original national-liberation rebellion.* By early 2006, this segment had lost most of its leaders and commanders, as well as popular support. Russian security officials noted that "this camp has practically ceased to exist as a military force," and that the remaining fighters were gradually withdrawing to peaceful safety in their villages.

2. *Islamist-Jihadist mujahedin.* The most militarily effective and constituting the long-term threat to Russia, this group was small in size—a few hundred, including foreign volunteers—but its mujahedin were vibrant, dedicated, well trained, and well financed. The jihadists were not interested in Chechen independence; while some were motivated by the vision of a caliphate throughout the northern Caucasus and southern Russia, most simply saw the jihad as a path to vengeance against Russia and the West, in the context of the worldwide jihad. Russian

security officials stressed that these jihadists represented "a real military force," serving as the "leading and defining lightning rod" for the Islamist-Jihadist movement throughout the entire region.

3. *The bandits.* The most violent group in Chechnya, Dagestan, and Ingushetiya. By the mid-2000s, this was also the only group that was constantly growing. These were cells of fighters, involved at one time or another in the liberation struggle or the jihad, who could no longer find a place in society and had no respect for law and order. They knew nothing but warfare and violence, and their "war" was a matter of robbery, largely for their own sustenance and enrichment. Some of their cells are involved with the Chechen Mafiya, but the majority are just local bandits.

The solidification of these three groups had been going on for quite some time. By the spring of 2006, however, it was irreversible.

The confluence of these two trends—the failed Chechenization of the conflict with Russia and the delinking of the anti-Russian fighters from society—were key to Moscow's success in reducing the Chechen conflict to its current, stable status as a stream of low-level terrorism, skirmishes, and banditry in the northern Caucasus. The near-dominant role of the Chechen, Dagestani, and Ingush security forces in this stage transformed the region. The bulk of the fighting was no longer between Chechen, Dagestani, or Ingush separatists and Russian troops, but rather between Chechen, Dagestani, and Ingush forces representing different views about the future of their nations. The growing numbers of pro-Russian forces, and the concurrent shrinking of the mujahedin, testified to the prevailing popular interest in normalizing life as part of the Russian Federation. Moscow simply resigned itself to maintaining an endurable balance between the ongoing low-level violence and the process of normalization.

The most important sign that the population had finally rejected the jihad was the marked increase and improvement in the flow of preventive intelligence to the security forces, which enabled security forces to prevent more terrorist strikes before they happened. One such incident took place in southern Russia, in early February 2006. Special forces launched a series of raids in the towns of Stavropol and Georgiyevsk and the village of Tukuy-Mekteb, in the Stavropol Krai Neftekumsk district, in order to prevent the launch of another Nalchik-type attack. Their target was a safe house in Tukuy-Mekteb where most of the operation's commanders and key fighters—all members of the Shelkovsky Jamaat (in northeastern Chechnya, near Dagestan)—were awaiting orders to strike. Some three hundred Russian troops, with armored vehicles and assault helicopters, stormed the house. In the long firefight that followed, eight mujahedin and seven troops were killed, and several mujahedin were captured. Although Islamist-Jihadist leaders would later claim success because they had inflicted casualties on the Russian forces, and several mujahedin succeeded in evading the raids and fleeing unharmed, the Russians had prevented a potentially disastrous strike—and the Shelkovsky Jamaat took several months to revive.

Although the Islamist-Jihadist leaders continued to claim achievements and growing popular support, several senior commanders began acknowledging difficulties and setbacks. In mid-February, Abdul-Khalim Sadulayev insisted that Chechnya was still the leading and dominant force of a regionwide jihad. "The Chechen Republic Ichkeria is a sovereign, independent, Islamic state based on the rule of law and created as a result of the self-determination of the Chechen people. The source of all decisions made is the Koran and the Sunna," Sadulayev asserted. He elaborated that "fulfilling their sacred duty before Allah, the Muslims of the Caucasus are uniting around the leadership of the ChRI and waging a national-liberation struggle to de-colonize the entire Caucasus." In early April, however, Sadulayev noted that "the greatest successes" of the jihad were not in Chechnya but in Dagestan, followed by Chechnya, Ingushetiya, Adygeya, Kabardino-Balkariya, and Karachayevo-Cherkessia. Sadulayev also acknowledged that the mujahedin had recently suffered significant casualties, including several field commanders—both locals and Arab volunteers.

In mid-April 2006, Dokka Umarov offered a bleaker summation of the state of the revolt. He acknowledged that shortages in funds and weapons had prevented the Chechens from supporting jihadist fronts outside their borders. In addition to the Chechen front, Umarov acknowledged only "three fronts: Nalchik, Ingushetia and Dagestan," rather than the half a dozen usually claimed by jihadist leaders. Nevertheless, he insisted that morale was high and that the revolt still could muster a punch. "We can at any time conduct a large military operation, it all depends on the political advantage for us. Apart from this, such an operation demands large financial and human resources," Umarov explained. He acknowledged that most operations amounted to hit-and-run attacks and sporadic bombings, and that they were being carried out by small groups in the remote mountains. He also noted that the jihadists could not accept volunteers "for physical and financial reasons," and because "mountain conditions are harsh, and not everyone can stand them." He further elaborated that dwindling popular support—which he attributed to the ramifications of growing repression by the Russian authorities—were driving away the mujahedin who could not endure the harsh conditions of the rugged mountains; because they "cannot live in the lowlands, they have to leave Chechnya."

Indeed, during the spring and early summer—the best fighting season in the Caucasus—the jihad was unable to seize the initiative. Sporadic ambushes, bombing strikes, and assassinations continued in Chechnya, Dagestan, and Ingushetiya, but they were little more than a nuisance. While some of these operations resulted in casualties among the civilian population and the Russian and local security forces, as well as some property damage, the jihadist terrorists proved incapable of reversing the trend toward peace. Even a few successful assassinations and unsuccessful assassination attempts—most notably the May 10 assassination of Ingush deputy interior minister Dzhabrail Kostoyev (by Amir Magas of Beslan fame, on behalf of Shamil Basayev)—failed to grab public attention.

Faced with its own increasing irrelevance, the jihadists in the northern Caucasus grew embittered and began taking revenge against the Chechen people for abandoning the cause.

The population, in response, turned on the rebels. Local leaders, both civic and religious, openly encouraged local youth to join the pro-Russia security services and the public at large to provide actionable intelligence to the security forces. This trend reached a new milestone in mid-June, when Russian security forces tracked down and killed Abdul-Khalim Sadulayev.

By the time Sadulayev was assassinated, the Islamist-Jihadist leaders were aware that the noose was tightening around them. In an interview with the Bulgarian weekly *Politika* (published in the June 9–15 issue and posted on chechenpress.org on June 17, the day Sadulayev was killed), Sadulayev himself declared that the Chechen leadership was ready for the imminent death of its leaders, including himself, and added that "there was someone [in command] before me. And if through the will of Allah I meet my death, there is someone to continue this task even better." He identified Dokka Umarov as his selected successor.

Sadulayev's assassination was a direct result of the grassroots rejection of the jihad. During the summer, rumors spread throughout the northern Caucasus that the Islamist-Jihadist leaders had resolved to break the deadlock by launching new strikes in both Chechnya and Russia in mid-July 2006, during the upcoming Group of Eight summit in St. Petersburg. The objective of these strikes would be to force the Chechnya issue onto the global stage, while humiliating Putin in the process. To the local population, these rumors suggested a reversal of the progress made in reconstruction—a prospect that was unacceptable to many. Scores of individuals came forward with intelligence information about the planned terror attacks in Chechnya and the whereabouts of the leaders behind them.

Acting on such tips, security forces started closing in on a section of Argun, twenty miles east of Grozny, where senior commanders were reportedly hiding. The break came when a Chechen mujahed confirmed to the security services that the operation's senior commander was Sadulayev, and provided them with the exact location of his safe house—all in exchange for 1,500 rubles ($50 to 55) to buy heroin. On the night of June 16, some three hundred Russian and fifty pro-Moscow Chechen special forces surrounded Sadulayev's safe house, and by midmorning they had launched an assault. Sadulayev was surprised to have been tracked down—he had only three or four bodyguards with him—but they put up tough resistance. After another long firefight, Sadulayev and one fighter were killed as special forces broke into the house, and the remaining mujahedin escaped. The Russian security forces also suffered casualties.

The Islamist-Jihadist leaders reacted to Sadulayev's assassination with characteristic defiance. "The death of even the most worthy will not weaken our Jihad," said a statement posted on chechenpress.info. "On the contrary, our martyrs encourage the fighters with their heroic example. . . . Today our enemies, not hiding their baseness, are celebrating. But it should be us, not them, who are celebrating, because our brother and leader Abdul-Khalim Sadulayev is in heaven, as is the will of Allah." They quickly announced that Dokka Umarov, Sadulayev's designated successor, had assumed power, and that all leaders and commanders—most notably Basayev—had already recognized Umarov's leadership.

On June 23, Umarov issued a message that essentially conceded the continued decline of the jihad in Chechnya. Discussing his objectives as a president, he stressed that he planned on "significantly expanding the area in which military operations are conducted." He also announced that "we are setting up a special unit consisting of the most experienced fighters with a single command and database" in order to fight not the Russians, but Chechens who abandoned the jihad and switched sides. "The task of the unit will be to kill the most odious national traitors from punitive units, as well as war criminals from occupational formations who have been sentenced by the Shariah court of the Chechen Republic of Ichkeria to capital punishment for their atrocities and murders. From now on, there will be no mercy to the killers of Chechens and other peoples of the North Caucasus, no matter where they are," Umarov declared, reflecting what has become the greatest threat to the Chechen cause.

On June 27, Umarov announced that he had named Basayev as both vice president and government chairman. He also reasserted his hard line vis-a-vis Moscow, reiterating what he had been saying for several months—that he "was no longer going to propose peace with Russia" and that the mujahedin could still "launch at any moment a large-scale military operation" anywhere in Russia. Umarov also stressed that he and Basayev were in agreement over the doctrine and strategy of the jihad.

Yet Shamil Basayev's tenure as vice president was short-lived. Just two weeks later, on July 10, Basayev was killed with several other rebels in an explosion while traveling through the Ingush village of Ekazhevo. Chechen sources called the blast an accident, but Russian security services took responsibility for the attack.

The problems of the jihad persisted. In a September 12 statement, Umarov continued his pattern of alluding to such troubles while insisting on extreme demands from Moscow. Appealing to the Chechen diaspora to help their brethren in the Caucasus. Umarov noted that "the Jihad had been continuing for 16 years already" and was taking its toll, but he urged "the Muslims abroad to unite and not to forget their brothers, the Caucasian Mujahedin, and to render them assistance." At the same time, Umarov reiterated that the Chechen Islamist-Jihadist leadership would not "offer any peace talks to the Russian invaders any more. If the Russians want to talk, they are to declare it publicly to show the seriousness of their intentions," he asserted. "I speak about this so that nobody has an impression that we are weak. And we want our brothers abroad to know about it," Umarov added. But the Chechen jihad needed help to survive, as Umarov conceded in his Ramadan message on September 25, in which he pleaded for foreign aid. "The Islamic Ummah should remember their brothers and sisters fighting in Jihad in the Caucasus. A good deed for a Muslim is the Dua'a [prayer] for Mujahedin and every kind of assistance to them: with property, weapons, word, support of Mujahedin and Shahideen families, help to injured and hapless," Umarov begged.

These shortages were also reflected in both internal writings and Internet postings from the Islamist-Jihadist community worldwide. One especially explicit, but not unique, message was an October 18 post on the website Ana al-Muslim entitled "A Reprimand from Chechnya." In this message, "Abu-Abdurrahman," apparently an Arab volunteer in

Chechnya, accused the Arab world of not doing its part in providing financial and other aid to the mujahedin in Chechnya. "Your mujahidin brothers in Chechnya are blaming you and they say that after the martyrdom of the leaders Khattab and Abu-al-Walid . . . the financial assistance coming from the Arab brothers has stopped and that they [the Arab brothers] are not sending aid the way they used to." Abu-Abdurrahman demanded action from his readers: "Distribute this reprimand everywhere. Distribute it in the mosque, the street, the forums, the schools, and the universities. Pray diligently for your brothers that Allah might have mercy on us." There were many similar posts in the months that followed.

Umarov's words constituted a strained effort to conceal the profound crisis the Chechen revolt was undergoing. In a June 18 interview with *Caucasus Times*, Apti Bisultanov, a former Chechen minister of social security, noted that "the greatest danger to the resistance forces is that their support base among the population is dwindling" because of the work of the pro-Moscow Chechen police and security forces. Under such circumstances, he argued, the only way to sustain the Chechen jihad politically was through new terrorist strikes on Russia and the West. It was the doctrine Basayev and the Islamist-Jihadist leaders had followed from the very beginning; after taking power in summer 2006, Umarov now realized—if he hadn't long before—that he had little choice.

Chapter 28

CENTER STAGE IN GLOBAL JIHAD

TODAY, THE BEST YARDSTICK FOR MEASURING THE IMPORTANCE of Chechnya within the worldwide Islamist-Jihadist movement is the growing preoccupation with Chechnya, and the higher profile of the Chechen "cause," in the statements of the key leaders of the movement—particularly Osama bin Laden and Ayman al-Zawahiri. Indeed, in audio and video statements starting in the summer of 2005, both leaders mentioned Chechnya as a leading part of the international jihadist movement. The first sign of this trend was a June 2005 "letter to the Muslim Ummah" announcing that Osama bin Laden was preparing for the next round of jihad, and instructing the localized Islamist-Jihadist leaders about the roles they would play. The letter emphasized the regional character of the main jihadist fronts, and instructed the local leaders urgently to recruit mujahedin in neighboring and adjacent areas—as Shamil Basayev was doing in the northern Caucasus at the time. "We want to make it clear that at present, Jihad is obligatory on all Muslims who are living around Palestine, Iraq, Afghanistan, and Chechnya. Then, it becomes the responsibility of the entire Ummah to wage Jihad," the letter explained—identifying Chechnya as one of the four most important jihad fronts worldwide.

From this point into mid-2006, the jihad in the Caucasus enjoyed a higher profile in messages from both bin Laden and Zawahiri. On September 1, 2005, after the July 7 bombing in the London Tube, Zawahiri issued a major videotape that was broadcast on Al-Jazeerah. "Like previous great raids in New York, Washington, and Madrid, this [London] raid has moved the battle to the land of the enemy after it kept moving the battle for many ages to our land, and after its armies and forces occupied our land in Chechnya, Afghanistan, Iraq, and Palestine. . . ." Zawahiri saw those four regional fronts serving as springboards for a new wave of strategically important strikes. "O people of the Crusader alliance, we have cautioned and warned you, but it seems that you want us to make you taste the horrors of death. So taste some of what you have made us taste," Zawahiri declared.

In a full version of the message posted on the Web in December, Zawahiri elaborated on his strategic vision, describing the escalating jihads in "occupied Muslim lands" as foundations for the next cycle of global jihad. Zawahiri was convinced that the world was at a crucial turning point in the struggle between the West and Islamdom, and attacked the "apostate, agent governments"—that is, the local governments resisting the jihadist onslaught—for cooperating with "the Crusader forces [that] are occupying Chechnya, Iraq, and Afghanistan, and bracing themselves for the next move." Zawahiri attacked local Muslim voices that had begun calling for abandoning the jihad in favor of normal life, demanding that they be penalized by "all Muslims."

Zawahiri expressed an abiding conviction that victory was imminent, claiming that the tides were turning in the four key jihadist theaters. "Look at what the mujahedin did for the Russians in Afghanistan and then in Chechnya. Look at what the mujahedin are doing for the Jews in Palestine. Look at what the mujahedin did to the Americans in Somalia and what the mujahedin are doing to them in Iraq and Afghanistan today," Zawahiri wrote, addressing the Ummah. At the same time, he admonished Muslims in those four theaters for seeking compromise with the local governments in order to normalize their own lives, instead of confronting the reality that there is no substitute for jihad. "If Jihad is the reason for losses and disasters, then show us what you have. Palestine has been occupied for more than eighty years, and the Islamic caliphate collapsed around that time. The apostate, agent governments have spread corruption in the countries of Islam, surrendered to Israel, and abandoned Palestine. Our oil and resources are being stolen and plundered. The crusader forces are occupying Chechnya, Iraq, and Afghanistan and are bracing themselves for the next move," Zawahiri lamented. On the other hand, he expressed faith that local jihadist forces would transform the four theaters into bastions of the next phase of jihad by reradicalizing the local populations. The keys to victory, he claimed, "are in our hands. We may choose to make sacrifices, win, be strengthened, and have successors, or we may choose to flee, be defeated, be deprived, and be replaced," Zawahiri concluded.

In a follow-up speech on March 5, 2006, Zawahiri warned of the dire ramifications of the unfolding democratic and economic reforms in the four contested lands. These reforms, Zawahiri explained, "expose to the Muslims the kind of freedom the Crusader West wants for us. It is the freedom of aggression on Islam and Muslims. Should these Crusaders seize our countries as they planned and are still planning, they will desecrate all that is sacred. . . . Their criminal, fearful plot will only be confronted and halted through the martyrdom and sacrifices of mujahedin in Palestine, Iraq, Afghanistan, and Chechnya. If it were not for those mujahedin, we could be in a more serious and more humiliating condition today." Reversing this all-encompassing global plot would take an all-out effort of the entire Muslim Ummah. "We must confront the enemy's conspiracy with a doctrinal, Jihadist plan, based on adhering to the rule of Shariah, rejecting the surrender agreements, continuing Jihad, and weakening its arsenal and economic system." It was a formula right out of the Chechen Jihadist playbook. "If [these] drastic measures are adopted," Zawahiri concluded, "then the land will be liberated, injustice will end, and the sanctities will be protected. However, if we sacrifice the rule of Shariah and bestow legitimacy on those who sell nations and sign the surrender agreements, in the hope of liberating the land, ending injustice, or protecting sanctities, then we will lose both religion and the present life and the land will remain occupied, injustice present, and sanctities violated."

In an April 13 video message commemorating the fourth anniversary of the battle for Tora Bora, Zawahiri stressed that Iraq was now emerging as the preeminent jihad front, in which local jihadist forces were promising to ignite a regional jihad that might deliver victory over the United States. Progress was being made in the other theaters as well, he claimed. "Muslim Ummah: The hope of your victory is being achieved at the hands of your mujahedin sons in Afghanistan, Iraq, Palestine, and Chechnya. O Nation of Islam: The arena of Jihad in Iraq is now the most important arena of Jihad in this age. The Ummah should support the heroic mujahedin of Iraq, who are fighting in the frontline to defend Islam's glory and dignity," Zawahiri declared. Zawahiri stressed the imperative of regional operations rather than localized jihad—following the logic of Basayev's shift from a Chechen-based jihad to one that embraced the wider Caucasus—as well as the shift from a self-liberation jihad to an anti-hostile power jihad. Zawahiri urged "Muslims in Turkey, al-Sham [Greater Syria] and the Arabian Peninsula: Take care of the mujahedin of Iraq. Support them with men, money, ammunition, and prayers. They are fighting at the gates of Palestine. We hope things will proceed well and they establish direct contact with the lions of Islam in the surroundings of Bayt al-Maqdis and achieve the greater conquest and the clear victory, God willing." In other words, Zawahiri argued, the primary mission of the jihadist forces in Iraq was the destruction of Israel, not the liberation of Iraq's Muslim population.

Nevertheless, on April 23, bin Laden rush-issued an audiotape aimed to correct any impression that the growing preoccupation with Iraq might have distracted the leaders' attention from the other jihadist fronts. Addressing the Western governments, bin Laden claimed that the entire Muslim world recognized the current state of affairs as a "Crusader war between the West and Islam," regardless of Western claims to the contrary. What matters, bin Laden stressed, are the actions of the Western forces, not the rhetoric of the governments. The West was guilty of "ridiculing people and holding them in contempt," he argued. "Your aircraft and tanks are destroying houses over the heads of our kinfolk and children in Palestine, Iraq, Afghanistan, Chechnya, and Pakistan. Meanwhile, you smile in our faces, saying: We are not hostile to Islam; we are hostile to terrorists, and we advocate peaceful coexistence and dialogue rather than a clash of civilizations." Highlighting what he considered the duplicity of the Western approach of waging war against Islam while pretending to want reconciliation with Muslims, he cited

Chechnya as a primary example. "What is the meaning of the [Western] silence over the Russian horrible crimes in Chechnya, and the lynching of Muslims and tearing their bodies by tying them between tracked vehicles? Despite this, the so-called civilized world blesses all this. In fact, they [the Western states] support this secretly," bin Laden explained. "This shows how great their rancor is. . . ."

In the summer of 2006, the true focus of the jihad was not Chechnya but Iraq and the Palestinian territories, where major resources and assets were being diverted by the leadership. Yet the leaders felt compelled not to alienate the other jihadist fronts—particularly the Caucasus, which they expected to play a significant role in the coming eruption. So they flattered mujahedin in the Caucasus by characterizing the flagging Chechen jihad (and other localized jihadist operations) as part of a newly consolidated global jihad under contemporary conditions.

This doctrine was first authoritatively articulated by Zawahiri in a video message posted on June 11, 2006. While placing the jihad in Palestine in its global context, Zawahiri singled out the significance of the Caucasus front in this context. Even mujahedin preoccupied with immediate local challenges, he noted, are fighting the global jihad in their way. "The cause of Palestine is one of the arenas of confrontation between the Islamic Nation and the Crusader-Zionist assault, and isolating the Jihad in Palestine from the Nation's Jihad against the Crusaders and their agents will only lead to the loss of religion and worldly life. I am not asking those who isolate the Jihad in Palestine to wage Jihad in Chechnya, for example. However, I call on every Muslim in Palestine to stand by the causes of his Nation in Chechnya, Iraq, Afghanistan, the Philippines, and Guantanamo even if it be with words, calls, or encouragement. Islam has never been nationalistic in its fighting in defense of political interest shackled by national unity. Rather, Islam was and still is Jihad for the sake of Allah to protect and spread the creed of monotheism." In this context, he hailed the contribution of the Chechen jihad to both the struggle against Russia and the global war against the United States. "May Allah grant long life to the lions of Islam in the mountains of lofty Chechnya who have rubbed Russia's pride in the dirt," Zawahiri declared. "Not only have the sacrifices of the mujahedin foiled the schemes of American against Muslims, but also got in the way of its crimes against humanity."

Zawahiri's last comment was an oblique reference to a major campaign already afoot—the eruption of fighting in mid-July, first in the Gaza Strip and then in southern Lebanon. Though the new violence was instigated and controlled by Iran, Zawahiri released a special video message on July 27 characterizing it as "Jihad for the liberation of Palestine, all Palestine, as well as every land that was a home for Islam, from Andalusia to Iraq. The whole world is an open field for us. As they attack us everywhere, we will attack them everywhere. As their armies got together to wage war on us, our Ummah will get together to fight them." Zawahiri expected the new surge to spread to foes of the Islamist-Jihadist movement everywhere, regardless of their connection to the Middle East crisis. "We must today target the Jewish and American interests everywhere. We must also target the interests of all the countries that took part in the aggression against the Muslims in Chechnya, Kashmir, Afghanistan, Iraq, Palestine, and Lebanon. According to the Shariah, all these governments and their people are fighting the Muslims," Zawahiri decreed.

Between the summers of 2005 and 2006, as this theological discourse was taking place, Tehran was formulating a bold new global strategy designed to preempt the building international pressure on Iran to end its nuclear program. The adoption of this strategy profoundly changed Iran's approach to the Caucasus, including its cooperation with the Islamist-Jihadist movement in the northern Caucasus.

Although Iran has long had close relations—both strategic and operational—with al Qaeda, and Iran-based senior commanders and operatives had shuttled back and forth to Chechnya and Georgia for years, Tehran took pains to avoid direct involvement with the Islamist-Jihadist movement there. There was a very pragmatic reason for their caution: Russia has become one of Tehran's primary sources of sophisticated weapon systems, military technologies, and nuclear technologies, second only to China. Moreover, Tehran needed Moscow's cooperation to stall and resist the U.S.-led Caspian Guard program, which was launched in 2003 to secure Western hegemony over oil resources in the Caspian Sea basin. Aware of Moscow's sensitivity about the Islamist-Jihadist movement in the Caucasus, Tehran steered clear of the region.

By the summer of 2005, however, Tehran found itself reexamining virtually its entire geopolitical posture. This election of Mahmoud Ahmadi-Nejad as president of Iran on June 24, 2005, energized Iran's war preparations. In early August, the new president appointed Ali Larijani, an IRGC brigadier-general and veteran of security and terrorism positions, as chairman of Iran's Supreme National Security Council, putting him in charge of Iran's nuclear policy. On August 16, Ahmadi-Nejad announced his new cabinet—composed entirely of hard-line Islamists. In the coming months, starting with a series of outrageous and threatening statements, Ahmadi-Nejad and Expediency Council Chairman Ali Akbar Hashemi-Rafsanjani launched a concerted effort designed to antagonize the international community, particularly the United States, and to invite foreign political attacks against Iran in a cynical bid to galvanize the largely disaffected Iranian population into supporting the regime at a time of national emergency. Among the strategic shifts was a reassessment of Iran's relations with Russia and of its support for the Islamist-Jihadist movement in the northern Caucasus. Tehran proved willing to capitalize on the Islamicization movement in Azerbaijan; Islamist-Jihadist operations in the northern Caucasus and Central Asia were now being supervised by senior al Qaeda commanders based in Tehran and northern Iran.

Back in spring 2005, as it prepared to implement the Working Strategy Until 2020, the Islamist-Jihadist leaders committed to building networks in Azerbaijan to link Central Asia and the Middle East with Russia's northern Caucasus. This was a major challenge: In the aftermath of 9/11, and increasingly after the October 2003 election of President Ilham Aliyev, Azerbaijani security services had cracked down on the myriad of Islamist-Jihadist organizations and charities operating in Azerbaijan. In late 2003, on President Aliyev's orders, Azerbaijani security services had targeted the Hizb-ul-Tahrir networks, particularly in the city of Gandzh, arresting numerous commanders and members. Several jihadist networks were later rolled back in similar fashion, and key commanders operating in or passing through Baku were arrested and rendered to their countries of origin with U.S. help. Several terrorist plots, including plans to attack the U.S. and Israeli embassies, were thwarted.

By early March 2005, the Azerbaijani security services had racked up an impressive counterterrorist track record, attacking several terrorist networks of the al-Jihad and the Islamic Army of the Caucasus, arresting more than thirty jihadists. Six of these "Azerbaijani Wahhabis" were sentenced to prison terms in February 2005 for "preparing terrorist acts and illegally possessing weapons." As well, fourteen terrorists belonging to the "radical Wahhabi movement of the North Caucasus" were detained in Azerbaijan and extradited to Russia, and more than twenty members of other foreign Islamist-Jihadist groups were arrested and extradited to other countries. Also important was the concentrated campaign to block the channels financing jihadist groups: The Azerbaijani security services shut down branches of six Baku-based Islamist charities, and forty-three people connected with the organizations—mostly Saudis—were expelled from Azerbaijan. Consequently, the Islamist-Jihadists all but aborted the use of Azerbaijan, except for humanitarian (especially medical) support for wounded Chechen and related mujahedin, shelter for the families of mujahedin (mainly commanders and notables), and emergency travel by a few commanders.

The supreme leaders' spring 2005 decision to rejuvenate their networks in Azerbaijan, then, was a major undertaking that reflected the growing importance of both the Caucasus front and Iran's new cooperative relationship. It was all the more significant considering that it came on the heels of a failed test run of one of the last remaining major sleeper networks in Baku—a network tasked mainly with trafficking key operatives and materiel to and from southern Russia. The network was run by Nurseem Abd al-Rahman (an Afghan national), and included an expert bomb maker and several veterans of Chechnya: Hammad Sadiga (a Jordanian) and Said-Husain Takayev, Alihan Balayev, and Suleiman Isayev (all Russian citizens from the Caucasus).

This network had first been noted by the Azerbaijani security services in the winter of 2004–2005, but appeared dormant at that time. In the spring of 2005, it was activated and ordered to prepare for a series of sabotage operations against various objectives in the greater Baku area. Toward this end, bomb specialist al-Idrisi asked for and received (via front entities in the United Kingdom and Germany) several specialized electronic systems and chemical materials he needed to construct a series of bombs. As they were finishing their work, however, the Azerbaijani security services lowered the boom. In late April, Baku announced the capture of a terrorist network "sponsored from abroad" that was planning terrorist acts "in the areas of compact dwelling of the foreigners" in the context of "the global Jihad." The Azerbaijani security forces captured stockpiles of high explosives, weapons, $80,000 U.S. in cash, cassette recordings of calls to jihad, and numerous passports of citizens of Iran, Tunisia, Morocco, Saudi Arabia, France, and Russia whose clandestine travel the network had recently facilitated.

Meanwhile, the Islamist-Jihadist supreme leaders were also working to bolster popular support for Islamist causes in the Sunni areas of predominantly Shiite Azerbaijan. In an effort to break free of their reliance on underground networks of isolated foreigners—like Nurseem Abd al-Rahman's network—the leaders now embraced the mobilizing power of the localized jamaats, which had succeeded so well in the northern Caucasus. In early 2005, militant Imams who had established and led underground jamaats in Dagestan were smuggled into northern Azerbaijan, particularly into the Lezghin-populated areas, in order to establish comparable jamaats in Azerbaijan. In April and May, a host of new jamaats were created in the Qusar, Quba, Xacmaz, Siyazan, and Balakan districts of northern Azerbaijan. One result of the emergence of these jamaats was that a growing number of believers started to perform their rites in Wahhabi "alternative" mosques that weren't licensed or subsidized by the Azerbaijani religious authorities. Just as they had in the northern Caucasus, the jamaats spread money around and mobilized their followers to exercise political influence even where their followers did not constitute a majority. "The spread of Wahhabism has reached threatening proportions in six Azerbaijani districts—Balakan, Zaqatala, Qax, Qabala, Saki and Qusar," Rovsan Novruzoglu, the director of the Azerbaijani Public Center to Combat International Terrorism, warned in late April 2005. Leaders of the underground jamaats even toyed with the idea of forming an Islamist party to take part in the November parliamentary elections. Although the idea never materialized, Islamist charities in Saudi Arabia, Egypt, Kuwait, and Jordan swiftly funneled $2.5 million U.S. "for election needs."

The theological center of Islamist activities was in the Abu-Bakr mosque in Baku. The Abu-Bakr mosque had had a long history of being a bastion of Wahhabi Islamism, and, during the security crackdown a few years before, it had been closed by authorities after about thirty of its parishioners were hauled in for "connection with the Chechen mujahedin," considered dangerous by the security authorities. The mosque was allowed to reopen after its leadership vowed not to get involved in Islamist politics. In the spring of 2005, however, the head of the Abu-Bakr mosque was Imam Qamat—"the torch-bearer of the Azerbaijani Wahhabis," in the words of an Azerbaijani senior security official. A charismatic rabble-rouser, Imam Qamat excelled "in stirring up passions." An Azerbaijani religious scholar noted that Imam Qamat "seems to have a talent for hypnotizing ordinary people who only have superficial knowledge of Islam and for distorting the very essence of the Koran."

Using funds from the UAE and elsewhere in the Middle East, the Islamists now expanded their presence and influence in the greater Baku area. Among what Novruzoglu called "secret Wahhabi offices," two were of singular importance. A bakery near Baku's "bes martaba" area on Fuzuli Street was taken over and transformed into "a secret meeting venue for the Wahhabis," where free or highly discounted food was distributed as a magnet for Baku's poor. Another office, located in the basement of a new skyscraper near the Insaatcilar metro station, was used for distribution of Islamist-Jihadist propaganda to the area's educated audience, created by the Islamists' own publishing house in Baku.

One of its first projects was the spring 2005 publication, in twenty thousand copies, of the book *Jihad,* by former Chechen president Zemlikhan Yandarbiyev. *Jihad* "calls on the Azerbaijani people to fight against Americans and western diplomats," Novruzoglu explained. The Islamists had their greatest success among the region's young, educated, and well-to-do: By the early summer, there was a noticeable increase in the number of university students consuming jihadist publications and attending Wahhabi mosques in the greater Baku area. A senior security official noted that these students "prefer[red] to talk more about politics than about the religion of Islam." They were also enticed by offers of fully funded overseas travel: Starting in late May, numerous students dropped out of university to go to Syria for what was described as Koranic studies in "Islamic learning centers" but was actually jihadist indoctrination, terrorist training, and an opportunity to wage the jihad in Iraq's Sunni Triangle. This program was lavishly funded by Islamist charities from the UAE.

By summer, the growing self-confidence of Baku's increasingly radicalized Islamist community was giving rise to an unprecedented rise in mob violence. Initially, the violence was directed at the mass media, which the Islamists accused of being un-Islamic. On June 8, a group of about twenty young Islamists attacked the Space TV channel station after it aired remarks deemed to be offensive to Islam. The next night, a larger group of enraged youth surrounded the apartment building where Ilhama Novruzova, the host of the "offensive" program, lived. The mob threatened to break in, and it took a sizeable police force to break through the mob and escort Novruzova and her children to a safe shelter elsewhere.

By mid-July, the Islamists were forming the beginnings of terrorist cells. Taking no chances, the Azerbaijani security forces swiftly stormed the safe house in the Novhani suburb of Baku, where one fledgling jihadist cell was hiding. In a brief but intense firefight, one trooper was wounded, seven jihadist terrorists were arrested, and a large number of components for constructing bombs were seized. Azerbaijan's Ministry of National Security (MNB) warned that "the radical Islamic movement's goal is to destabilize the situation before the upcoming November parliamentary elections."

Emboldened by their initial success in Azerbaijan, the Islamist-Jihadist leaders expanded the push in the summer. The surge was managed from southern Dagestan, based on the lessons and experience of recent Islamicization drives throughout the northern Caucasus. A stream of Islamist preachers and charity officials flooded into the area from Iran, Afghanistan, Pakistan, Saudi Arabia, Bahrain, Jordan, and Egypt. Provided with lavish funds, they established numerous community centers, including "alternative" mosques and distribution points for humanitarian aid. Working with young Azerbaijani Islamists and a few activists from the northern Caucasus, the foreign Islamists reached out to the poorest areas of Azerbaijan, especially the refugee camps for the displaced population from Nagorno-Karabakh, concentrating on the local youth.

This intense recruitment and funding effort led many secular Shiites, most of them youngsters, to convert to Sunni Islamism. The foreign Imams and charity workers were also able to promise access to a growing number of scholarships for "advanced Islamic studies" in the Middle East and Pakistan. With no prospects for higher education or employment in Azerbaijan, many underprivileged youth readily adopted Sunni Islamism as a vehicle for studying overseas and perhaps getting a job in the Gulf states. Though they were aware of these foreign preachers' and charity workers' hidden agendas, Azerbaijani security services were reluctant to suppress them because of the popularity of—and need for—their charity and educational work. Rafael Gasanov, a Baku political scientist, noted in mid-July that "the Wahhabis have turned into a force capable of affecting the political life of the country."

In the summer of 2005, Tehran started examining the idea of launching an "oil war" as part of its new approach to resolving the quagmire in the Middle East. One of the very first things Ahmadi-Nejad did after taking office was to ask for a detailed briefing on Iran's strategic capabilities and options throughout the greater Middle East. He instructed that special attention be paid to Tehran's capabilities to strike Azerbaijan and destroy its energy infrastructure. Even before receiving the specific instruction from Ahmadi-Nejad, the Iranian government had embarked on a major program designed to enable strikes against Azerbaijan that would destroy its energy infrastructure—both resources and transportation. Ahmadi-Nejad warmly endorsed the idea of an oil war, instructing the defense and security establishment to accelerate their war preparations.

In early August, Ali Akbar Hashemi-Rafsanjani, Iran's most influential strategist, articulated the importance of the oil war strategy, noting that "maintaining energy security in the world is not possible without Iran." Because of Iran's ability to influence the global energy economy, Hashemi-Rafsanjani explained, "the world is dependent on us and Iran can no longer be ignored in international calculations." To date, Iran "has graciously avoided taking any adventurous approach and has remained committed to international and regional security," he added, but this approach might not last forever.

In their late June briefings for Ahmadi-Nejad and the Iranian uppermost leadership, the chiefs of the military and intelligence services distinguished between readily available capabilities (such as ballistic missile strikes) and emerging options—including a myriad of covert operations using Tehran-controlled and -sponsored jihadist movements. While Tehran prioritized the covert option, it decided to deter Baku in the meantime by floating the idea that it was considering missile strikes. This task was entrusted to Ahmadi-Nejad confidant Jelal Muhammedi, an Azerbaijan expert and longtime clandestine liaison between Tehran and "friendly" media and intellectuals in Baku. "Iran plans to deliver a preventive missile strike on the territory of Azerbaijan," Jelal Muhammedi informed his Azerbaijani contacts in late June. He warned of "a sharp deterioration of Iran-Azerbaijan attitudes" in the near future because of Baku's political tilt toward the United States in the brewing crisis over Iran's nuclear program. He warned that Iran would not tolerate the possibility of a U.S. force deployment to Azerbaijan, and would most likely deliver a preemptive missile strike rather than risk the presence of U.S. forces on its border with Azerbaijan. Muhammedi urged his interlocutors to think "how Baku will look after two missiles strike the area," and reiterated that Iran's combined missile and air power would "allow Iran to deal a blazing blow on the large cities of Azerbaijan."

Iran, however, was anxious to limit the danger of U.S. retaliation and ensure its own ability to gain future control over Azerbaijan's energy resources. Hence, even as it was making its missile threats, Iran embarked on a major program to foment rebellion within Azerbaijan by exploiting indigenous religious and ethnic factions. Ahmadi-Nejad instructed Iranian intelligence to build pressure on Azerbaijan, by deniable covert means, to convince Baku that any cooperation with the U.S. would be dangerous to Azerbaijan's stability and well-being. Tehran envisaged an evolving secret war between the Azerbaijani and the Iranian secret services. If Baku failed to heed Tehran's message, it could expect destabilization within its borders, in the form of a violent jihadist insurrection—thus making any U.S. presence in Azerbaijan practically impossible.

The population of Azerbaijan is complex. Ethnic Azerbaijanis constitute about ninety percent of the population, but the remainder is split among roughly twenty national minorities. Three of these—the Lezghins, Talyshs, and Kurds—are the indigenous inhabitants of ancient Azerbaijan, and radicalized ethnocentrist elements in their midst were susceptible to Islamist-Jihadist influence.

Iran's initial instinct was to use the Talyshs and Kurds—Shiite elements along Iran's border with its ally Armenia—to fight its secret war with Azerbaijan. Indeed, Iranian intelligence had long sponsored two clandestine militant organizations in Azerbaijan—the HizbAllah and the JaishAllah, both of which had bases in the minority zones and vast underground networks of Azerbaijani jihadists throughout the country. The MNB knew that both organizations were operating "in connection with the Iranian special services"—albeit mainly against Iranian opposition figures, particularly activists of the Iranian-Azerbaijani separatist movement who had fled to Baku. In mid-July 2005, Iranian intelligence had offered reinforcement to both HizbAllah and JaishAllah throughout Azerbaijan. Yet Tehran dampened the extent of these activities—fearing that the incitement of jihadist ethnocentrist militancy might backfire and spread across the border back into Iran.

Far more promising—and less threatening—to Iran were the jihadist networks of Sunni Lezghins, which were already being sponsored and influenced by the Islamist-Jihadist leadership, Iran's allies in other Jihadist theaters. Clandestine and subversive operations involving Lezghin and Azerbaijani Sunni Islamists were best run from Dagestan. By now tensions were so high in the Caucasus that, as one Russian senior security official put it, "one match will be enough to explode the whole of the Russian Caucasus." The spread of jihadist insurrection from Dagestan across the border into northeast Azerbaijan would appear more like a natural spread of regional violence than like an Iran-sponsored gambit. Using Russian territory as a base for destabilizing Azerbaijan also gave Iran a fig leaf of deniability, since they always point to the fact that Iran had long avoided cooperation with the Islamist-

Jihadist trend in Russia's Caucasus. Indeed, in the summer of 2005, Tehran's main challenge was convincing the Islamist-Jihadist supreme leaders to cooperate with Iran after years of disconnection.

To convince the Islamists that they were in earnest about supporting the jihad, Tehran relied on the presence of several jihadist leaders in Iran—primarily Saad bin Laden, Sayf-al Adel (one of Zawahiri's closest confidants and strategic advisers), Shawqi al-Islambuli (who had worked in Chechnya), and Abu-Muhammad al-Misri (who had cooperated with Iranian intelligence in East Africa in the late 1990s, including in the bombing of the U.S. embassies in Kenya and Tanzania). With Baku working overtime to battle the Islamist-Jihadist networks in Azerbaijan, the Islamist-Jihadist leaders needed all the help they could get to sustain operations, and they ultimately agreed to work with Iranian intelligence.

Meanwhile, throughout the summer and fall of 2005, a small group of senior experts from Iran's intelligence operation, its Revolutionary Guards Corps, and its oil ministry worked under the direct supervision of Hashemi-Rafsanjani to formulate the detailed strategy for Iran's overall oil war. Hashemi-Rafsanjani's recommendations were adopted by the Iranian leadership in a session run by Ayatollah Khamenei in late November or early December 2005. Having all but exhausted its attempts at political maneuvering, the Iranian leaders decided to embark on a strategic confrontation with the West, using oil as its primary weapon. The first test of this strategy came in mid-January 2006, when the latest round of Western pressure on Iran's nuclear program led to a series of threats to hike prices on Iranian oil supplies or cut them off altogether. "Any possible sanctions from the West could possibly, by disturbing Iran's political and economic situation, raise oil prices beyond levels the West expects," Iran's economy minister, Davoud Danesh-Jafari, told Iran's state radio. Ahmadi-Nejad went further, threatening the West: "You [the West] need us more than we need you. All of you today need the Iranian nation." These threats caused rifts among the United States, Western Europe, Russia, and China regarding "the merits" of sanctions against Iran—rifts that persist today.

By the winter of 2005–2006, Tehran had formulated a three-tier strategy for its oil war. Tier One, the "Core" of the strategy, involved disrupting the production and transportation of oil and gas in the areas immediately surrounding Iran—that is, the Middle East, the Caucasus, and Central Asia. About half of the world's oil and gas production comes from these regions. The instruments available to Tehran to implement its plans ranged from overt and covert acts of war by Iranian forces to a myriad of terrorist strikes and covert operations by a web of both Shiite and Sunni Islamist-Jihadist groups. To the north, the Iranians planned to sabotage the vast energy infrastructure in the Caucasus, the Caspian Sea, and Central Asia. Toward this end, Tehran foresaw cooperation with the myriad of Chechen and Central Asian Islamist-Jihadist forces affiliated with Osama bin Laden. Any Iranian cooperation with the Chechens was bound to alienate Moscow, at present Iran's staunchest supporter in the nuclear face-off; the mere existence of these plans suggested that Tehran considered striking in the Caucasus more important than retaining its friendship with Moscow.

In early March 2006, the Iranian strategy was endorsed by Ayman al-Zawahiri in an audio message that identified the oil war as the first and most important jihadist objective. "The First Front is that we must continue to cause maximum economic damage to the cross worshipers using strikes that make the enemy bleed for decades. We have seen the effect of the attacks on New York, Washington D.C., Madrid, and London. We must deprive the cross worshipers from our oil and we must practice a full economic boycott against . . . all countries which participated or helped in the fight against Islam," Zawahiri instructed.

On March 8, in the scope of a few hours, Ahmadi-Nejad and several senior officials promised to inflict "harm and pain" upon the United States if the international pressure on Iran's nuclear program continued. "The United States has the power to cause harm and pain," read a statement delivered to the IAEA in Vienna. "But the United States is also susceptible to harm and pain. So if that is the path that the U.S. wishes to choose, let the ball roll." The next day, Ahmadi-Nejad raised the ante. "They [the West] know that they are not capable of inflicting the slightest blow on the Iranian nation because they need the Iranian nation," he declared. "They will suffer more and they are vulnerable," he added without elaborating. Tehran seemed to have concluded that its ability to play the West through doublespeak and convoluted negotiations was coming to an end, and that a bold strategic gambit—such as a concentrated assault on the West's oil economy—would be required to keep the West off balance.

There were growing indications that Iran and its allies, most notably Osama bin Laden's terrorist elite, were aiming higher than just its usual terrorist strikes. Tehran was contemplating a major strategic gambit that could have a dire and lasting effect on Western society, such as a concentrated assault on the West's oil economy.

In the first week of March, senior Iranian intelligence officials met with Chechen emissaries in the UAE to discuss an increase of Iranian aid to Chechen and other Islamist-Jihadist movements in the Caucasus (specifically Azerbaijani jihadist forces) and the conduct of joint sabotage operations against energy targets in Azerbaijan and southern Russia. Significantly, the Iranians managed to convince the initially skeptical Chechens of Tehran's commitment to launch strikes in the Caucasus and the Caspian Sea basin. With this commitment in hand, the Islamist-Jihadist leaders moved quickly to implement their part of the strategy. Within a week, dozens of Islamist terrorists—highly trained graduates of the fighting in Iraq under Zarqawi—left Iraq to facilitate the escalation of jihad in the Caucasus and Afghanistan/Pakistan. In mid-March, the Pakistanis captured a commander, newly arrived from Iraq, who provided details about their training and escalation orders. He told his Pakistani interrogators that the Zarqawi-trained terrorists—from all parts of the Caucasus (including Azerbaijan)—had been dispatched to form a new group called the Jaish-ul-Islam al-Kawkazi, the Islamic Army of the Caucasus. With the scent of new violence in the air, the Azerbaijani security services raided a few houses associated with Wahhabi groups and recovered weapons and explosives. On the night of March 11, one such raid on a house on Salatin Asgarova Street in central Baku ended with an exchange of automatic fire between police and members of the Wahhabi group. Surprised by the intensity of the jihadists' fire, the police were unable to storm the house, and the armed jihadists managed to escape.

Later that month, the office of Iran's Supreme Leader, Ayatollah Khamenei, issued instructions to Iranian intelligence concerning the forthcoming clash with the West. The intelligence services were assigned to prevent Azerbaijan from being used in military actions against Iran, using "various measures . . . from influencing public opinion through mass media, up to conduct of diversions [by Iranian intelligence and special forces] and acts of sabotage-terrorism [by Islamist-Jihadist terrorists] against American and governmental facilities, and also attempts to destabilize the republic with the help of radical Islamic elements and the Talysh national minority." In Armenia, the main task of Iranian intelligence was "preventing in every possible way the reorientation of the republic [of Armenia] to the West. This can be promoted by activating contacts with the local mass media and politicians, and also by strengthening the pro-Iranian lobby in the business and military circles of the republic."

A couple of days later, a small delegation of senior Pasdaran intelligence officials, led by a confidant of Ahmadi-Nejad, arrived in the Armenian capital of Yerevan for secret talks with Armenian intelligence officials about the growing crisis with the United States. Yet Yerevan's behavior was highly unusual, and troubling for Western interests: Normally, the Armenian authorities inform its friends and supporters in both Moscow and London of such sensitive visits from Tehran, so that they can shield Yerevan against Washington's displeasure. This time, however, the Armenians chose not to inform their friends about the Iranian visit—and adamantly refused to answer questions from London and Moscow. On March 17 or 18, the Iranian military attaché in Yerevan, Colonel Bizhan Hamzeil Hashame, was transferred from the military to Khamenei's office and ordered to help plan for a confrontation with Azerbaijan—particularly the launch of a terrorism campaign in the context of Iran's oil war.

On the morning of April 9, Ayatollah Khamenei convened an emergency meeting with the leaders of Iranian intelligence and security establishment—including Ahmadi-Nejad and Hashemi-Rafsanjani—to discuss the sudden cancellation of the negotiations with the United States over Iraq. The meeting signaled that Iran was ready to launch its oil war—including sponsoring terrorism against energy-related objectives in Azerbaijan—while abandoning ongoing contacts with the United States.

Until a couple of days beforehand, the Iranian leaders had still been intrigued by the possibility of a high-level direct dialogue with the United States over the future of Iraq. Although a March 24 preparatory meeting between the two nations in Zurich had ended in a deadlock, Tehran had planned on sending senior officials to a Baghdad meeting scheduled for April 8. One scheduled topic on the agenda for that meeting was the future government in Baghdad. In Zurich, the Iranian representatives specifically told the Americans that Ibrahim al-Jaafari was Tehran's favored candidate for prime minister of Iraq, and demanded that no decision on his future be made until after the Baghdad meeting. After a telling visit to Baghdad by Condoleezza Rice and the U.K.'s Jack Straw, however, Tehran decided that their evident effort to unseat Jaafari meant that the United States was intentionally reneging on its representations in Zurich. When the Iranian team preparing for the Baghdad summit raised the subject with their American counterparts on April 7–8, the United States abruptly cancelled the Baghdad meeting. Concurrently, major U.S. marine units in Najaf, Karbalah, and other key Shiite centers were involved in clashes with local Shiite militias thought to be sponsored by Iran, and several pro-Iran clerics and officials were arrested.

Khamenei and the Iranian leadership considered this turn of events a major affront to Tehran and a proof of American perfidy. Khamenei told the April 9 meeting that Tehran could not permit the United States to challenge Iran's vital interests without proper retaliation. The possibility of launching an oil war in the Middle East and the Caucasus was raised as one preferred option. The other leading option was to have the HizbAllah provoke Israel into a major confrontation along the Israeli-Lebanese border, which might even escalate into an Israeli-Syrian war with the potential to incite and radicalize the entire Arab world against the United States—thus creating a viable excuse for the oil war. The meeting adjourned without a decision, and Khamenei asked the attendants to submit their detailed ideas on retaliation in a few days.

Ultimately, Tehran chose to have the HizbAllah trigger a flare-up on the Israeli-Lebanese border, killing eight Israeli soldiers and kidnapping two soldiers on July 12, 2006, in an attack that led to a lengthy war in southern Lebanon. During the weekend of August 11–12, the chiefs of the Iranian Revolutionary Guards and Iranian intelligence, as well as senior military officers and strategy experts, briefed the top Iranian leaders—including Khamenei, Ahmadi-Nejad, Hashemi-Rafsanjani, and national security chief Ali Larijani—about the state of the war in Lebanon, the continuing face-off with the West over nuclear development, and their recommendations for short- and long-term Iranian policy.

The meeting reached a decision to expand its confrontation with the "U.S.-Israeli alliance," in Iraq and Lebanon, to divert attention from the nuclear issue while making strategic strides for Tehran. The fighting must continue for at least three to four months, it was resolved—to extend through the U.S. congressional elections and into the winter. Instructions to that effect were transmitted on the morning of August 13 to the senior Iranian officials assigned to Nasrallah, Assad, and Sadr, so that these leaders could be prepared for the next phase, regardless of the impending UNSC Resolution 1701 calling for a cease-fire.

In their briefing on long-term strategic issues, the Iranian leaders were warned that "a U.S.-led new bloc with Azerbaijan, Georgia, Turkey, and Israel" was emerging, one that was designed to regulate and control the flow of oil and gas from Central Asia and the Caspian Sea basin to the West. (Tehran saw little hope in Saudi efforts to win over Turkey and wean it away from a U.S.-Israeli alliance against Iran. Tehran was convinced that the brewing Iranian-Turkish confrontation over the Kurds and Turkomans—and the control of the oil—of northern Iraq would ultimately determine Ankara's policy.) The development of this bloc, made possible by the activation of the Baku-Tblisi-Ceyhan (BTC) oil pipeline on July 13, could markedly reduce the West's dependence on Middle East oil. In turn, it would also reduce Iran's ability to threaten disruption of oil supplies from the Middle East and undercut Iran's plans to dominate the oil resources of the region. With the growing likelihood of a U.S. strike on Iran, the briefers stressed the importance of forestalling the West's ability to weather the retaliatory strike against the oil resources of the Middle East and of ensuring "a very cold winter" and "debilitating economic crisis" in Europe.

Moreover, the briefers noted, the long-term success of the BTC pipeline, including a planned Ceyhan-Ashkelon underwater extension to facilitate oil deliveries to India via the Red Sea, would have far-reaching global ramifications beyond simply isolating Iran. The briefers stressed that "this design is intended to weaken Russia's role in Central Asia and cut off China from Central Asian oil resources." If Iran were to disrupt this U.S.-directed "conspiracy," it would serve the vital interests of Russia and China, thus enticing both Moscow and Beijing to reciprocate by supporting Tehran at the UN and in other international forums.

The Iranian briefers called for a preemptive defeat of this new bloc by striking at the energy infrastructure of Azerbaijan, assuring the Iranian leaders that they had standing plans for both cross-border missile strikes from Iran and deniable covert terrorist and sabotage operations to be launched from Russia-Georgia (using Chechen and Azerbaijani-Lezghin assets) and/or Armenia (using Kurdish and Azerbaijani-Talysh assets). The leaders instructed the chiefs of the Revolutionary Guards and intelligence to review the plans and return in a couple of days with possible timetables. The chiefs were convinced that their leaders might authorize immediate strikes at the next meeting, scheduled for August 16–17.

Enter the Chechen Islamist-Jihadists. As a few expert terrorist cells were deployed to the region, the Islamist leadership prepared a series of public announcements explaining the nvolvement of Chechen fighters in attacks on the West's energy infrastructure. In a brief video message recorded around August 20, Umarov condemned "the double-standard policy of so-called Western democracies that stay blind [to] Russian-led genocide, atrocities and bloody murders of Chechen civilians in order to keep their energy supplies intact. This shows the real face of the Western democratic system," Umarov stressed. Umarov's mention of the West's energy needs came out of the blue; it's difficult to see it as anything but a justification for what was to come.

As the target date for its first strikes, Tehran had selected the night of August 21, the date of Muhammad's surge to heaven from Al-Aqsa in Jerusalem. A few days before, Iran launched nationwide military exercises that served as cover for national mobilization and massive troop movements. Meanwhile, on August 12–13, HizbAllah forces began to collapse under Israeli bombardment and shelling. With their casualties mounting, and their supplies of rockets, ammunition, water, food, and medical stuff depleted, most of the HizbAllah forward units would have had to surrender had the fighting continued for ten more days—just past the August 22 target date. On August 14, Sheikh Nasrallah accepted UNSC resolution 1701 and a cease-fire went into effect. "The cease-fire acted as a life vest for the organization," a HizbAllah officer acknowledged. HizbAllah proved unable to fulfill its part of the plan by distracting Israel with strikes on the Lebanese front, and its failure sparked anxiety in Tehran. There were fears that the anticipated eruption might backfire—that Israel would be able to retaliate against any Iranian strike or provocation. Tehran notified Damascus to cease preparations for further operations and shelved its own plans for the August 21 eruption.

The prospect of sabotaging the energy infrastructure in Azerbaijan remained tempting to Tehran, but even there things weren't as straightforward as Ahmadi-Nejad would have liked. The Azerbaijanis' success in capturing and neutralizing Islamist-Jihadist cells, and in capturing and/or scaring away Iranian-Azerbaijani terrorists and intelligence operatives, gave Iranian intelligence pause. With Shamil Basayev dead and Abu-Hafs on the run, the Chechen Islamist-Jihadist leadership was only reluctantly prepared to commit its remaining elite resources—the best expert terrorists still operational—to Iran-sponsored operations in Azerbaijan. And although a few detachments deployed to the vicinity of the Russian-Azerbaijani border, the leaders nevertheless feared that the capture of any of these expert terrorists by Azerbaijani security would lead to the rolling back of the few networks left intact in the northern Caucasus.

And so Iran's grand strategic initiative fizzled almost as quickly as it had emerged. Tehran shelved its plans to strike. The Islamist-Jihadist leaders were relieved. And the threat of an Iran-sponsored oil war subsided, at least for now.

By mid-2006, Chechen mujahedin had largely left the Caucasus behind and moved on to other jihadist fronts throughout the world, mainly in Asia and Africa. They wandered into these hot spots in quest of both "action" and hope—and often attracted notice for their bravery and military skill. In July–August 2006, during Israel's war with the HizbAllah in southern Lebanon, a group of snipers—all Russian emigrants who had fought in Chechnya and Afghanistan—joined the fighters of the 51st Battalion of the Golani Brigade in some of the fiercest battles of the conflict. From the bunkers and fortified positions they encountered to the tactics employed by the HizbAllah squads of "tunnel warriors," the experience recalled their years spent fighting in the first and second Chechen wars. Russian voices (with Caucasian accents) were heard sporadically on the HizbAllah and Iranian communications networks, though the origin of these transmissions proved impossible to pinpoint.

British and other NATO special forces had similar encounters during their assault in southern Afghanistan—particularly in the Helmand valley. In late August, British intelligence experts noted that foreign fighters were appearing among the Taliban forces in southern Afghanistan. "So far the Taliban have been drawing very heavily on the locals in the south, [but] this is beginning to change," a senior British intelligence official said. "More by way of foreign fighters . . . Jihadists, are beginning to appear." Asked which of these foreign fighters posed the greatest threat to the British forces, the official replied that "Chechens get mentioned, among others." In the fall, Chechen fighters were reported in the ranks of the Taliban fighting in eastern Afghanistan. In Paktia, Nuristan, and Qandahar provinces, NATO forces encountered "fighters from Chechnya, Pakistan, Turkey and Yemen," another senior official reported. Chechen mujahedin were even reported in the Horn of Africa and the Far East, having abandoned the apparently lost cause of Chechnya in favor of still-active jihadist fronts elsewhere in the world.

Chapter 29

END GAME

BY THE FALL OF 2006, ANY THREAT THAT THE WESTERN POWERS would rally to the Chechen separatist cause had evaporated. The Chechen people—exhausted by the war, dejected by the radicalization and Islamicization of the resistance—appeared interested only in returning to relatively normal life. The flow of Russian reconstruction and development funds helped breed support, or at least passive tolerance, for the pro-Moscow authorities.

At the same time, there is concern among security experts and senior officials—not just in Chechnya, but also throughout the Caucasus and in the Kremlin—that the problem of Chechnya may "not [have] been resolved," but was merely "frozen." As far as Putin's Kremlin is concerned, strategic victory was attained when the West, particularly the United States, abandoned any possibility of supporting independence or other special status for Chechnya. With no more pressure from the West, the Kremlin can take its time trying to put an end to the sporadic violence that persists to this day. Moscow has managed to reduce the level of hostilities; though it continues to tolerate protracted low-level violence, it has reduced pressure on the civilian population and expedited the return of near-normal life.

The Russian effort is helped by the absence of rebel leaders with any ability to win public trust and confidence. Basayev's assassination removed the last of the giants from the scene. Despite his leading role from the very beginning of the Chechen separatist revolt, Dokka Umarov lacks the popular touch, and his inner circle's decision to favor the

Islamist-Jihadist movement over Chechen nationalism has tarnished his reputation and harmed public trust. Umarov's regime announced a series of regional undertakings and aggressive threats, while it was actually isolated in the remote and frozen mountains of southern Chechnya and Georgia—another tactic that did nothing for public confidence. When Umarov announced on September 30, 2006, that "due to changes in operational situation and in order to improve military coordination," he was promoting several senior officers and reorganizing the "fronts" of the jihad not just in the northern Caucasus, but also on "the Volga Front" and "the Urals Front" within Russia, it was clear that he and the Islamist-Jihadists were on the defensive—no matter how many reorganizations of "General Staff" Umarov announced.

Moreover, the exodus of nationalists from the ranks of the revolt continued. Many fighters simply went home with their guns and resolved to make the best of life as it was. Their hostility to the Russian and pro-Russian Chechen authorities notwithstanding, these fighters and the civilian population that supported them—the backbone of the original Chechen revolt—have increasingly fought the jihadists to preserve their newly restored stability. While the amnesty supported by Kadyrov made things easier for the former rebels and their families, the shift in sympathies among the Chechen population at large—not just the fighters—was more than just a reaction to the lure of amnesty and decent income: It was evidence of the profound transformation of Chechen society.

The key to this shift was the rise of competent Chechen security forces—GRU-controlled special forces, FSB-controlled intelligence and border guard units, and a myriad of security forces reporting first to Chechen Republic president Alu Alkhanov, and then to Ramzan Kadyrov, who took office as prime minister with Alkhanov's departure in March 2006 and was named president in February 2007. Kadyrov has expressed confidence that "the Chechen security forces are able to fully supervise the situation in the republic." Their handling of the bulk of the raids and mop-up operations—though ruthless and, at times, extrajudicial—were no longer manifestations of Russian occupation, but rather domestic law enforcement operations (and internal power struggles). The security forces, made up of predominantly ethnic Chechen forces under Chechen command, were instrumental in reducing popular support for the last vestiges of rebel activity within the republic's borders.

By 2006, the counterinsurgency efforts in Chechnya had taken an intriguing new shape—as a popular struggle to save the indigenous Islam of the Chechens from the jaws of the Arabization-Islamicization movement launched in the 1990s by the Chechen neo-Sallafite jihadists (so-called Wahhabists) and their Arab allies. The majority of the pro-Russia Chechen security forces are adherents of the brotherhood of Kunta Haji Tariqat, which is related to the Qadiriya brotherhoods of Sufi Islam. The members of the Kunta Haji Tariqat have historically constituted the most nationalist teips in Chechnya, and thus contributed a significant component of the rebel forces during the First Chechen War. But they considered the jihadists' Islamicization campaign a threat to the indigenous character of Chechnya. When the Second Chechen War exploded in summer 1999, the majority of Kunta Haji Tariqat, led by Akhmad Kadyrov (then the mufti of Chechnya), sided with Russia over the Islamist-Jihadists. Kadyrov later explained why he and his father shifted their allegiance: "We didn't [change sides]," he stressed. "We were never against Russia, and we were never for Russia. We were always with the Chechen people. And they changed their minds." By then, the Kunta Haji Tariqat had joined the brotherhood of Arsanovs, which is related to the Naqshbandi Tariqat of Sufi Islam, and which has always sided with the Russian authorities.

From an operational point of view, the key to understanding the success and growing legitimacy of the Chechen security forces is that the vast majority of their personnel—including very prominent senior commanders—are former nationalist rebels who had changed sides because they felt that the Islamist-Jihadist leaders had betrayed the real Chechen "cause," and that they must be defeated in order to save Chechnya. Most of these commanders and fighters resented the unilateral forfeiting of de facto independence represented by the 1996 Khasavyurt Agreement, and particularly the Islamist provocations in the summer of 1999 that led to the Second Chechen War. They point to the fall of 1999—when Akhmad Kadyrov, the revered ideologist of the Chechen independence struggle, left the resistance and joined the

Russians "to save Chechnya"—as the most important turning point in the character of the rebellion. (Kadyrov's son, Ramzan Kadyrov, was responsible for building the republic's Security Services as he climbed through the nation's command and political structure.)

Today's Chechen security forces consist largely of fighters attached to the Ministry of Interior, most of them former separatist fighters who received amnesty in exchange for returning to the fold. Most of the commanders are associated with the Kadyrov clan. The most elite units are the special forces battalions known as Sever (North), under the command of Alibek Delimkhanov, who switched sides in 1999, and Yug (South), under the command of Muslim Ilyasov, who switched sides in 2003. The most combat-effective and ruthless units—the Chechen special-task police unit (OMON)—are under the command of Ruslan Alkhanov, a former rebel commander in the first war. OMON commander Arthur Akhmadov was Aslan Maskhadov's chief of staff. Highly specialized security operations are carried out by a group led by Movladi Baisarov, another former rebel commander, and the Oil Regiment, which protects oil pipelines and infrastructure, is under the command of another former rebel commander, Adam Delimkhanov. A veteran commander from the Second Chechen War named Alambek Yasayev now commands a special regiment of police (PMON) named after Akhmad Kadyrov. Hairudi Visengeriyev, a top military commander under Dudayev and Maskhadov, became the new secretary of the Security Council in December 2005. The effectiveness and loyalty of these commanders, and their subordinates and troops, demonstrate the profound grassroots rejection of the jihadist cause.

At the core of every successful counterintelligence campaign are strategic intelligence networks, which take intelligence work beyond occasional tips about tactical occurrences to offer broader and deeper intelligence reports. Such strategic networks generally take years to mature. In the case of Chechnya, Russian intelligence services began having major breakthroughs in their recruitment efforts following the outbreak of the Second Chechen War. At first, there were fundamental difficulties, as one "former Russian intelligence serviceman" explained in a late October 2002 interview with Interfax. "The work with the agent network in the Oriental countries was always characterized by difficulties of ethnic, linguistic, cultural type. These problems are even more difficult in Chechnya's case. The militants trust only to the Chechens, and only to those Chechens with whom they were brought up . . . and whom they know for decades. Therefore, it is an extremely hard mission to infiltrate a prepared external informer in their surroundings." Yet the number of Chechens willing to cooperate with Russian intelligence to help save Chechnya from the Islamists grew throughout the early years of the twenty-first century, and Russian intelligence agents slowly built a reserve of trust and cooperation with them. Efforts to build a successful intelligence network that were launched late in 1999 were bearing fruit in 2005–2006.

Moscow also improved its intelligence cooperation with several countries, most notably Israel, helping Russian intelligence infiltrate and penetrate the predominantly Arab and Circassian foreign volunteers arriving from the Middle East. By mid-2006, the Russian secret services—the Federal Security Service (FSB) and the Military Intelligence service (GRU)—had recruited roughly one thousand quality agents in and around Chechnya. Moscow still suffers from friction among its various intelligence services. Anatoly Safonov, a former deputy director of the FSB and the Russian president's special representative for International Cooperation in Fighting Terrorism, stressed this point in a June 2005 interview with *Nezavisimaya Gazeta*. Asked how many Russian ministries and services were engaged in the fight against terrorism, Safonov acknowledged that "it is impossible to count them. Often, there is a problem with interaction: for example, when either several contesting organizations act in the same field, or no one act[s] at all." Safonov, and other senior intelligence officials, candidly acknowledge that Russian intelligence still has much room for improvement.

Nevertheless, the data provided by these sources, as well as technical intelligence, were of great use to the Russians in their suppression of the jihadist movement. These networks worked hand-in-glove with the special forces and security services, which proved capable of swiftly capitalizing on intelligence to conduct highly effective preventive and preemptive operations. At this, Russian elite forces are putting the remaining jihadist forces on the defensive in the remote mountainous corners of Chechnya and across the border in Georgia—reducing their ability to interact with and influence the civilian population.

As with the Chechen security forces, the most effective Kremlin-controlled GRU special forces—two SPETSNAZ battalions known as Vostok (East) and Zapad (West)—are composed of Chechens. Both battalions report directly to the Russian General Staff, but only one battalion is commanded by ex-rebels. The other is commanded by Russian Chechens who have always fought on Moscow's side. This battalion is used for cross-border raids. These Kremlin-controlled forces help respond to strategic and regional threats not only in Chechnya, but also in Dagestan and Ingushetiya.

The Vostok Battalion operates mainly in remote southeastern Chechnya and across the Dagestani border, conducting deep reconnaissance and focused strikes in areas considered rebel safe havens. The troops of the Vostok Battalion—all of them former rebels—are credited with some of the greatest successes among Russian security forces in recent years. The battalion's commander is Sulim Yamadayev, one of the five Yamadayev brothers, among the most important commanders to have switched over to the Russian sides. During the First Chechen War, the Yamadayev brothers were the commanders of Gudermes, Chechnya's second-largest city. In the fall of 1999, however, they switched over to the Russian side—along with their entire force and most of the Yamadayev clan—because they disagreed with the Islamists' decision to provoke the war, and were determined to save their home city from destruction. Soon thereafter, they established a special forces company using former rebel fighters to hunt down mujahedin in the Gudermes area. In 2003, the GRU assumed supervision of the Yamadayevs' force, expanding it to a full battalion. In March 2003, Dzhabrail Yamadayev, the first commander of the Vostok Battalion, was blown up in his own house, and his brother Sulim assumed command. Another Yamadayev brother—Khalid—was elected to the Russian State Duma from the ruling party, Edinaya Rossia.

The Zapad Battalion formally operates in the western parts of Chechnya and across the border in neighboring Ingushetiya and Northern Ossetia. Its declared missions are the same as those of the Vostok Battalion, but Zapad is the only Chechen special force that has practically no former rebels in its ranks. Its personnel come from the Nadterechny area of northwestern Chechnya—a strongly pro-Russian area that refused to join the rebellion in the early 1990s. The battalion's commander, Said-Magomed Kakiyev, is a veteran of Russian special forces who hails from this region. Kakiyev and the entire battalion are all devout followers of the local Sufi Tariqat and believe that the Sufi socioreligious teachings constitute the key to the survival of the Chechens. Kakiyev describes himself as a "soldier of Islam," but a "fierce opponent of Wahhabism" who has decried the spread of neo-Sallafite influences in the northern Caucasus as the greatest threat to Sufi Islam and Chechen national identity. In his battalion's mission "to annihilate camps and bases of the militants in the mountainous part of the [Chechen] republic [and] to carry out search operations," Kakiyev acknowledged that his forces also raid mujahedin safe havens in neighboring republics. But his battalion's most important mission is the infiltration of Georgia's Pankisi Gorge region, to collect intelligence and conduct covert operations against the mujahedin camps and supply lines there. Although Moscow has remained silent about these strikes, senior security officials in the Ministry of Interior of Georgia have repeatedly complained that "Chechen members of the GRU have indeed participated in secret operations in territory of the republic" largely aimed against "Chechen refugee camps."

The collective impact of the successes of the Zapad Battalion in Georgia and the targeted killing of key Islamist-Jihadist commanders—most notably Shamil Basayev and Abu-Hafs—emboldened the Kremlin. With the Chechen jihadist leaders focusing on planning new terrorist strikes within Russia and against Russia's Western allies, the Kremlin started contemplating preemptive strikes against mujahedin objectives, and at safe havens and supply lines in central Asia. In the late summer, Kadyrov openly discussed using Chechen special forces for long-reach strikes, asserting that Chechen forces would soon be permitted to operate outside Chechen borders in support of regional stability. (Along that line, Kakiyev's SPETSNAZ forces continued their raids on terrorist objectives in Georgia.)

And while the Kremlin remained focused on conducting raids within the former Soviet Union, the escalation of Israeli HizbAllah fighting along the Israeli-Lebanese border gave Moscow an unexpected opportunity to test its new assertive doctrine. In late September, the Lebanese government accepted Moscow's offer to send a detachment of combat engineers and construction troops to help rebuild bridges and roads destroyed by the Israeli air force. Ramzan Kadyrov had already suggested to Moscow that Chechen troops might be used as peacekeepers in Lebanon. The Kremlin seized on the idea, and in early October it announced that it was deploying "elements" of the 42nd motorized rifle division—which is permanently deployed in Chechnya—to provide security for construction and engineering units in Lebanon. Moscow also mentioned that these security forces would compose two platoons—one each from the division's Vostok and Zapad Battalions. "The[se] servicemen have vast combat experience which can be used to guard our battalion effectively," explained Russia's deputy prime minister and defense minister, Sergei Ivanov.

On October 3, roughly 150 noncombat troops from the 100th Bridge Battalion landed at Beirut International Airport. They were the first of about 550 Russian troops to arrive in Lebanon. The Kremlin characterized the deployment as largely political: Ivanov told the Russian media that the deployment of the Chechen troops was intended to "improve Moscow's image in the Arab and Muslim world." The media largely accepted this description: An editorial in the daily *Kommersant* noted that "Al Qaeda puts Chechnya, along with Iraq and Afghanistan, on the list of regions where Christian-Muslim war is still going on. Apparently, by sending 54 Chechen soldiers to Lebanon, Moscow is trying to change the attitude to its policy in Chechnya."

When the two SPETSNAZ platoons landed quietly in Sidon, however, it was clear that their small camp, headquarters, and surveillance posts—some two hundred meters from the water in the port of Sidon—were basically operating in a war zone. Russian security officials acknowledged that "some Russians will probably have to come face to face with HizbAllah militants who control southern Lebanon." Deploying the Chechen battalions into Lebanon, in truth, was a first step in the

Kremlin's decision to use the GRU's Chechen SPETSNAZ outside Russia in the war against terrorism. The primary mission of the SPETSNAZ was to hunt down and kill Chechen Islamist-Jihadists hiding in southern Lebanon, mainly in Palestinian refugee camps in the Sidon area. The GRU's main target was Usbat al-Ansar (the League of Partisans), the local representative of al Qaeda, which operated in the Ein al-Hilweh refugee camp near Sidon. The Usbat al-Ansar consists of several hundred Palestinian and Lebanese mujahedin who had fought in Chechnya, Afghanistan, Bosnia, and Kashmir, as well as foreign mujahedin, many of them Chechens, who found refuge in the southern Lebanon camps. The SPETSNAZ were also tasked with collecting intelligence about the Arab instructors and organizers who sponsor and sustain the networks throughout the northern Caucasus, and with disrupting their work if possible. Some of these Arab-run facilities—including "charities"—are located not far from where the Russian troops were doing their construction work, fixing bridges and roads.

Today, the situation in Chechnya, and the Caucasus as a whole, is at a crucial turning point.

On one hand, the Chechenization and Islamicization movement of the wartime years has been all but defeated. After abandoning the revolt, the public resigned itself to the fact that any return to normal life and recovery, at least in the near term, must be under the rule of Moscow and its local allies. Rebuilding and recovery in Chechnya has progressed at an impressive rate, and grassroots optimism has returned. It would be easy to conclude that Russian strong-arm tactics compelled the population into ending the violence, but the optimism and commitment to rebuilding Chechnya appear to be genuine—and in them lies the core of Moscow's success. Indeed, the real victory over the Islamist-Jihadist movement in Chechnya belonged not to Moscow but to the people: Though they had supported the quest for national liberation, the vast majority of Chechens had refused to abandon their heritage and indigenous traditions in favor of the radical Islamicization and Arabization that their leaders insisted they adopt as the price of freedom.

The statistics tell a story of their own: Today, there are fewer than one thousand rebel fighters of all stripes in Chechnya—but between six and seven thousand former rebel fighters now populate the pro-Moscow Chechen security forces.

On the other hand, however, the hard core of the Islamicization and radicalization movement still has a certain momentum in the North Caucasus—buoyed by the zeal of the most persistent jihadist leaders and preying upon the frustrations of those who have missed out on the social and economic development flourishing around them. The Islamist-Jihadist movement has shifted its goals from mass recruitment to the cultivation of smaller numbers of dedicated mujahedin—and they continue to find a small stream of individuals willing to offer themselves as suicide terrorists. As the frustration of the Islamist-Jihadists grows and their self-imposed isolation only intensifies their radical sensibilities, their readiness to inflict carnage on society remains a permanent threat.

There is a limited amount that a modern state, founded on law and order, can do to combat such a threat. The very same factors of economic development and sociopolitical normalization that defeated the spread of popular Chechenization only further radicalize those at the fringes, driving them into the open arms of hardcore Islamist-Jihadists. The radicalization and alienation of Muslim youth in the face of modernity appear to be all but permanent factors in contemporary society, but in Chechnya the Russian government can claim an impressive, though still incomplete, victory in keeping its wrath at bay—even as it awaits the further terrorists that are almost inevitable.

The West, particularly the United States, has much to learn from Russia's experience in the Caucasus.

In the last five years, the Bush administration's war on terrorism has devolved into horrific quagmires in both Iraq and Afghanistan. The steady stream of anti-U.S. attacks and grassroots fratricidal carnage continue to escalate, despite—or, more accurately, because of—the immense efforts and sacrifices of the U.S. armed forces. At the core of this failure is the intensifying alienation of ever bigger segments of Iraqi and Afghan society—sending the youth into the fold of the armed resistance, which by 2007 comprised everything from localized ethnocentrist elements to global Islamist-Jihadist forces.

Washington would be wise to follow the crucial lessons of Russia's performance in resolving the Chechen conflict. The U.S. government must realize that the demands of national security and world order no longer allow for the population to have what it really wants—namely, a return to the politically amorphous Islamic way of life, based around localized ethnocentric frameworks (tribes, clans, etc.). The only approach that will bring stability to the situation, in practical terms, is to persuade the population to face the reality of the modern world and compromise accordingly.

The United States needs to find ways to show the two populations it claims to have liberated from the yoke of oppression—the people of Afghanistan and those of Iraq—that they must either accept a compromised version of their indigenous way of life, as the Chechens have under Russian authority, or face an unacceptable level of persistent violence and suppression—being caught in the crossfire between ruthless and radicalized insurgencies and the superior firepower of modern military powers. Instead, thus far, the United States has attempted to impose a centralized unitary state, based on Western democratic principles, on the ethnically and religiously diverse, and mutually hostile, populations of both nations. Simply put, the people of Iraq and Afghanistan are convinced that the United States cannot, or will not, permit them the kind of society they aspire to be—and they are not afraid to continue fighting the U.S. occupation of their lands, honor, and heritage. So where is the incentive to stop fighting and compromise?

Moreover, even as the fighting drags on, the Islamist-Jihadists continue their appeal to the population by providing social services to people in need, providing religious-ideological fiber to hold on to as the social order collapses around them, and by constantly proving their own commitment through their martyrdom missions against the occupation forces. The process of Chechenization, finally rejected in its nation of origin, triumphs among the people the United States is attempting to liberate.

Washington's inexplicable blunders are not limited to these two violent countries. The United States has repeatedly demonstrated shortsighted ignorance in other parts of the Muslim world—initiating what must be considered self-defeating missions in other hotbeds of Muslim foment. In the spring of 2005, for example, the United States was the driving force behind the overthrow of Askar Akayev, the president of Kyrgyzstan, even though he had successfully adopted and instilled the region's Epos of Manas in an effort to counter the Islamicization spreading through Central Asia. Now radical Islamism-Jihadism is growing unhindered in an area that sits on energy reserves second only to those of the Middle East. Similarly, the United States is pressuring the president of Azerbaijan, Ilham Aliyev, to implement Western-style "democratic reforms" as he strives to sustain a secular moderate state by relying on the traditions of the Silk Road—when Azerbaijan, a state with a predominantly Shiite population on the border of Islamic Iran, is the West's gateway to the energy resources of Central Asia and the Caspian Sea basin. The United States has embarked on several similar "initiatives" throughout the Hub of Islam—in north and central Africa, on the fringes of the Middle East, and in Central Asia. These areas share certain unfortunate common denominators: growing Islamist-Jihadist activity among the grassroots population, and the beginning of the Chechenization process, with both society and government becoming alienated from the West. If this trend continues, the United States armed forces will find themselves "liberating" more Iraqs and Afghanistans, at a terrible cost in both blood and reputation, even as U.S. policy allows the Islamist-Jihadist movement to flourish where the indigenous population was inclined to reject it.

In a time when the United States has failed to gain any lasting traction in its battle with Islamist-Jihadist terrorism, its leaders would do well to look to Russia and learn from its experience and expertise in Chechnya if only because, unlike the United States, Russia has succeeded in reversing the spread of Islamism-Jihadism and restoring stability, progress, and relative peace.

POSTSCRIPT

THE TWO DOMINANT TRENDS IN THE JIHAD IN THE CAUCASUS—normalization and recovery in Chechnya, and further integration of the jihadist movement into regional and global causes at the expense of Chechnya—have been even more pronounced between February and August 2007, the first political half year of Ramzan Kadyrov's rule over Chechnya.

In Chechnya itself, this period saw a major consolidation of the Moscow-dominated government. The Kadyrov administration, with all its imperfections, has been accepted as legitimate by the majority of people. The vast majority of Chechens are most interested in rebuilding their lives, return to normalization, and guaranteeing their personal security—all of which, they are now convinced, only Kadyrov can deliver.

The other side of this coin is the continued stifling of the revolt in Chechnya itself for lack of popular support. This development is so pronounced that rebel leaders can no longer ignore or deny it. Communiques and messages issued by the Chechen leadership no longer discuss an independent Chechnya as a near-term objective and, instead, stress long-term jihad as an Islamic obligation irrespective of the prospects of victory. Even when the situation in Chechnya is now addressed, the emphasis is on the current fighting being but a phase in a centuries-long jihad that is likely to continue for centuries to come before Allah provides victory. Moreover, having failed to reverse normalization in Chechnya, the leaders now emphasize their intentions to mount further terror campaigns throughout the Caucasus region as a whole (not just within Russia's northern Caucasus). The jihadist leaders insist on having monopoly over the "correct" Islam by urging their followers to assassinate pro-government Muslim clergymen—which has since happened throughout the Caucasus and southern Russia.

Moreover, even the declared objective of this regional jihad has been transformed during this period into an integral component of the global jihadist cause—rather than an instrument for attaining freedom for the Caucasus. The global jihadist terminology was adopted fully as well. The jamaat in Dagestan, for example, echoed the Shield Fatwa, declaring that all infidels in the Caucasus were now "lawful military targets for the mujahedin" regardless of who they are or what they do. Similarly, projected attacks on the 2014 Sochi Winter Olympics, jihadist communiques assert, are described as efforts to punish "countries fighting against Islam and Muslims" all over the world, not just Russia. Another communique urged attacks on synagogues throughout the Caucasus in order to punish Israel for "endangering the Al-Aqsa Mosque" in Jerusalem. In this context, Azerbaijan has been getting increased attention from the jihadist leadership in the Caucasus, both because of Baku's growing role and effectiveness in the war on terrorism and because of Tehran's growing determination to subvert Azerbaijan—using Islamist-Jihadist networks toward that end—so that it cannot serve as an anti-Iran base in a future confrontation with the U.S.-led West.

The first half of 2007 also saw a further expansion of the involvement of both Chechen and "Chechen" mujahedin in jihadist fronts all over the world. With the all-Caucasus efforts now defeated, the jihadist supreme leadership highlighted the growing impact of Chechen commanders and mujahedin in far-flung major theaters as an expression of their continued participation in the global Islamist-Jihadist movement.

The ascent of Ramzan Kadyrov and the consolidation of his presidency marked a true watershed. In mid-February, immediately after Kadyrov's thirtieth birthday (the minimum age for holding the office), Russian president Vladimir Putin named him president of Chechnya. He was sworn in on April 5, completing the process of Moscow's handing over responsibility for Chechnya to an administration predominantly comprising ethnic Chechens. "I want to achieve a peaceful Chechnya within the Russian Federation," Kadyrov declared after his inauguration in Gudermes.

By mid-August, Kadyrov was recognized and legitimized by the Muslim world. He traveled to Mecca to perform the *Umrah* (a minor, off-season Hajj pilgrimage) and was accorded all the honors of a head of state. Most important, Kadyrov was granted an audience by the far-seeing King Abdallah of Saudi Arabia, who recognizes that the revival of traditional values and social heritage by indigenous Muslim peoples—whether mountain peoples in the Caucasus or the Bedouin tribes on the Arabian Peninsula—is the sole defense against militant Islamism-Jihadism. Kadyrov's reception signaled King Abdallah's support for Kadyrov's approach to the continued rebuilding and stabilizing of Chechnya.

Before he took office, Kadyrov expressed his vision of the evolution of the war in Chechnya in an interview with RFE/RL. He described his own transformation from fighting in the anti-Russia revolt to being Putin's man in Chechnya as a response to the need to reverse the Islamist-Jihadist impact on his people—even if this required recognizing Russian rule. "I took up arms first and foremost because I'm a Chechen. The Wahhabist devils who came to us at the beginning said that we should be not Chechens, but Arabs, mujahedin, Afghans. . . . Then the others [foreign jihadists] came and tormented the Chechen people. We're the ones who wouldn't tolerate the torment of the Chechen people, and stood up against it," Kadyrov explained. Indeed, as the Chechen population was cooperating with the authorities and ejecting the Islamists, Kadyrov ordered the relaxation of the security regime—essentially a reduction in the abuse of human rights in Chechnya. In late May, Oleg Orlov, an expert with the Russian human rights center Memorial, corroborated the decline in abductions in Chechnya since the beginning of 2007. He attributed the trend to orders issued by Kadyrov back in January. "Those orders were presumably part of a broader campaign to craft a new, more benevolent and less threatening image of the younger Kadyrov," Orlov noted.

Kadyrov has always considered the economic recovery of Chechnya and the return to normal life to be both the key to the future of Chechnya and his own greatest challenge. In his interview with RFE/RL, he highlighted the extent of the recent economic development and recovery. "Chechnya, which has been destroyed over the last fifteen years, now has eighty-six educational institutions, more than a hundred hospitals, and many roads that have been restored. There is now a gas pipeline running to the Sharoy district, which never had gas before. There are 712 buildings in Grozny that have been repaired, which equals more than 24,000 apartments that will be ready by May 1. Thousands of sick people have been sent to hospitals." Even hostile Western observers acknowledged the rapid and thorough rebuilding of Chechnya and the accelerated return to normalcy under Kadyrov.

A group of British correspondents who visited Grozny in early March were taken to see the new rows of apartment blocks in the city and its huge new mosque. One of the journalists, who'd been to Grozny in early 2006, was "stunned" by the changes. "Last year this whole street was still a bomb site," he said. "Grozny is being rebuilt at a frenetic pace." Another British journalist summed up Kdyrov's logic that "the more young Chechen men working on building sites, the fewer that will be tempted to pick up a gun against Russia. But while Moscow pays the bills, running Chechnya is today left to Ramzan Kadyrov and his band of former rebels." And Kadyrov considers the current effort only the beginning of a profound transformation of Chechnya. "According to our program, Grozny should be completely restored in the next year. The cement factory should begin operations this year. There will be brick and reinforced concrete factories in Grozny. There will be new oil refineries using new technologies. It is becoming clear if that Chechens reach accord and, like others, become their own masters, then the federal center will work for our benefit," Kadyrov told RFE/RL. With growing population support, Kadyrov predicted in August 2007 that the bulk of the remaining violence would be resolved by year-end.

Ultimately, the Kremlin's staunch support for the consolidation of Kadyrov's presidency, and the continued rebuilding of Chechnya, is a manifestation of Putin's growing emphasis on the strategic importance of the Caucasus for Russia. The Kremlin pays close attention to the evolving energy policy vis-a-vis Europe—particularly the transportation of gas and oil via the region. The emergence of Azerbaijan as "the Spigot of Europe" is being watched very closely by friends and foes alike. While the Kremlin is seeking to reach a compromise with the Aliyev administration to ensure the sustenance of Russia's influence in Europe without alienating Baku and the states of Central Asia, led by Tehran, the Islamist-Jihadist movement is seeking to improve capabilities to strike and destroy the vast energy infrastructure throughout the Caucasus in order to bring Europe to its knees and pain Russia. The Islamist-Jihadist forces throughout the northern Caucasus, not just Chechnya, are the primary instrument for waging the jihadists' war on oil and gas. Meanwhile, Azerbaijan has emerged as the unsung hero of the war on terrorism—particularly in the tumultuous Caucasus. Under the leadership of President Ilham Aliyev, Azerbaijan has retained its moderate Islamic character, sustaining democratic reforms and Western modernization while resolutely fighting the spread of Islamism and jihadism.

Between February and August 2007, as the Russians consolidated their hold on Chechnya, the resistance continued its transformation from a Chechen force into an Arab jihadist movement aimed at the Caucasus as a whole.

By early 2007, the Russian-Chechen proactive operations of 2004–2006 were having a discernible impact on the Islamist-Jihadist movement. According to an FSB senior official, seventeen al Qaeda emissaries were killed and seven were arrested during this period. Several dozen foreign mujahedin were also killed and many others captured or forced to flee across the Georgian border. In 2007, the official noted, the Russian security services would "focus on neutralizing figures such as Dokka Umarov, Rappani Khalilov, and Achemez Gochiyayev." "They have committed many crimes and should be arrested or, if they offer resistance, eliminated," he stated.

Concurrently, the Islamist-Jihadist leadership made yet another effort to bolster the operational capabilities of the Chechen resistance. New Arab commanders were elevated. At the core of the new command structure were some thirty foreign mujahedin. Most significant was the rise, in early 2007, of a senior Algerian mujahed known only as Yasser to the position of effective commander of the jihadist elite forces in Chechnya—effectively Abu-Hafs's original command. Abu-Hafs's declared successor—the Sudanese Commander Muhannad—is of lower rank and posture in the jihadist movement. Another key commander to be elevated in early 2007 is a Jordanian mujahed known as Saif al-Islam. He has been responsible for the flow of funds from foreign Islamist organizations and charities to Umarov's command cell. With the dwindling popular support, Umarov is presently completely dependent on foreign funds—from Islamist contributions to shares in drug trade. Moreover, Russian intelligence unearthed new documents proving close cooperation between most senior Chechen leaders and the intelligence services of states supporting jihadist causes. Some of these relations go back to contacts and modalities of cooperation established and sustained by Shamil Basayev.

The Islamist-Jihadist camp, by this point, was almost defunct. On the eve of the fighting season in early 2007, Chechen resistance was barely functioning. According to General-Colonel Arkady Yedelev, the Russian deputy minister of interior who headed the FSB Operation and Coordination Department in North Caucasus in 2001–2004, "The total number of the anti-Russian armed troops in the territory of this republic is about 450 men. They are divided into forty-six groups." According to the Russian security services, "several dozen" foreign mujahedin were still fighting on the Chechen side. According to jihadist sources, their numbers ranged between one hundred and two hundred. According to Russian intelligence, most of these mujahedin hailed from Arab countries, Turkey, and Muslim emigre communities in Western Europe. There were also few mujahedin from Central Asia, Afghanistan, and Pakistan. There were, however, a growing number of recent converts from Europe—not only the usual flow of Western Europeans, but also growing numbers of Ukrainians, Estonians, and Belorussians. As well, there was a discernible increase in the number of mujahedin from the southern Caucasus—mainly Georgia and Azerbaijan—preparing to take the all-Caucasus jihad back to their home countries.

Most significant, however, was the movement's effort to put a theological face on the decision to prioritize the global jihad over the situation in the Caucasus. A study published in mid-February 2007 by the Islamic Center of Strategic Research and Political Technologies and distributed by Chechen Islamist circles claimed to be "an attempt of generalization and analysis of the world Islamic experience, and the Movement for Islamic Renaissance in the Caucasus within the context of ideology, strategy and tactics of the restoration of the Shariah in the Caucasus." The study asserted that the jihad in Chechnya could not be furthered in isolation from the prevailing trends in global jihad. "We will not be able to work out the right and appropriate strategy in the cause of the Shariah Renaissance without clear understanding of the present situation in the world, without taking into consideration the Muslim world fighting experience and peculiarities of the Caucasian region."

The study argued that the number one factor hindering the ascent and triumph of the jihad all over the world was "the absence of Islamic Center"—namely, a new caliphate that will serve as a viable source of guidance and support for all far-flung fronts. This assertion led the authors to raise hitherto sacrilegious questions regarding the jihad in Chechnya and the Caucasus. "In order to establish the Shariah State we strive to drive away Russian occupants from our soil. But a question that may cause bewilderment at the start arises: 'Is it necessary to hurry up to turn out the invaders?' Are we ready for the peaceful period? After all, our first aim is the liberation of our Muslim consciousness, which has long been subjected to ideological occupation." The mere raising of these issues was a reflection of the grim realities of the Caucasus jihad. The authors were clearly aware that the recent attempts at establishing a Muslim state stretching beyond the Caucasus had failed because of the Islamists' inability to cope with the aspects of modernity the people expect and demand. An all-Islamic center capable of providing guidance for such quandaries is a precondition for success, the study noted, and such a center does not exist at this time.

The study stressed that the experiences of the late twentieth century repeatedly showed that the mere banishment of the infidels from areas inhabited predominantly by Muslims was not sufficient for the establishment of a Muslim state. The authors noted that earlier attempts by the Chechen Islamist leadership to learn from the experience of Islamist luminaries had come to naught. "In 1992 [a] Chechen delegation flew to Khartoum, where it met and discussed these questions with the Islamic leadership of Sudan. We had repeated discussions of this problem with the leaders of the party 'Ikhvanul-muslimin' [Ikhwan al-Muslimin—the Muslim Brotherhood] as well." Chechen efforts to learn from other Islamic causes and jihad movements were not successful either.

The war in Chechnya, the authors stated, was a milestone event "on the path of the Caucasus Islamic Revival," and one that must not be examined in isolation from global dynamics. "The struggle of Muslims of Ichkeria isn't weakening, but owing to the Most High Allah is spreading being supported by Mujahedin of the Caucasus." The study warned against the lure of ending the hostilities and establishing a modern state—which the authors considered "Jahilijjah" (*Jahiliyah* meaning "barbarity"). The establishment of such a Jahiliyah state was even more dangerous than the continuation of Russian occupation and oppression, they argued, because people would tend to accept the emergence of a non-Muslim state as the realization of their aspirations and then stop fighting for the establishment of a real Islamic state.

The authors acknowledged the profound gap between the leadership's objectives and the grassroots aspirations of the people. They explained that "the main common aim—throwing down the colonial yoke—united everyone" during the early phase of the jihad. Consequently, many Chechens failed to see the profound difference between the ultimate objective of the genuine jihadists—a Shariah state—and that of the rest of the fighters and leaders: a viable modern state, possibly within the framework of the Russian Federation. Moreover, the authors note, many Chechens fought solely for the liberation of the Chechens when, for the leadership, the jihad has always been for the liberation of the entire Muslim Caucasus. Therefore, for the jihadists, no agreement regarding the future of Chechnya—such as the Khasavyurt agreement—warranted the cessation of hostilities.

Since the mid-1990s, the jihadist leadership had been trying to remedy the situation. "Everything that was not for the sake of Allah collapsed, showing its groundlessness and uselessness. What remained was the idea of Freedom, which didn't conflict with Islam. . . . And it was possible to establish formally Shariah in the provinces which were under the control of Mujahideen." The previous efforts to consolidate a Chechen state—a reference to the post-Khasavyurt efforts—were meaningless because the would-be Chechen state was based on nationalist principles and sought support and recognition from "kaffir" Western Europe, the study explained. The leadership decided against "copying state institutions of unbelievers" because "the structure and form of these institutions are meant to support their religion, their laws; to support the laws of Taghut." Adoption of values and moralities of the "kaffir" West only undermines the standing and undertaking of the jihad. Similarly, the tenets of state building must be uncompromising. "Only providing that the Jahilijjah system is completely broken, Shariah state can be built. Taking something from Allah and something from Taghut we will build something . . . , but it will not be Shariah."

The study authors called for the jihad to break contacts with the West—even with otherwise sympathetic movements and states that declined to endorse spectacular terrorism like the attacks at Nord Ost and Beslan—and celebrated the role of Khattab's camps in formulating the Islamist state-building doctrine with international jihadist character. "The entire atmosphere [of the camp] was saturated with the spirit of Qur'an, the spirit of Jihad," they observed. "Here one did not even dream about any other laws, but for the laws of Shariah. Mujahedin from all over Caucasus and Muslims from other countries longing for taking the path of Allah gathered here." The authors emphasized the central role played by the graduates of the jihadist training system in Chechnya in the transformation of the Chechen jihad into a Caucasian and later global jihad, but acknowledged its ultimate failure. "The major part of them [Khattab-trained mujahedin] became Shahids [martyrs], but we can see that a new generation of Mujahedin is coming to take their place. In this way the greatest mercy of Allah to the Islamic state is being revealed."

The authors concluded with a stark reminder of the real objective of jihad in the Caucasus. "It is impossible to combine Shariah with institutions of the Roman law. It is impossible to observe both the orders of Allah and 'International law' at one time." The message is simple: It is imperative to give precedence to establishing the Shariah center through global jihad before it is viable to once again attempt to establish a Shariah state in the Caucasus, let alone only in Chechnya.

The study's grim conclusions regarding the state of the jihad in Chechnya were reflected on the battlefield. What fighting remained was low-key—localized clashes and skirmishes, mainly ambushes and counterambushes by special forces. In late February, Dokka Umarov acknowledged that the "situation at the fronts of Russian-Caucasian war continues to remain stressed," and emphasized that there was no longer a Chechen war. "Don't despair and don't stop fighting," he urged his mujahedin. "Fight on the path of God, be God-fearing, do good deeds, be righteous, obey the laws of God and refrain from things that He has forbidden, and then the Almighty will grant you victory, Inshallah!"

In early March, Umarov released an "interview" distributed by Kavkazcenter.com in which he seemed to change his tune somewhat, claiming that "the war continues." But he acknowledged both enduring problems and new limitations, explaining that the Chechens "reorganize our battle formations in the correspondence with the new tactics of waging of war in this stage." He granted that they faced acute logistical shortages and noted that the jihadists "cannot provide [sufficent] weapons, but in every way possible we try to solve this problem." Umarov adopted a philosophical outlook: "We don't doubt the success and the inevitable defeat of the Russian Empire, which, in the final analysis, compulsorily will lead to the expulsion of occupiers from the Caucasus." But he conceded his inability to determine and dictate the pace of the war. "We don't hurry anywhere. For believing Muslim Jihad is a blessing from Allah, anyone can't succeed himself in it participating," he added.

Umarov vigorously sought to deny that the majority of mujahedin had crossed over to Kadyrov's side. As he pointed out, the majority of the fighters during the first Chechen war—who constitute the majority of Kadyrov's men—were not even mujahedin, for "the first war was only preparation for the second"—that is, not a jihad in its own right. Umarov acknowledged that there were mujahedin in the first war and even at the beginning of the second who had since changed sides, but he insisted that "[the] number is very small." He argued that the new vision of an all-Caucasus jihad was the best way to defeat Russia in the long run. "Russia in every possible way attempted to limit war [by keeping it within] the framework of Chechen territory. . . . However, the war has its rules, and Most High has its plan, different from Moscow." Umarov considered this expansion of the jihad to reflect a return to the region's historical contours. "The peoples of North Caucasus have experience [as a] joint state . . . during the times of Sheikh Mansur, Imam Shamil, and [the] Mountain Republic, and the North-Caucasian Emirate of Sheikh Uzun-Hadji. The basis of this association was always Islam, and in the case of the Mountain Republic, [it was the] idea of all Caucasus unity and decolonization."

In late March, Amir Muhannad, the commander of the foreign volunteers, was also drafted to reassure the supporters of the Chechen cause. In a report posted on Qoqaz.net, Muhannad "assured the Chechens that the plans of the Mujahideen and their preparations are moving according to a program that has been established and the rumors that the Russians are spreading that the Chechen resistance is finished has no basis in reality." The prevailing reports about Kadyrov's growing popularity, Muhannad insisted, were lies disseminated by a frustrated Moscow "perplexed by the determination of the mujahedin in Chechnya and their steadfastness in spite of the death and injury they have faced." He also addressed the growing fear of abandonment by the global jihad movement, stressing that "the Muslim Ummah must not fail their Chechen brothers." On June 6, 2007, the Chechen jihadist media began referring to Muhannad as "the Military Vice-Commander of Chechanya"—echoing the title Khattab held at the height of his relations with Basayev.

Yet the jihadists' much-heralded "spring offensive" fizzled. Even the most optimistic reports from mujahedin circles portray an endless series of ambushes and raids. More often than not, these engagements took place over the same battlefields Khattab and Basayev had been fighting since the First Chechen War—a reminder that the jihadists had gained no ground in all that time. A March 23, 2007, jihadist report of a major ambush by the forces of Amir Khair Ullah, the military commander of the western front, noted that their position "has witnessed many major operations carried out by the Mujahedin against the Russians from the very beginning of the war until today. This was a favorite location of Commander Ibn Khattab, may Allah have mercy on him, who was the first to carry out operations on this road during the first Chechen war." For more than a decade, in other words, the mujahedin had failed to dislodge the Russians from the area.

The mujahedin remained on the defensive throughout the spring. Their detachments were pursued and cornered by relentless forays of special forces—particularly the GRU's Vostok (East) and Zapad (West) Battalions. Rebel forces encountered were usually outnumbered and outgunned by the Chechen special forces. Although many of the captured and surrendering fighters acknowledged that theirs was "a hopeless cause," security officials, particularly former Chechen rebels, had healthy respect for the determination and endurance of their enemies. "The war is not over," Colonel Sulim Yamadayev, the commander of the Vostok Battalion, said in late March. "The war is far from over. What we are facing now is basically a classic partisan war and my prognosis is that it will last two, three, maybe even five more years." At the same time, the security forces were constantly on the offensive at the jihadist heartland, forcing the mujahedin to abandon their sanctuaries and withdraw deeper and deeper into the forbidding and inhospitable mountains. "We are coming under fire in this area all the time," noted Magomet, one of Yamadayev's company commanders. "There are a lot of bad guys around here—this is Wahhabi Central." Concurrent battle reports from jihadist commanders were portraying their forces as being "surrounded and pounded" by Chechen and Russian security forces with superior fighting skills and firepower. The mujahedin's ability to evade and withdraw with little or no casualties was now considered achievement.

Even where the mujahedin were capable of inflicting tactical setbacks, their own forces were so strained that they were incapable of capitalizing on these successes. On April 27, a combined mujahedin force led by Amir Ramzan, together with forces of Amirs Aslam Bek and Umar, conducted what Amir Ramzan called "a major operation in the area of Shatoi" in southern Chechnya, an isolated region of little relevance to the greater Chechen population. The mujahedin set an ambush along the local main road to serve as a decoy for the Russian and Chechen security forces. As anticipated, the security forces reacted swiftly and in no time dislodged the ambush and started pushing the mujahedin deeper and deeper into the rough terrain. This time, however, the main mujahedin forces were at the ready and, a couple of hours later, were able to surround their pursuers in a series of counterambushes. The security forces called for reinforcements, and about an hour later, three Mi-8 assault helicopters carrying GRU Special Forces arrived on the scene. Another mujahedin detachment—the Saif Allah group, which includes many foreign mujahedin—fired an SA-7 and brought down an Mi-8 helicopter. Four crew members and thirteen GRU troops onboard were killed. But the Russian and Chechen forces brought in more reinforcements, as well as artillery and air support, faster than the mujahedin were able to disengage. The security forces surrounded the area, resumed pushing and fracturing the mujahedin forces, bombarded the mujahedin's isolated pockets, and eventually forced the survivors to withdraw and melt into the rugged mountains. Though the Russian and Chechen security forces suffered casualties, these could be easily replenished. The mujahedin forces involved, however, were so devastated that by August 2007 they had not yet regrouped as a fighting unit.

Meanwhile, a close examination of the jihadist high command shows the growing success of the Russian and Chechen special forces in tracking down and eliminating the surviving commanders of the Chechen jihad. In late March, Chechen forces surrounded the house in Gudermes where the commander of the northeastern front, Amir Takhir (real name Takhir Batayev), was hiding. He was killed after a brief firefight. Around the same time, Commander Rustam Bashayev was located and eliminated in Khasavyurt. In early April, the Russians located and killed the military commander of the western front, Amir

Khair Ullah (real name: Suleiman Imurzayev). In late April, Amir Muslim was named the new commander of the northeastern front, replacing Amir Takhir. In mid-May, special forces intercepted a raid under Amir Muslim's command. He was badly wounded but could not be evacuated from his remote hideaway, and he died in late July. In early June, Commander Aslam Bek (real name Aslambek Vaduyev) was tapped to succeed the late Khair Ullah. In late July, Amir Daud was appointed new commander of the northeastern front.

A new series of clashes in late May and early June, predominantly in the remote mountains and isolated villages, saw the Chechen security forces—supported by Russian intelligence and special forces—inflict heavy losses on the mujahedin, and the pace of defections accelerated. The jihadist forces did manage some localized successes—mainly successful ambushes in which several troops were killed and wounded. In late July, the jihadists scored a momentary success in the Vedeno area: When Russian and Chechen special forces surrounded a mujahedin force on its way to lay an ambush, they were attacked from the rear by another mujahedin force that happened to be nearby and noticed the battle. The Russian and Chechen special forces suffered heavier casualties than usual, though the number was debatable: The Russians acknowledged six fatalities, while the jihadists claimed to have killed thirty Russians. But this clash, like most others, ended with the special forces calling in air and fire support, which dispersed the mujahedin forces and inflicted additional casualties. Flare-ups like these had little effect on the overall strategic environment. The jihad remained on the defensive and largely isolated from the population—both physically and ideologically.

By mid-2007, Umarov was trying once more to shake up the jihadist high command, elevating the jihadist terrorist and foreign mujahedin commanders. In late April, Umarov reintegrated the famous Sharia Guards, disbanded in 1998, into the CRI Army. General Abdul Malik Mezhidov, the last commander of the Guards before 1998, was reinstalled as commander—reversing one of Maskhadov's key steps toward legitimate statehood and adherence to the Khasavyurt Agreement. In mid-July, Umarov announced a new high command, appointing Amir

Muhannad as First Deputy Military Amir, Amir Tarkhan (real name Tarkhan Gaziyev) as Second Deputy Military Amir, and Amir Aslam Bek (real name Aslambek Vaduyev) as Deputy Military Amir of the CRI Armed Forces.

Umarov then nominated Amir Magas (real name Akhmed Yevloyev) to Shamil Basayev's old post as Military Amir of the CRI Armed Forces—a significant move, since Magas is not a Chechen but an Ingush. Indeed, Magas had never held a command position in Chechnya itself; He first rose to prominence in June 2004 when he led a large-scale raid on Ingushetia. In October 2006, he was made commander of the Caucasian Front—responsible for areas west of Chechnya in Ingushetia, North Ossetia, Kabardino-Balkariya, Karachayevo-Cherkessia, Krasnodar, and Stavropol. The nomination of Magas as Basayev's successor reflects more than the shift of the focus of the jihad away from Chechnya, a trend that began under the leadership of Chechen commanders. Rather, it was a stark reflection of Umarov's inability to find qualified Chechen commanders, after years of Russian-Chechen counterattack and the continued defection of commanders and fighters.

By late July, then, the so-called Chechen jihad was being directed by an Ingush commander and a Sudanese deputy—and Umarov himself was being referred to as CRI President and Amir of the Mojahedin of the Caucasus. These developments did not sit well with the fragments of the Chechen population that were still committed to the anti-Russia rebellion. Umarov was feeling pressure to demonstrate his commitment to Chechen affairs, and on August 6, Kavkaz-Tsentr published a sensational report of a secret visit he paid to Grozny, escorted by his personnel guard and a special group of mujahedin. According to the report, Umarov met with Amir Abubakr Basayev, commander of the central front, and several middle-rank commanders to discuss the next phase of the jihad. The report also claimed that Umarov "visited several districts of the capital and met local residents" in flagrant defiance of Kadyrov's ostensible control over Grozny. Yet no pictures were published of this purportedly public appearance—an odd omission in the era of miniature electronic cameras—and no independent confirmation of the visit was to be found. The fizzling of the story was yet another manifestation of the shift of focus away from Chechnya and into the northen Caucasus.

The most dangerous trend of 2007 was the further radicalization of the jamaats throughout the northern Caucasus. Though these groups had shrunk in size, their remaining ranks were growing more dedicated and zealous—and harder for the Russian security services to pin down. Chechen jihadist leaders considered the jamaats their best instrument to counter the decline of the jihad in Chechnya. In early March, Umarov issued a message to "the Muslims of the Caucasus" urging them "to step up the Jihad . . . against the Russian occupying forces and the pro-Moscow Chechen regime" rather than the authorities in their own republics. Umarov called for "setting up combat groups and spreading the word among Muslims who have not yet joined the Jihad," and urged "those Muslims fighting on the Russian side to return to Islam." This was a clear role reversal—where once the Chechen jihad had been the primary sponsor of the all-Caucasus jihad, now it was reduced to appealing to the jihadists of the Caucasus for help in a time of need.

The burgeoning Nogai Jamaat, because of its distinct traits, serves as a good example of the overall emerging trend. The Nogai are a small population of Mongolian origin (about seventy thousand strong) dwelling in northern Chechnya and southern Russia. Involved in the anti-Russia struggle from the beginning, they lived close to the key military axes of advance into Chechnya and suffered heavy losses during both Chechen wars. The surviving Nogai mujahedin withdrew with the Chechen forces to the Grozny area, where they fought under Basayev. In the late 1990s several Nogai mujahedin received advanced training in Khattab's camps, and as the fighting subsided the Nogai commanders were dispatched to resurrect the Nogai Jamaat along professional lines. By 2007, the Nogai Jamaat had assumed a leading role in the Caucasus jihadist movement, despite its small size—only a few dozen members, including some thirty to forty veteran mujahedin and commanders. The jamaat is divided into several small cells of three to five men each. Members of each cell live in close proximity so that they can supervise and assist each other. The Nogai's distinctive Mongolian features allowed them to specialize in terrorist operations within Russia, where they can pass as Central Asians and Mongolians.

In early February, Russian security forces began moving against the jamaats networks. Their intelligence sources discovered that the cells in various republics were cooperating operationally, as confirmed by the presence of operatives from hitherto unrelated jamaats in the same cars and safe houses. At the same time, there was a reduction in the quantity and quality of operations in the North Caucasus because of the enhanced security regime. The jamaats' own communiques reflected the trend: Where once it would have targeted a senior police official himself, the Dagestani Jamaat now hailed the assassination of his driver/bodyguard as a great achievement.

Yet the jihadists still set their aims high. In late February, the jihadist media reported that "the press-service of [the] Dagestani *Jamaat* 'Shariah' [has] warned that the mujahedin will make it impossible [for the Russians] to conduct so those called Olympic Games in Sochi in 2014. . . . the Daghestan mujahedin [have] also warned that the provocative actions of Zionist authorities in Jerusalem against the sacred mosque Al-Aqsa [may elicit] appropriate reaction and reciprocal actions from the side of the mujahedin."

In pursuit of these threats, the jihadists did attempt a few terrorist strikes in early 2007. All of these attempts were intercepted and prevented by the Russian security services, who had superior penetration of the jihadist networks and had been receiving warnings from civilians throughout the northern Caucasus and from Caucasian communities in Moscow and other cities. In late February, one such attempt was averted at the last minute when police arrested twenty-nine-year-old Farid Magomedov of Makhachkala as he was waiting for a bus on Marshal Zhukov Avenue in central Moscow. He was carrying an improvised explosive device equivalent to 500 grams of TNT. The structure of the bomb—encased in a solid plastic container filled with buckshot, chopped nails, and other metal objects—was similar to bombs used by the jihadists in London and Madrid in recent years. Magomedov was also carrying a remote control for the bomb, suggesting that he wasn't planning a suicide mission. Shocked by his apprehension in the middle of the street, Magomedov was neutralized before he could activate the bomb and later acknowledged to authorities that he was an expert bomb maker responsible for building most of the antipersonnel bombs used by the jihadists in Dagestan.

In late March, the Dagestani Jamaat 'Shariah' issued a statement confirming that it saw its goals as "the restoration of Islamic state on the territory of Caucasus." The communique stressed that its relationship with the Chechen jihadist leadership was in the context of regional objectives. "*Jamaat* 'Shariah' has sworn [allegiance] to Dokka Umarov, who is Amir of the Caucasian Muslims," the communique clarified. "Our primary task is [the] liberation of the Muslim land from Russian occupation and the construction of Shariah state." The jamaat expressed no obligation to any political arrangement in Chechnya, and pledged that its resistance would stop only after "the wiping out of occupation forces from the land of Caucasus and the restoration of Islamic state on the Muslims' land." Yet the jamaat leaders were also realistic enough to concede that negotiations with the Russians might prove inevitable. "The question of negotiations with Russian infidels will be solved by Amir of the Muslims of Caucasus and the oppressed Muslims of Russia, by Dokka Umarov, and by Shura of Muslims," it suggested. "But Allah knows better . . ."

The few attacks there were that summer were isolated and inconclusive, with one exception. On the night of July 5, near the Ingush village of Troitskaya, a large armed group attacked elements of the 503rd Motorized Rifle Regiment garrisoned near the village. For an hour and a half, the base was subjected to sustained attack with mortars, grenade launchers, and large-caliber machine guns. The Russian authorities acknowledged "light" casualties, but the jihadists claimed that at least thirteen Russian troops were killed and about twenty-five wounded when mortar shells scored direct hits on living quarters. Their report also claimed that the attack was carried out by a "70-strong rebel detachment" and that four of them suffered light wounds in the battle.

Official Moscow registered this as evidence of a newly threatening jihadist resurgence. "The separatists in the North Caucasus are regrouping," warned Viktor Ilyukhin, the deputy chairman of the State Duma's Security Committee, on July 10. He noted that the jihadists in Dagestan, Adygeya, and Krasnodar Krai were planning new strikes in an effort to "cause maximum military, strategic, economic, and political damage to the Russian state." As a response, the Kremlin assigned Federal Interior Minister Rashid Nurgaliyev to shore up the security situation in the Caucasus. Police General Mikhail Sukhodolsky was nominated Nurgaliyev's deputy. The Kremlin was concentrating on preventive measures aimed to contain the resurgence of terrorist activities, primarily of political significance, to prevent strikes on the growing oil exports from the port of Novorossiisk in Krasnodar Krai on the Black Sea coast, and to prevent interference with the preparations for the parliamentary and presidential elections in 2008 and the Sochi Winter Olympic Games in 2014. Nurgaliyev was aware of the challenge he faced: "Separatist leaders have plans to destabilize the social and political situation not only in Chechnya, but in the whole North Caucasus," he warned in a late July meeting with his senior officers, adding that the jihadist leaders had already "started recruiting new members more actively, especially young people," for the conduct of such terrorist strikes.

The jihadists were also shifting their focus away from Chechnya and the neighboring republics of Ingushetiya and Dagestan to the farther reaches of the northern Caucasus, where a long period of quiet had led the Russians to reduce their security presence. Significantly, the jihadists had finally changed their approach, avoiding popular outreach and Islamicization efforts. Instead, they concentrated on establishing a few very small clandestine cells operating in complete isolation from society. The jihadists established a network of "mountain camps" and "guerrilla bases" in Dagestan, Ingushetiya, Kabardino-Balkariya, and Karachayevo-Cherkessia. From these bases, mujahedin "diversion squads" were preparing to strike out in the Stavropol and Krasnodar regions. The jihadists also sustained a viable elite terrorist force in southern Chechnya and northern Georgia, giving them the option of striking within Russia as well as launching support operations in Europe. As of late August 2007, Amir Rabbani of the Dagestani Jamaat 'Shariah' was warning of a new escalation, reporting that jihadist commanders from all over the northern Caucasus had recently met to discuss a new recruitment drive throughout the region. Several operations new operations for the autumn-winter period were also discussed.

On August 19, 2007, the jihadist leadership posted an ideological tract on Kavkaz-Tsentr.com entitled "Problems of Islamic State-Building." Recapping the evolution of the Chechen jihad in the context of global developments, the release explicitly equated the jihadist struggle in the Middle East (not just Iraq), and the Iran-led War on Oil, with the struggle in Chechnya and the rest of the northern Caucasus. The study argued that the collapse of the Chechen Sharia state and the success of the Kadyrov regime were tied directly to the rise and fall of the Taliban regime in Afghanistan. In its effort to gain legitimacy, the leaders contended, the Taliban had compromised the tenets of Islam in favor of international norms, weakening their resolve and shattering their grassroots support, leaving them helpless to resist when the Americans struck Afghanistan. The parallel with the rise and collapse of the Chechen Sharia state was obvious.

The study also stressed the importance of the teachings of Sheikh al-Islam Ibn-Taymiyyah concerning "false Muslims," individuals and institutions who considered themselves Muslim but rejected the jihadist interpretation of the Sharia. Ibn-Taymiyyah considered these "false Muslims" to be the worst enemies of true Islam, and called for them all—including children, women, and the elderly—to be exterminated. The study pointed to Ibn-Taymiyyah's teachings as the solution for the current situation in Chechnya: The key to the country's salvation, it argued, lay in jihad not only against the Kadyrov administration, but also against Chechen civilians who wanted a return to normalcy and stability.

Yet this August 2007 statement also made a profound leap of judgment regarding the Persian Gulf and the southern Caucasus, where Azerbaijan was the sole "false Muslim" state. Led by Saudi Arabia, the report charged, the Gulf states were "false Muslim" states, which obstructed the efforts of their youth to go to Iraq and fight the jihad, while adapting policies designed to please the United States. The leaders called for these states to be attacked with "all available means, including nuclear weapons if available," in order to eradicate these "false Muslim" regimes—a call to arms that legitimized ongoing preparations that were already accelerating through much of 2007.

This convergence of strategic interests between the jihadist movement in the Caucasus and Mahdist Iran put Azerbaijan squarely in its sights. By 2007, President Ilham Aliyev's forward-looking policy had turned his nation into the burgeoning "Spigot of Europe," the crucial new player in the global long-term production and transportation of both oil and gas—particularly to Western Europe. Keenly aware of the growing importance of the Caucasus to the supply of energy to Europe, the Aliyev administration saw its own potential to play a key role in helping to diversify Europe's energy sources—a goal with grand strategic implications. Aliyev's government propelled a rapid expansion of a web of gas and oil pipelines via the Caucasus, as well as the rejuvenation of the Silk Road as the main transportation artery across Europe and Asia. Moreover, Aliyev has maintained that Azerbaijan can succeed in this pivotal role only by maintaining its moderate Islamic character while pursuing further Western modernization.

Little wonder, therefore, that as of this writing the jihadists were making new efforts to infiltrate Azerbaijan, targeting this new energy infrastructure while planning to extort the West—mainly Europe—for support. Operating out of Chechnya and Dagestan, with lavish funding from Saudi Arabian and other Gulf states "charities," the jihadists have intensified their recruitment efforts, using a campaign to spread neo-Salafi Sunni Islam (popularly known as "alternative Islam") as a cover for recruiting small clandestine networks for launching terrorist strikes—a method borrowed from the northern Caucasus jamaats. In summer 2007, the most popular center of alternative Islam in Azerbaijan was the Abu-Bakr Mosque in Baku. On average, some seven thousand to ten thousand worshipers attend the Friday prayers and sermons, while during key holidays attendance reaches twelve thousand to fifteen thousand worshipers—an impressive number, considering that the entire Salafi/neo-Salafi community in Azerbaijan is estimated at twenty-five thousand. Hadji Gamet Suleymanov, the mosque's Saudi-educated young imam, stresses his outreach programs to help the disadvantaged and denies that the Mosque and community are involved in "politics." Yet his Friday sermons harp regularly on socioeconomic issues such as poverty, corruption, and social injustice. Meanwhile, the more radicalized young worshipers and their guests from the northern Caucasus and the Arabian Peninsula are allowed to continue uninterrupted, and in recent months the Azerbaijani National Security Ministry has picked up signs that jamaat-style activities are expanding in Baku, and have arrested several "Wahhabi" networks and cells that were planning terrorist acts.

This impending escalation in Azerbaijan is not unique. The growing pressure on Tehran raises the specter of a regional eruption; meanwhile Iran and Syria are pushing the Middle East toward explosion against the United States (in Iraq and the Arabian Peninsula) and Israel—a potential conflagration that could spread into the Caucasus in the context of Iran's War on Oil. The Iranians and their jihadist allies have been relying on the Chechen diaspora communities and affiliated Sufi networks as sources of operatives and support networks. The first half of 2007 saw a noticeable increase in the radicalization of the significant Chechen diaspora communities in Jordan, Azerbaijan, Turkey, and Kazakhstan. Veterans of the jihad in Chechnya—both native sons and migrants from the Caucasus—assumed greater prominence in community affairs and organized cells and networks, and the pattern was also spreading to other Sufi communities in the Middle East, even where there was no significant Chechen population.

The Chechen jihadist movement also had a profound impact on the war in Iraq, in the form of the Naqshbandi Sufi armed/terrorist resistance in Iraq. The Sufis had largely stayed out of the fighting as a distinct organized entity, until they were influenced by the Sufi-turned-jihadist Chechens to join the war. The Naqshbandi jihadist movement first appeared on January 17, 2007, when a communique of the Army of the Men of the Naqshbandi Order [Jaysh Rijal Al-Tariqah Al-Naqshbandiyah] claimed several attacks against U.S. targets in Iraq, including videos showing four of the attacks.

On February 7, the Army of the Men of the Naqshbandi Order issued a lengthy follow-up statement (backdated to December 30,

2006) explaining their background. Though their theological and ideological orientation seemed to blend elements of Nationalist, Ba'athist, and Sunni-Jihadist sentiment, the Naqshbandi jihadists made it clear that their sole objective was fighting the U.S. occupation and its Iraqi supporters, and that they would never fight other Iraqi patriotic entities regardless of religion or ideology. The Naqshbandi jihadists vowed to "target the head of the serpent, the infidel occupiers, not the Iraqi citizens. Our hands never got smeared with Iraqi blood. This army will continue to fight the infidel occupiers to foil all their evil plots until we expel the last infidel and those sectarian surrogates who assist their Zionist masters in carrying out their plans in attacking Islam and tearing apart our beloved Iraq." The Army stressed its solidarity with Palestinian and other jihadist causes—a signal it reiterated on the second anniversary of the death (in U.S. custody) of Palestinian terrorist leader Muhammad Abbas (a.k.a. Abu-al-Abbas). This document was followed by a highly detailed manifesto on April 18 called "Creed and Methodology," tracing the role of the Naqshbandi Sufi Order through the history of Islamic revolts against foreign occupiers, highlighting the inspirational role of "the revolution of Imam Shamil al-Naqshabandi in Dagestan and Chechnya against Tsarist Russia," and promising to "fight for the unity of the people and land of Iraq and to preserve its Arab and Islamic identity" against Westerners and false Muslims alike.

The Naqshbandi jihadists also began taking retroactive credit for jihadist strikes against the United States going back to December 2003—operations in which their members had participated at the time but which were claimed by mainstream jihadist entities, usually affiliated with al Qaeda. Many of these claims included videos that show the use of Chechen-style bombs, sophisticated rockets, and other weapons contributed by Chechen and "Chechen" fighters. These attacks—including numerous strikes on the Green Zone and an attempt on U.S. Secretary of Defense Robert Gates during a visit to a remote U.S. base on April 19, 2007—highlighted the jihadists' ongoing exploitation of intelligence sources, a fact they trumpeted in their public statements. Another communique announced the formation of a Naqshbandi jihadist "Black Widow" unit of at least sixteen women training for martyrdom operations. Echoing the Chechen theme, the leading female mujahed claimed in the video that her husband was "martyred" and that her brothers "were detained"; she claimed she could find solace only by expressing her "eagerness for martyrdom" to avenge the crimes and humiliation she suffered.

But the most permanent impact of the Chechen jihad has been in spreading its guerrilla ethos throughout the worldwide jihadist movement—an influence clearly reflected in the Army of the Men of the Naqshbandi Order. In an August 20, 2007, statement, they offered a revealing description of the organization and tactics of their army. The Naqshbandi jihadists are organized in "small Jihadist combat groups" of seven to ten mujahedin. To ensure centralized control, "all groups in each region have a regional Amir and the commanders in each governorate are connected with the Amir of the governorate, and all Amirs of the governorates are militarily connected with the [national] command council." The communique stressed the importance of solid training in warfare and modern weaponry, which is conducted in specialized camps. The communique also described the Army's goal as a "long-term attrition war"—a convenient way of shifting the focus away from the present day, when they have little to show for their trouble.

The communique also pointed to its reliance on Chechen-style operational planning procedures that distinguish between centrally controlled spectacular operations and ongoing localized activities. "The [Army's] big and qualitative operations planning are centralized," it explained, but they depend on "decentralization in planning and carrying out the field Jihadist operations . . . to ensure the fast reaction that is required in attack and retreat battles." The Army of the Men of the Naqshbandi Order continues to pay close attention to Chechen-style intelligence operations: "[d]epending on the principle of secrecy and security in planning and carrying out Jihadist operations from one side and penetrating the American enemy intelligence on the other to acquire precise information from the ranks of the enemy." Taken together, the communique offers a portrait of the the Army of the Men of the Naqshbandi Order as a clear derivation of the Chechen jihadist elite forces model, adapted to conditions in Iraq.

The continued flow of Chechen and "Chechen" expert mujahedin also continued to make itself felt on the Iraqi battlefield. The summer of 2007 saw the introduction of new "home bombs"—a classic Chechen weapon used effectively against the Russians in the Second Chechen War, and one that drastically reduced the U.S. forces' chances in urban warfare. Their appearance in Iraq came after a new wave of expert terrorists reached the Middle East from the Caucasus, from the late winter through the spring of 2007. The terrorists booby-trapped houses the mujahedin expected the enemy to search and then positioned snipers to hit additional enemy troops when they rushed in to help their injured comrades. The key to this tactic was meticulous planning, anticipating the moves of enemy troops and building on those expecations to calculate a fiendishly effective series of fuses, bombs, and sniper attacks. "House bombs can be some of the most difficult explosives to detect because of the myriad ways they can be activated," U.S. commanders noted. "The blast can be set off by a trip wire, a pressure plate or a remote device." The appearance of home bombing appears to herald a new round of modernization and improvement among the jihadist forces. "The growing use of house bombs is part of a larger pattern of more complex and coordinated attacks against US forces by al-Qaeda in Iraq," explained a senior U.S. military official in Iraq in early August.

Chechen jihadists also continued to take part in other jihad fronts in the Middle East. In northern Lebanon, a few Chechens fought with the Fatah al-Islam group in the lengthy battle in the Nahr al-Bared refugee camp. On June 24, for example, Lebanese troops killed six Islamist terrorists in a clash with a group of jihadists who were trying to strike at the Abu Samra district of nearby Tripoli. According to Lebanese intelligence, documents found on the bodies identified them as "three Saudis, a Russian from Chechnya, and two Lebanese" who also held European citizenships. A few additional Chechens were killed in the ensuing siege of Nahr al-Bared, and a couple were captured and extradited to Russia.

No less important were the Chechen presence and leading roles in the fighting in the Afghan-Pakistani tribal areas and in southern-central Afghanistan—particularly the Lashkar-Gah area and the Helmand Valley—between February and August 2007.

The presence and combat effectiveness of Chechen and other foreign mujahedin were noted immediately after fighting resumed in February 2007. On February 11, Assadullah Wafa, the governor of Helmand Province, attributed the recent escalation to the arrival of expert fighters who had crossed over from Pakistan. "I cannot tell you an exact figure, but according to our information around seven hundred al Qaeda terrorists have come to the Sangeen and Kajaki districts from Waziristan [in Pakistan]," Wafa said. "They are mostly Chechen fighters, Chinese, Uzbeks, and Pakistanis." The flow of foreign mujahedin accelerated in March as the fighting in the Pakistani border region of Waziristan escalated. One Afghan tribal leader in the Helmand area reported that local Taliban commanders "had already approached" the Uzbek-led foreign mujahedin in Waziristan and urged them to join the Taliban-led jihad in Afghanistan in order to "reinvigorate their campaign of violence against NATO troops." The foreign mujahedin were offered safe passage to Kunar, Paktia, and Helmand Provinces so that they could spearhead the Taliban's spring offensive.

Meanwhile, under mounting U.S. pressure, Pakistani forces launched a new offensive in the South Waziristan tribal zone bordering Afghanistan. In mid-March, an ex-Pakistani Taliban commander, Mullah Maulvi Nazir, ordered all foreign militants to disarm or leave the area. The local Ahmadzai Wazir tribes of Wana, who were sheltering the foreign mujahedin—mostly from Uzbekistan, Turkmenistan, and Kazakhstan—predictably refused, citing the tenets of the Pushtunwali sacred tribal code, and vowed to protect their guests. On March 19, Islamabad unleashed a mercenary tribal force backed by the Army's artillery and helicopter gunships on the tribal forces and their guests. The Pakistani forces were pitted against a force of mujahedin ten thousand strong—composed predominantly of the region's Pushtun tribesmen and reinforced by a hard core of more than a thousand foreign mujahedin led by the Uzbek commander Tahir Yuldashev (locally known as Qari Farooq).

Information about the casualties demonstrates that the foreign mujahedin conducted most of the fighting. On March 23, Ali Muhammad Jan Aurakzai, the governor of North West Frontier Province, reported that roughly 160 mujahedin, "including 130 Uzbeks and Chechens," were killed in the first offensive. "Another sixty-two foreign fighters were arrested during the clashes, including Chechens, Uzbeks, and other foreigners," Aurakzai added. By March 31, the Pakistani military completed yet another offensive sweep against mujahedin encampments. Pakistani interior minister Aftab Sherpao reported that "fierce fighting between tribesmen and foreign militants in the tribal region killed fifty-two people, including forty-five Uzbek and Chechen rebels." Fighting continued to escalate in early April with the foreign mujahedin suffering most of the losses.

By mid-April, as fighting escalated, many of Nazir's tribal forces had second thoughts about siding with Islamabad against their erstwhile Pushtun and mujahedin allies. It didn't help matters that the heavy and indiscriminate bombing and shelling by the Pakistani Army hit numerous innocent civilians—including families of Nazir's fighters. On April 20, Mullah Maulvi Nazir formally changed sides, embracing the cause of the "oppressed people" from Central Asia and all other al Qaeda mujahedin seeking shelter in the tribal lands—including Osama bin Laden. "If he comes here and wants to live according to tribal traditions, then we can provide protection to him because we support oppressed people," Nazir announced. He added that the "thousands of al Qaeda–linked foreign fighters, including Uzbeks, Chechens, and Arabs," who had fled to Pakistan's tribal lands after the jihad in Afghanistan, could return for as long as they respected the Pushtunwali. Nazir expressed his regrets over the roughly three hundred foreign mujahedin and forty-odd tribal fighters killed in the battles.

However, only a few hundred foreign mujahedin returned to the Wana area in South Waziristan. Many moved north to bolst
mujahedin communities in North Waziristan. By late J
Shah, the former security chief for Pakistan's tri
that about two thousand foreign mujahedi
Chechen, and Tajik fighters, along with a s
and organizers"—were being sheltered in fou
of Mir Ali and in territory west of Datta Khel,
took over a few local mud fortresses and perfe

training young Pakistanis and Afghans as suicide bombers and quality fighters and then dispatched them to reinforce the Taliban forces fighting NATO in Afghanistan.

The Taliban capitalized on the arrival of highly trained foreign mujahedin reinforcements to launch a new spring offensive in early May. After a year of relative calm, Taliban forces, supported by units of Chechen and Uzbek elite fighters, mounted an effort to retake the strategic Baylough Bowl from U.S.-led Afghan forces. The foreign mujahedin outperformed the local Taliban forces, exploiting the mountainous terrain to evade the heavy bombardment by U.S. and British fighter jets. Moreover, they planted in the dust tracks numerous sophisticated mines and improvised explosive devices—making it impossible for U.S. and Afghan forces to use their vehicles or heavy weapons. In mid-June, Lieutenant Colonel Rob Walker, the commander of the Canadian battle group in the Shahwali Kot district north of Kandahar City, reported that "Chechen and Arab fighters have flooded" his area of operations and forced an escalation in the fighting. "The insurgents decided they are going to mass within that area, and there are a lot of foreign fighters there—Chechens and Arabs," Walker explained. "It's a bit of a sanctuary, so ISAF [the NATO-led International Security Assistance Force] has decided we need to go up there to confront them, so that's where the fighting is."

The fight around Baylough was still raging in mid-July. With plenty of air and long-range firepower at their disposal, the U.S. and Afghan forces were still fighting to reach and hold the controlling high ground. Squads and fighting teams of elite mujahedin fighters—especially Chechens and Uzbeks, but also Tajik and Turkmen fighters—were constantly maneuvering in the area. They were ambushing the U.S. and Afghan forces on the move, as well as attacking mountaintop disposi-
tions held by Afghan forces. The foreign fighters were better trained
n the Taliban, and they sported military-style uniforms, chest rigs,
armor, and modern Russian-origin weapons including sniper rifles
rtars. "The U.S. troops uniformly hate the insurgents, yet show a
respect for their tenacity and steadily maturing tactics, which
tern-style ambushes and outflanking moves," one visiting

military expert noted. Of the few hundred mujahedin from the Caucasus and Central Asia, U.S. intelligence singled out the dozen or so Chechen fighters, with their mountain-warfare experience, for their fighting skills, tactics, and tenacity. The Chechens managed to dislodge U.S.-led Afghan forces from mountain-peak positions they were trying to hold—usually in daring nighttime attacks. U.S. intelligence brought special teams to track the Chechens' radio transmissions, and Afghan National Police patrols, led by U.S. Special Forces, were sent out to spot them after reported attacks and encounters. Whenever the Chechens were located, the U.S. would direct mortar and artillery fire and airpower at them—but with little success. The Chechens would melt into the rugged mountains, only to reappear in yet another daring attack.

Meanwhile, the elite foreign fighters were spreading throughout Afghanistan—markedly improving the military capabilities of the Taliban forces. NATO forces now frequently reported "encountering complex Taliban ambushes and other maneuvers that increasingly remind them of their own training and tactics," as one British intelligence officer noted. U.S. military reports noted that the foreign mujahedin were equipped with modern weapons, efficient uniforms, body armor, and possibly night-vision devices. In early July, General Dan McNeill, ISAF Commander in Afghanistan, cited evidence that "the militants are being coached by experienced outsiders." Senior Taliban officials readily acknowledged the major impact of the imported fighters on their escalating jihad. "We are always looking for new ways to learn and defeat our enemies; we are fighting with modern forces and have to use modern tactics," Taliban official Zabiyullah Mujahed explained. "Since we are fighting Western troops we have to copy tactics used in other parts of the world, like those used in Iraq." He singled out the contribution made by "Chechens who learned their skills fighting the Russian army in the North Caucasus."

In August, the arrival of "well-trained Jihadists from around the world" in the Helmand Valley tilted the war against the British forces. As the battle escalated, long convoys of Toyota Land Cruisers started racing from Pakistan—carrying Pakistani and foreign mujahedin, weapons, equipment, and supplies. The Helmand campaign had become "one of the most difficult campaigns in British military history," a senior defense official acknowledged. "You have to ask whether British troops should have been sent here in the first place; our presence has only succeeded in attracting trouble," opined another British senior officer. Beyond their expertise in tactics, organization, command and control, and logistics, the Chechens also excelled at reconstituting Taliban units after seemingly debilitating encounters with British forces. "Defeated in the morning, the insurgents strengthened overnight," observed one British officer on site after a devastating raid. The experienced Chechen fighters made a lasting impression on the British: "They look smarter than the Taliban. Most now wear the Pakistan dishdash. Their fire and weaponry is getting more accurate all the time. Someone is training these guys," the British officer explained. The Chechen and other foreign mujahedin also brought with them shoulder-fired SAMs, heavy long-range machine guns with armor-piercing ammunition, and several types of mortar. Chechen-style roadside bombs and expertly concealed mines were taking a heavy toll on the British and Afghan traffic throughout the area. In late August, a new team of Chechens—identified through the monitoring of their initial communications—brought what they called "hi-tech stuff" to the theater, including diverse electronic warfare systems that helped in both monitoring and jamming British communications.

Confronting eroding popular support for the Karzai regime, the constantly improving fighting capabilities of jihadist and tribal forces—bolstered by the Chechen and other foreign mujahedin—and the insufficient NATO forces, it's little wonder that senior NATO officers were heard to suggest that it could take decades before the jihadist forces (popularly lumped together as "the Taliban") are finally defeated. "I am sure that Afghanistan will be a better place in twenty years," Major General Bruno Kasdorf, the highest-ranking German officer at ISAF headquarters in Kabul, predicted in early September 2007.

By summer 2007, the Chechens' ongoing contribution to jihadist fronts and causes around the world was having an aggregate impact. Chechnya had taken a seemingly permanent role as the renewable resource of jihadist inspiration, know-how, and manpower to jihadist movements worldwide. The work of Dara Shukoh, the famous Assamese "poet of Jihad" and the theological leader of the jihadist movement in India's Assam state, seemed newly apropos: "Chechnya is a leader of Jihad in the Muslim world!" Shukoh declared. "The Chechnian history itself speaks of its brave and courageous fight against the Russian invaders. The legends of Dzhokhar Dudaev, Aslan Maskhadov, Shamil Basayev, Abdul-Halim Sadulayev, and now Dokka Umarov are famous all over the Muslim World."

The transformation of violence in the Caucasus from a popular revolt into a small but zealous and secretive Islamist-Jihadist web of networks will continue. While no longer capable of attracting widespread public support or sustained mobilization, the jihadist jamaats still constitute a major threat. Driven by despair and vindictiveness, these tiny cells are ready, much like their jihadist soulmates in the West, to strike out and terrorize the civilian population. The growing importance of the Caucasus as both a transportation artery and an energy avenue from East to West will only increase the importance of the Caucasus for the jihadist movement. Beyond giving them access to hit targets of vital importance for the West—the energy and transportation infrastructure—the Caucasus is also their primary springboard for reaching Central and South Asia, the greater Middle East, and Europe.

Moreover, the growing importance of the Chechen jihadists to the worldwide movement, regardless of their failure at home, makes Chechnya a mythological symbol for the Islamist-Jihadist movement—a symbol they still dream of reclaiming from the Russians. And so the global jihadist movement will continue to invest in the Caucasus, deploying quality fighters and launching new terrorist strikes. Though the region's intelligence services, most notably those of Russia and Azerbaijan, have thwarted most such attempts, they cannot be expected to have one hundred percent success. And so—despite the failure of Chechenization—the destabilized and radicalized Caucasus will remain a focus of world attention, a region whose future may be either peaceful or violent, constructive or cataclysmic.

A NOTE ON SOURCES AND METHODS

Chechen Jihad addresses a very specific issue—the interaction between the revolt in Chechnya, and subsequently the Northern Caucasus, and the international Islamist-Jihadist movement (popularly known as al Qaeda, although in reality it is a far more complex and multifaceted phenomenon). This book is not an exhaustive history of the wars in Chechnya. It does not address operations by the Russian military or the rebel forces, except where relevant; nor does it address the state of the civilian population, beyond its effect on the jihad. Many other authors and scholars have already addressed these issues.

My reason for focusing on the relationship between the revolt in Chechnya and the Islamist-Jihadist movement is that the subject has been largely ignored by most Western authors and scholars of Chechnya and the northern Caucasus. Many have chosen to concentrate on more obvious aspects of the conflicts, examining the country's political/military history and/or the plight of its civilian population, while others—mostly supporters of the Chechen "cause"—have apparently sidestepped the jihadist story for ideological reasons, fearing the association of their pet cause with the jihadist bogeymen (particularly after 9/11).

The Chechen-Jihadist relationship goes far beyond its obvious evidence—the presence of Arab and other foreign volunteers in Chechnya—to include the role of veterans and Chechens in other jihadist fronts, as well as the impact of the combat experience accumulated in Chechnya on the jihadist movement worldwide. Although reports of Chechen mujahedin appearing in other theaters around the world frequently pop up in the popular media, no effort has been made to separate the facts from the myths and to analyze the phenomenon and its implications for the West. Moreover, although the international jihadist movement and its sponsoring states have repeatedly threatened to sabotage and otherwise harm the flow of oil and gas to the West in times of crisis—and although the Caucasus is one of the most crucial arteries of energy supplies to the West—there has been no real study of the potential relationship between the wars in Chechnya and the energy security of the West.

Chechen Jihad concentrates on these important, yet hitherto unknown, aspects of the crisis in the Caucasus.

This book was made possible by access to unique and diverse sources. The key to writing a comprehensive, factual, and objective analysis of the challenging subject at hand is access to original source material—both written and oral—from all sides of the conflict. This book is unique because of the use of Islamist-Jihadist material, Russian material (including captured documents), and diverse material from all areas affected by the Chechen jihad. Only such a wide array of sources allows a comprehensive study of the subject from all sides.

The vast majority of the material in this book comes from standard sources—documents and media reports from the pertinent countries and organizations (which are detailed below). But it is also informed by unique sources that put things in perspective, providing both the overall framework and the distinct details that helped give the story the human touch. After a professional career spanning more than thirty years—sixteen of them as director of the Task Force on Terrorism and Unconventional Warfare of the U.S. House of Representatives—I have been able to draw on friendships with several world leaders, senior officials, and leading experts from around the world. Many of these friendships span well over a decade, and some go back more than a quarter of a century. As I was researching and writing this book after leaving Congress, these friends (some of whom are mentioned in the acknowledgments) have provided extensive help, shared their own knowledge and expertise, and fostered my understanding of their countries' relevant national institutions. It is the help of my numerous friends with both knowledge and access that makes *Chechen Jihad* the special book that it is.

The primary sources I used—including original source material and documents from diverse open sources, as well as from the pertinent governments—can be divided roughly into three major categories:

(1) indigenous Islamist-Jihadist sources from the Caucasus (including Chechnya); (2) indigenous Islamist-Jihadist sources from the Middle East and Central Asia; and (3) indigenous source material from the countries that are fighting Islamist-Jihadist terrorism on their own territory, in their own regions, and worldwide. These open, though not necessarily easily available, sources were bolstered by extensive unique data collected and/or captured by the intelligence and security services of the governments involved (most notably Russia, but many others as well).

Ultimately, however, the key to writing this book lay in the methodology involved in using such sources. Writing any book-length treatment of still-unfolding events is a challenge, and *Chechen Jihad* covers both overt political and theological dynamics and discrete violent undertakings, the events they inspired, and the activities they caused—the ramifications of roads taken and not taken. The leaders involved—political, military, and terrorist—all had a great deal at stake in the outcome. And this reality has direct bearing on the commonly available public record. Throughout the Chechen conflict, its leaders and commanders have used the media (overtly and covertly) to manipulate unfolding political dynamics, to influence the record of their actions, and to control their historical legacy. The truth, and the objective historical record, have frequently been sacrificed on the altars of short-term political expediency and long-term history. And, of course, one must always concede that other unknown dynamics—clandestine contacts, covert operations, political arm-twisting, and handshake agreements—are lurking behind the documents and sources, cloaked in darkness to even the most dogged researcher.

Hence, in writing *Chechen Jihad*, I relied as much as possible on contemporary indigenous sources—that is, material collected from, as well as engagements and communications with, senior officials and professional staff directly involved in the events described. However, the human memory is far from perfect. With so much in the balance—from the fate of peoples to the careers and legacies of leaders—individuals' memories tend to tilt in their own favor. Hence, whenever possible, I relied on minutes and written reports, as well as contemporary documents and "nonpapers," prepared by and for officials and staff themselves for their own use. No less important were the documents and other material captured by various intelligence and security services that I was privy to study and use. As well, I relied heavily both on the vibrant media of all the parties involved—broadcast, electronic (Internet), and printed—as well as on a huge number of personal contacts on all sides.

In recent years, specialized electronic news services and online periodicals have become an indispensable source of timely reporting. Officials and experts alike use these elite outlets as both sources of their own knowledge and primary vehicles for disseminating their own take on events. As such, these newsletters and periodicals constitute a contemporary record of events against which the more private communications can be checked. Of the large quantity of news services, three merit singling out for their unequaled quality: The Middle East News Line (MENL), with its chief editor Steve Rodan, is an indispensable resource for news and analysis on regional security matters. The Johnson's Russia List, compiled and edited by David Johnson of the World Security Institute in Washington, D.C., is a unique and indispensable resource for news and analysis concerning Russia. Comprehensive news and insightful analysis of global security developments is provided by the Global Information System (GIS), with its chief editor Gregory R. Copley. (I must stress that the GIS is objectively an excellent resource, even though I have written several stories for it.)

From the very beginning, the wars in Chechnya elicited a most active and diverse electronic media. Supporters of the Chechen rebels and the Islamist-Jihadist camp established numerous websites in several languages. Of singular importance are the various editions of Movladi Udugov's Kavkazcenter.com (also known as Kavkaz-Tsenter) and Chechenpress, which, like the revolt in Chechnya, has evolved into an Islamist-Jihadist organ. On the other side, Dr. Paul Murphy's RETWA (Russia-Eurasia Terror Watch) website is an excellent resource for news and analysis of the conflict. Regarding the security situation in the Caucasus and jihadist terrorism, the various electronic periodicals of the Jamestown Foundation—particularly *Chechnya Weekly* and *Global Terrorism Analysis*—are especially noteworthy. While frequently opinionated, they nevertheless include a wealth of well-researched information and thoughtful analysis. Also of significance is Yigal Carmon's Middle East Media Research Institute (MEMRI), a most valuable source of annotated translations and analysis of media from the entire Middle East and the Muslim world as a whole. Giora Shamis's DEBKAfile is also noteworthy because of the unique scoops and intriguing analysis of Middle East and Caucasus affairs they repeatedly publish.

Beyond the large array of extensive indigenous source material from the Caucasus, Russia, and the Middle East I've obtained and acquired, additional primary original source material for the book came from numerous West European countries, Central and South Asia, and other parts of the Muslim world. Moreover, I have had extensive conversations and other communications with numerous government officials, diplomats, "spooks" of many stripes, senior security and defense officials, as well as terrorists, militia commanders, émigrés, and otherwise involved individuals from all sides of the conflicts' tangled web of loyalties and associations. These unique sources supplement the large quantities of open sources—primarily regional media—that by themselves provide a wealth of data and documentation. This open source material includes wire-service reports by local and international news agencies; numerous articles from local newspapers, periodicals, and newsletters; numerous articles from newspapers, periodicals, and newsletters of the diverse émigré communities—mainly from the Caucasus, South Asia, and the Arab world—in Western Europe; numerous articles from newspapers, periodicals, newsletters, and academic journals in the United States, Europe, Russia, and so on; transcripts of broadcasts by the local electronic media (mostly translated by the U.S. government's excellent Open Source Center [formerly the Foreign Broadcast Information Service, or FBIS]); and huge quantities of original source material retrieved through the Internet. For background information, a unique collection of primary sources—plus original publications, documents, and reports—developed over more than three decades of intensive research was also consulted. This wide range of sources constitutes a unique database for expert analysis regarding the subjects in question.

Although I have stressed above the numerous friends and colleagues in high places who have been of great help in writing *Chechen*

Jihad, they are not the only human sources who were instrumental to my work. Indeed, the dry definition of "human sources" does not do justice to the actual people involved. Throughout the last thirty-odd years, numerous people have made tremendous contributions to my knowledge and understanding in two major ways.

1. *Many hundreds, if not thousands, of people* from all over the world talked to me, communicated otherwise, and sent stuff from obscure places, at times at a risk to life and liberty. Special thanks to those who patiently told me things and answered what must have been countless overly detailed and/or just plain dumb questions. Thanks to those who sought, acquired, and delivered piles of documents and other material in "funny" languages and illegible scripts. Many of these individuals live and operate "on the other side," and have communicated and provided material at great risk to themselves and their families, because they care about their own countries and peoples. Others, usually members of "the other camp," have communicated because they want to make sure we understand what they stand for and believe in. Theirs was not an easy task either.

2. *The periodicals, newspapers, bulletins, newsletters, communiqués,* and other written material that pour in from remote and forbidding regions of the world tell part of the story, but not all of it. Their quality varies from the absurd to the excellent, as do their reliability and pertinence. But they are all important, for in their wide diversity they constitute an accurate reflection of the area's colorful and vibrant civilization. To help decipher and comprehend the nuances woven into them, I relied on native speakers who patiently translated and explained the multiple layers of meanings and innuendo in the rich, often flowery, and fascinating languages of the Muslim East. Thanks to all the translators and readers who worked with me over the years, teaching me how to "read" the stuff even when I thought I knew the language. (Well, I knew the alphabet.)

Despite the diversity and multitude of the sources I used, and despite my frequent use of published material, precise source-noting is inadvisable in writing about such volatile matters—in large part out of concern for the safety and survival of the human sources. As a rule, the moment a critical work is published, hostile counterintelligence and security organs launch relentless efforts to discover and silence the human sources still in their midst. Whenever such an individual is exposed, that individual, along with his/her family, are punished severely—usually by means of torture and death—in order to deter others. Using "anonymous sources" or "officials" as specific entries in an otherwise academic-style source-noting is not sufficient to protect most human sources—particularly those providing access to most sensitive inside information. Source notes that detail what material was acquired from which human sources only helps hostile counterintelligence and security organs to narrow the scope of their search, better identify the institutions from which the leaks came, and ultimately hunt down the human sources themselves. It has been my professional experience—both as the director of the Congressional Task Force on Terrorism and Unconventional Warfare and as a published author—that when confronted with a monolithic text in which the specific sources have been blurred, the hostile counterintelligence and security organs find it virtually impossible to narrow down their searches and thus stifle the human sources.

To these brave individuals, who provide crucial and distinct information at great risk to themselves and their loved ones, we owe certain fundamental protections. The omission of precise source notes is the least one can do.

ACKNOWLEDGMENTS

Chechen Jihad could not have been written without the help of longtime friends throughout the world—most of our friendships and mutual trust spanning well over a decade, and some going back more than a quarter of a century—who provided the most crucial information included in this book. No less important were the contributions of the many individuals who offered their knowledge and unique source material on a wholly confidential basis, as described in the Note on Sources and Methods. To many of these, I extend my special gratitude for their warm hospitality as well.

Embarking on this undertaking, I discovered just how blessed I've been with good friends, whose help made this book possible. A few can be recognized by name.

Special thanks to Dr. Roman Murashkovsky, of Moscow University Touro, for his friendship and the unwavering support that made this undertaking possible. Roman was among the very first to suggest that I write about the jihad in Chechnya and the Caucasus. He encouraged and helped the project from the very beginning. While I was writing and researching in Russia, Moscow University Touro was more than a home away from home. Roman also helped with introductions to his numerous contacts and associates. As well, special thanks to Dr. Renee Lekach, the rector of Moscow University Touro, for her hospitality, help, and encouragement.

For more than a decade, Major General Evgeniy Nikitenko, of the Russian National Security Council, has been first and foremost a great family friend and a soulmate. Evgeniy is one of the leading experts on the insurgency and jihad in the Caucasus, and his encyclopedic knowledge is second to none. Our long discussions about Chechnya and the Caucasus over the years have been crucial to my comprehension of the region, its intricacies, and subtleties. Thanks to the extended Nikitenko family for their hospitality in Moscow.

Not only is Askar Akayev the first president (1991–2005) and founder of modern Kyrgyzstan, but history will recognize him, because of his work on the National Epos *Manas*, as the father of post-Soviet Central Asia, who saved the region from succumbing to the lures of either jihadism or militant chauvinism. Over the years, we've become close friends. I treasure our numerous long conversations, in which he opened my eyes to the intricacies and subtleties of Central Asia. President Akayev also introduced me to many of his colleagues—introductions that led to the broadening of my knowledge. Thanks for the hospitality and help in both Byshkek and Moscow.

Gregory R. Copley, president of the International Strategic Studies Association, author of the highly acclaimed book *The Art of Victory: Strategies for Personal Success and Global Survival in a Changing World*, and a friend for a quarter of a century, shared his vast knowledge and understanding of world affairs. Always ready to lend an ear to my doubts, he constantly helped me with his sound judgment. Special thanks to Pamela von Gruber, publisher of *Defense & Foreign Affairs: Strategic Policy.*

In Morocco, my very dear friends Ambassador Hassan Abouyoub, the chief foreign policy adviser to King Mohammed VI, and Abdeljebbar Azzaoui, the Director of Counter-Intelligence and Counter-Terrorism (CSDN), provided both warm hospitality and invaluable help. Thanks for the unique insights into the recruitment of terrorists and their travel in Western Europe, North Africa, and such jihadist fronts as the Caucasus and Iraq.

Dr. Darko Trifunovic, of the Security Studies Department of the University of Belgrade, has been a great friend and generous host in Belgrade and Banja Luka. Darko is also the world's leading expert on jihadism in the Balkans, and I benefited greatly from his vast knowledge and understanding. Special thanks for providing me with the Ali Hammad manuscript.

Most of my Russian interlocutors cannot be named because of their continued involvement with the Kremlin and Russia's various security forces—the FSB, the SVR, the GRU, the MVD, the Armed Forces, and so on. This anonymity need not diminish my recognition of their excellence and expertise, as well as my gratitude for all their help and support. A few in Russia can be recognized by name and thanked for the many years of both friendship and professional interaction: Special thanks go to Vladimir B. Rushailo, presently the executive secretary of the Commonwealth of Independent States and formerly the national security adviser to President Putin and minister of the interior (responsible for internal security), for a very long and fruitful cooperation; and thanks to General Aleksandr N. Malinovskiy. Special thanks to General Anatoly S. Kulikov, presently the chairman of the State Duma's Security Committee and formerly the minister of the interior, for sharing his extensive personal experience in, and vast knowledge of, the war in Chechnya, and thanks to his colleagues at WAAF. Special thanks to Ambassador Anatoly E. Safonov, President Putin's special representative for international cooperation in combating terrorism, for our fascinating and thought-provoking conversations. Special thanks to General Valery L. Manilov, member of the Federation Council of Russia, and formerly deputy chief of staff of the Russian Armed Forces, for his erudite analysis of the military situation in the Caucasus. Special thanks to Mrs. Svetlana Miranyuk, director-general of RIA-Novosti, and to her assistant Leonid Burmistrov, for helping this book project in so many ways. And, last but not least, special thanks to Aleksandr A. Sharavin of the Institute of Military and Political Analysis for the many interesting conversations about the security posture in the Caucasus, and for the excellent maps of Chechnya.

Unfortunately, given the subject of this book and my decision not to address other issues pertaining to the Russian-Chechen wars (mainly alleged human rights abuses) because they are not relevant to the book's subject, my interlocutors from Chechnya and other parts of the North Caucasus did not agree to be identified even when the meetings were on neutral grounds, such as European capitals or in Central Asia. Nevertheless, their articulation of the Chechen cause and their in-depth analysis of the situation in the Caucasus were invaluable even when we disagreed. Special thanks to my Jordanian, all Circassian and Chechen, and Pakistani friends and colleagues who helped arrange many of these meetings—a most challenging task, given the subject I was interested in.

Regrettably, the organizers of a most memorable and significant meeting in Washington in the autumn of 1997 still insist on the anonymity of my interlocutor.

Professor Murray Kahl was there with me and for me throughout the chaotic experience of writing a book on still-unfolding events. He helped by reading and commenting on early drafts, locating data on the Internet, "holding my hand" as my computers kept crashing, and just being a good friend.

Other friends rallied as well. Steve Rodan, the director of the Middle East Newsline, was most generous with his vast knowledge and deep understanding of Middle East security affairs. Yoichiro Kawai was exceptionally generous with his impressions and interview material accumulated during many years. Guido Olimpio also shared a lot of material from his unique sources. Dr. Marc Ellenbogen, president of the Prague Society, and I had stimulating and thought-provoking conversations that helped me see things in the right perspective. The indefatigable "Jacques" did what only he can. Dr. Assad Homayoun shared his unique insight and knowledge of Iran. Rosanne Klass kept my files bursting with clippings.

Daniel Bial, my tireless agent, took care of business as I kept typing. Thanks to Judith Regan and Calvert Morgan, who committed to the book from the very beginning. Thanks to the great team at HarperCollins, including Jonathan Burnham, Kathy Schneider, Cindy Achar, John Jusino, Christine Boyd, and Campbell Wharton for their masterful work and their contribution to the book's success.

Last but not least, thanks to my father, Shmuel, for helping with German sources; to my mother, Siona, for helping with French sources and for the flow of clippings from Israel; and to my wife, Lena, for translating and helping with the Russian sources. As well, hugs and kisses to Lena and Masha for enduring my hectic typing and the loud jazz playing into the wee hours and for their love. And special pats for Max, for being there.

INDEX

Abu-al-Walid
  death of, 272
  jihad in Iraq, support for, 306
  Khattab assassination, role in, 228–29
  martyrdom operations, role in, 259, 261
  power and authority of, 230, 231, 234
Abu-Hafs al-Urduni
  death of, 383
  G-8 Summit attack plan, xv, xvi–xvii, xviii
  global jihad, escalation of, 291–92
  power and authority of, 272, 276
Abu-Idris, Amir Abdallah Shamil. *See* Basayev, Shamil
Abu-Musab al-Zarqawi, 209, 238, 240, 254, 311–12, 313, 316–17
Afghanistan
  Anaconda Operation, 214–17
  Chechenization of, 387
  Chechen mujahedin in, 205–6, 323–31, 376, 413–18
  contributions of to jihadist movement, 218–20
  Kunar Province operation, 326–31
  Kunduz seige, 205–6
  quagmire of war in, 386–87
  Taliban and, 199–203, 326, 330
  Tora Bora, battle for, 210–13
  training of mujahedin in, 37–38, 60–61, 159, 203
  UN sanctions declaration against, 148
  U.S. attacks on, 100–101
  Uzbek mujahedin in, 205–6
Afghan mujahedin, 25–28, 38, 58–59, 217–18
Ahmadi-Nejad, Mahmoud, 360, 368, 369–70
Albania, 114–15
Algerian operatives, 254–56
Alkhan-Kala operation, 224
Amirat, Youssuf. *See* Abu-Hafs al-Urduni
Anaconda Operation, 214–17
al-Aqsa Operation, 193–94
Arabization campaign, 50, 61–62, 379
Arafat, Yasser, 190
Armed Islamic Movement, 38
Armenia, 25–27, 371
Arsanov, Vakha, 72, 100–101
Autonomous Soviet Socialist Republic (A.S.S.R.), 17
Azerbaijan
  Afghan mujahedin in, 25–28
  Armenian control of, 26
  attacks from Iran, 365–67
  Baku Islamist network, importance of, 101–2
  British intervention in, 15
  Chechen conflict, resistance to involvement in, 174–78
  Chechen support for, 176, 178
  commitment to struggle against Islamist-Jihadists, 28, 360–61
  as false Muslims, 408–9
  German interest in, 16
  Islamist-Jihadists' networks in, 360–68, 370–71, 409–10
  jamaats in, 362–63

Azerbaijan *(cont'd)*
  jihad against, Iranian support for, 367–68, 390
  oil pipeline through, 174–76, 373–74, 393, 409
  oil pipeline through, sabotage operations against, 369, 372, 374, 375
  population of, 367
  supply and logistical-support routes through, 169–74, 178
  U.S.-Azerbaijan relationship, 366–67, 388, 390
  U.S.-led block to control oil and gas resources, 373–74

Badran, Fuaz, 196–97
Balkans, 1–2, 3, 275, 280, 310
Barayev, Arbi Alaa-Eddin, 223–24
Basayev, Shamil
  Beaulieu summit, 131–32
  Beslan school seige, 281, 282, 285–86
  bombing attack plan for Russia, 241–45
  break with Maskhadov, 119
  cease-fire agreement with Russia, 296–97
  challenges to Maskhadov's authority, 298–99, 300
  committment to jihad of, 189
  Dagestan, invasion of, 106, 111, 133–34
  death of, xvii, 352, 383
  drug smuggling by, 29
  Dubrovka Theater seige, ix, 242–43, 252–53
  foreign workers, terrorism threat against, 108
  G-8 Summit attack plan, ix–x, xi–xiii, xiv, xv
  global jihad, escalation of, 293–94
  government role of, 69, 71, 73–74, 96, 317, 351–52
  jihad, escalation of, 303
  Khattab's influence on, 41
  leadership role of, ix, 43, 148
  Magomedov assassination, role in, 229–30
  martyrdom operations, role in, 259, 260, 261, 266–67
  on Moscow bombings, 136, 139, 243
  Nalchik operations statement, 343–44
  nuclear terrorism incident, 49
  power and authority of, 35–37, 54, 231, 233–34
  power-sharing agreement, 233
  raids and sabotage operations of, 53–54
  responsibility claims by, 183, 292–93
  strong-arm strategy of, 294
  suicide terrorists announcement, 138
  support for Palestinian Intifadah, 191–92, 279
  terrorism denouncement, 136
  terrorist strikes against Russia, 44–45, 46, 85, 94, 163–64, 231–32
  U.S. and Russia as enemies, 119–20
  wounding of, 156, 225
Basayev, Shirvani, 29, 35, 187
Beslan school seige, ix, xv, 274, 280–90, 291, 292–93
bin Laden, Osama
  asylum negotiations, 109, 111
  evacuation from Afghanistan, 206–7
  headquarters of in Afghanistan, 200, 201–2, 205
  health of, 157–59
  importance of Islamist-Jihadists to, 101–2
  Islam, war between West and, 357–58
  jihad, management of, 41, 94–95, 354
  jihad, support for, 132–33, 140
  jihad against U.S., 151–52, 305–6
  as leader of HizbAllah International, 57
  leadership role in jihadist movement, 148
  master plan for fighting in Iraq, 311
  nuclear suitcase bombs for, 102
  safe haven in Pakistan for, 207
  Shield Fatwa, 297, 318, 319
  terrorism against Russia, support for, 143–44
Black Widows, 257–59, 262–70, 277–78, 308–9, 411–12
bombs and bombings. *See also* martyrdom operations
  Beslan school seige, 289
  bomb-making technology, 279
  instructions for, distribution of, 332
  nuclear suitcase bombs, 102–5
  passenger jet bombings, 274, 277–79, 290
  Russia, operations in, 135–38, 139, 405–6

shoe bombs, 279
training for martyr-bombers, 261
Uzbekistan operations, 112
Vladikavkaz operations, 113
Budennovsk raid, 44–45
Buinaksk raid, 75–82, 83, 85

caliphate in the Middle East, 309–12, 317
A Campaign with the Turks in Asia (Duncan), 10
Caucasian common market, 83–84
Caucasus
"All-Caucasus Security Structures," 315–16
Crimean War theater, 8–12
Islamicization of culture of, 50, 121–23, 232
Islamist's claim to, 1
jihad call by Tagayev, 32–34
jihadist front in, 2, 405–8, 419
liberation of, 134, 177–78
martyrdom operations in, 265–66
radicalization of Islam in, 339
regional jihad, focus of in, 334
revolt against Russia, support for, 15–17
Russian "liberation," resistance to, 14–15
Russification of, 20
Russification of, reversal of, 33, 42
self-identities, importance of, 13
Shamil and struggle against Russia, 7–13, 17–19
strategic importance of, 392–93
terrorist strikes in, 112–13
U.S. intervention in, 71
Central Asia
Chechen population exiled to, 17
Livo's goals for, 207–9
U.S.-led block to control oil and gas resources, 373–74
Chechenization
active areas of conversion by, 3–4
concept of, 2–3
failure of, 386, 419
funding for, 3
pan-Turkism and, 4–5
process of, 3, 5, 386–88
success of, 294
Chechen jihad
call for, 32–34, 113
collapse of, 376, 408
committment to, 189
effectiveness of, decline in, 335–36, 345–53, 394–404
escalation of, 233–35, 275–77, 290–94, 302–4, 353, 405–8
fatwa in support of, 144–45, 149–51
funding for, 167–74, 352–53
global jihad, escalation to, 143–52, 305–6, 323, 390, 395–404
importance of to jihad movement, 95, 101–2, 354–55, 358–59
influence of foreign mujahedin on, 48, 160–61, 184–86, 232–33
internal fighting in, 336–37
Iraq war, impact on, 410–12
nationalist character of, 232
planning for, 121
popular support for, loss of, 232–33, 271–72, 273–74, 276, 295–96, 334–35, 337–38, 347–50, 353, 385–86, 389
regional jihad, escalation to, 42–43, 54, 57–61, 82–83, 176–77, 299–300, 389–90
reorganization of, 378, 402–3
revenge against population, 349
Russian crackdown on, 223–28, 234–35
support for, 35, 38–39, 40–41, 44, 117, 307–8, 335
training for, 36–38
Chechen Mafiya
activities of, 23, 28–31
bandits involved in, 347
dialysis machine, smuggling of, 158
financial support for jihad, 167–68
nuclear suitcase bombs through, 102–5
Chechen mujahedin
Afghanistan, operations in, 205–6, 323–31, 376, 413–18
amnesty for, 378, 380
Anaconda Operation, 214–17
Asia and Africa, operations in, 376
Azerbaijan energy resources, operations against, 374–76
bond with Islamist-Jihadist leadership, 157–59
characteristics of force divisions, 346–47
definition of, 4

Chechen mujahedin *(cont'd)*
disenchantment of, 161, 183–84, 225–27, 234–35, 255–56, 336–37
effectiveness of, 348–49, 398–99, 412–19
fierce fighting by, 205–6, 213, 326
global jihad, role in, 4
IED construction by, 306–7
Iraq, operations in, 413
Kosovo, operations in, 115–16
Lebanon, operations in, 220–22, 413
Livo role of, 208–9
organization of, 42–43
Pakistan, operations in, 323–31, 413–18
raids and sabotage operations of, 62–64, 130, 186–87
recruitment of, 335
Second Chechen War, 156–57
skills of, 4
as support for Palestinian Intifadah, 192–96, 279
Tora Bora, battle for, 210–13
training of, 4, 60, 67–68, 124–27, 204
training of Palestinians by, 195–97
Chechnya
Alkhan-Kala operation, 224
"All-Caucasus Security Structures," 315–16
Arabization of culture of, 50, 61–62, 379
Arab support, failure to gain, 68, 70–71
beginning of crisis in, 21
cease-fire agreement with Russia, 41–42, 46–47, 54–58
contributions of to jihadist movement, 218–20
counterfeit currency operations, 30–31
drug smuggling in, 29
economy, destruction and disruption of, 127–28, 167–68
education of population of, 20–21
employment of population, 20–21
exile of population to Central Asia, 17
First Chechen War, 28, 259, 379
humanitarian work in, 63–64
independence declaration by, 21–22, 68
independence negotiations with Russia, 71–73, 131–32
independence of, 56, 315–16, 377
Islamicization of, 38, 39, 47–48, 50–51, 61–62, 122–23, 313–14
Islamicization of, resistance to, 379, 385–86
Islamicization of, reversal of, 95–96
Islamic movement, base for, 64
jihad call by Tagayev, 32–34
jihadist efforts to Islamicize the conflict, 28–31
martyrdom operations in, 261–63
negotiations with Russia, 22–24, 66, 84–86, 131–32, 187–88
pacification policy of Russia for, 12
peace and stability in, 334
private armies, disbanding of, 70
private armies, training of, 124–27
pro-Russian attitudes in, 271–73, 295, 334, 378–81, 391
radicalization of Islam in, 339
reconstruction of, 377–78, 385–86, 392
role in Working Strategy, 310
Russian raids on, 47–50, 76
Second Chechen War, 137–38, 155–57, 259, 379
security forces and counterinsurgency efforts, 378–83, 386, 393, 401–2
self-identities, importance of, 13, 385–86
Shamil and struggle against Russia, 7–13, 17–19
Sharia practice and enforcement, 62, 70, 73, 74, 106–7, 313–14
terrorist-sponsoring nation, transformation to, 94–95, 108–11
unification with Dagestan, 121–22
violence in by mujahedin, 107–8
war cabinet, 317–18
war stance of, 56–57
weapons for, 29–30, 124
withdrawal of Russian troops from, 55, 297, 337
chemical and biological weapons
French recovery of, 255
training camps for use of, 238–39, 240–41, 253–56
Turkey, attack on, 238–39
WMD strategy, 141–43, 293
Churchill, Winston, 15, 16
Confederation of Caucasian People, 90

Confederation of Mountain People (KNK), 37, 90
Congress of Peoples of Ichkeria and Dagestan, 111, 115
Congress of Peoples of Ichkeria and Dagestan, Second, 120–23, 131
Congress of the Islamic Nation, 89–90
counterfeit currency operations, 30–31
Crimean War, 8–12

Dagestan
  Buinaksk raid, 75–82, 83, 85
  characteristics of force divisions, 346–47
  fighting in, 40, 349
  importance of to jihad movement, 82–83
  invasion of by Islamic Army of Dagestan, 133–34
  Islamicization of culture of, 87–89, 121–22, 134
  jihad movement in, 32–34, 81, 405–8
  Makhachkala riots and strikes, 93–94, 97, 98
  regional jihad, escalation of in, 338–40
  Russian raids on, 51, 76
  unification with Chechnya, 121–22
drug smuggling operations, 29, 200
Dubrovka Theater seige, ix, 242–43, 245–53
Dudayev, Dzhokar
  cease-fire agreement with Russia, 41–42
  death of, 54
  funding network of, 167
  Islam as premise for war, 25, 34–35
  negotiations with Kremlin by, 23–24
  power and authority of, 21–22, 44, 49
  Russification of, 21
  stance against Russia, Islamic pressure on, 48
  support for, 22–23
  weapons agreement with Pakistan, 29–30

European Union, 318, 331–33

Fatah, HAMAS cooperation with, 198
fatwa declarations
  on martyrdom operations, 179–83, 280, 308–9
  Shield Fatwa, 297, 318, 319, 390
  in support of Chechen jihad, 144–45, 149–51
foreign workers, terrorism threat against, 108
France, 15–16, 254–55, 332–33
French-Algerian operatives, 254–56

Gaza Strip, 359
Georgia
  Abkhaz war, 36–37
  Afghan mujahedin covert operations against, 27
  Chechen support for, 176, 178
  jihadist training camp in, 236, 237–39, 240, 253–56
  support for jihad, 128
  U.S.-led block to control oil and gas resources, 373–74
  visas for mujahedin, 139
Germany, 16–17, 332
Great Britain
  Caucasian common market, 83–84
  chemical and biological attack plan, 255–56
  French-Algerian operatives in, 254, 255–56
  jihad, call for support for, 146, 147–48
  London attacks, 319–21, 354–55
  support for anti-Soviet popular revolt, 15–16
Group of 8 (G-8) Summit attack
  decoy operation as diversion, xi–xv
  jihadist plan for, ix–xi, 350
  weapons for strike, xv–xvii
Grozny, Chechnya
  fighting in, 22, 39
  martyrdom operations in, 261–63
  military action against, 24
  oil and petroleum resources, 20
  raids and sabotage operations in, 53, 54
  Second Chechen War, 155–57
Gudermes, Chechnya, 49

HAMAS
  Chechen mujahedin connection to, 197–98
  Fatah cooperation with, 198
  jihad against Israel, 317–18
  martyrdom operations, 308–9
  suicide bombings, responsibility for, 197

HAMAS *(cont'd)*
support for Chechen jihad, 307–8
support for Islamist-Jihadists, 279–80
Hammad, Ali, 41, 52–53
Hamzah, Mustafa, 209
Harakat-ul-Ansar, 60
Har Dov bombing, 195
HizbAllah International
in Azerbaijan, 367
bin Laden as leader of, 57
Chechnya, activity in, 58
Israeli-Lebanese border confrontation, 372–73, 375, 376, 384–85
training by, 60
Hizb-i-Islami, 25, 35, 37–38
Holy War, or How to Become Immortal (Tagayev), 117–19
hostage taking
Beslan school seige, ix, xv, 274, 280–90, 291, 292–93
Dubrovka Theater seige, ix, 242–43, 245–53
jihadist preference for, x
Penza hostage incident, 63
Hurricane Katrina, 321–22
Husseini, Hajj Amin al-, 16

Ibn al-Khattab
assassination of, 227–28
asylum negotiations, 109, 110
Buinaksk raid, 75–82
Dagestan, invasion of, 106, 111, 133, 134
inspection of positions by, 88
leadership role in jihadist movement, 93, 148
Moscow bombings, support for, 136, 137
near-miss attack on, 70
power and authority of, 40–41, 67, 184, 225
private army of, 125–27
raids and sabotage operations of, 53–54, 66–67
responsibility claims by, 183
terrorist strikes against Russia, threat of, 163–64
terrorist training by, 67–68, 77, 125–27
WMD strategy, 142
wounding of, 225

Imamat, 90
improvised explosive devices (IEDs), 306–7
Indonesia, 3
Ingushetiya (Inguishetia)
ambush operations in, xvii, 91–92
characteristics of force divisions, 346–47
mujahedin fighting in, 349
regional jihad, escalation of in, 339–40
Russian raids on, 76
terrorism operations in, 283–85
International Committee of the Red Cross (ICRC) incident, 63–64, 66–67
Intifadah, 190–98, 279
Iran
as advocate for Islamist-Jihadists, 28
Azerbaijan, attacks on, 365–67
Azerbaijan, secret war against, 367–68, 390
Baghdad summit on Iraq, 372
counterfeit currency operations, 30–31
global strategy development, 359–60
Islamist-Jihadists, avoidance of, 359
Islamist-Jihadists command cell in, 209
nuclear program of, 368, 369
oil war strategy, 365–66, 368–76
role in Working Strategy, 310
Russian-Iranian relationship, 359–60
support for Chechen jihad, 38, 44
training of mujahedin in, 61
Iraq
Baghdad summit on, 372
bin Laden's master plan for fighting, 311
Chechenization of, 3, 387
Chechen mujahedin in, 413
Islamist-Jihadists operating in, xiii
jihad against U.S. in, 209, 305–7, 317–18, 410–12
jihad front in, 356–58
jihadist training camp in, 237–40
al Qaeda's goal in, 317
quagmire of war in, 386–87
regional jihad, call for, 357
Russian diplomats kidnapping, xii–xv
Islam
false Muslims, 408–9
global war within, 1–2, 6
as premise for Chechen war, 25
war between West and, 355–59

Islambuli, Muhammad Shawqi al-, 208–9
Islamic Army of Dagestan, 133–34
Islamicization campaign
in Caucasus, 50, 121–23, 232
in Chechnya, 38, 39, 47–48, 50–51, 61–62, 122–23, 313–14
in Chechnya, resistance to, 379, 385–86
in Chechnya, reversal of, 95–96
in Dagestan, 87–89, 121–22, 134
jihadist efforts to Islamicize the conflict, 28–31
Islamic Movement of Uzbekistan (IMU), 164–66
Islamic Partisans League, 154–55
Islamist caliphate, 309–12, 317
Islamist-Jihadists
anti-Semitism of, 100, 190–91
assassination campaign of, 272–73
Azerbaijan networks, 360–68, 370–71, 409–10
Caucasus, importance of, 419
challenges to Chechen government, 99–100, 106–7
characteristics of force divisions, 346–47
Chechen conflict, influence over, 28–31, 54, 232–33
Chechen jihad, integration with, 4, 95
Chechnya leadership role, 45–47
communication networks, 322, 331–32
Dagestan humanitarian activities, 97–98
effectiveness of, decline in, 345–53
expansion networks of, 209
financial and logistical support systems, 127–28, 167–74
foreign workers, terrorism threat against, 108
G-8 Summit attack plan, ix–xix, 350
guerrilla warfare strategy, 139
HAMAS support for, 279–80
importance of to bin Laden, 101
Iraq operations, xiii
Palestinian Intifadah, support for, 190–98, 279
passports and visas for, 130, 139
power and influence of, 95–100, 120–23, 386
right to determine fate of Caucasus, 177–78
security breach, 140
sleeper agents in Moscow, 237
support for, 128
terrorist strikes against Russia by, 44–47, 91–94
threat from Russian intelligence to, xvi–xix
training camps, 236, 237–41, 253–56
WMD strategy, 141–43
Islamophobia, 314–15
Israel
Great Ramadan Offensive, 316–17
Israeli-Lebanese border confrontation, 372–73, 375, 376, 384–85
jihad against, 317–18
U.S.-led block to control oil and gas resources, 373–74

JaishAllah, 367
jamaats
in Azerbaijan, 362–63
definition of, 110
importance of, 299–300, 303–4, 334
in Kabardino-Balkariya, 340, 344–45
radicalization of, 404–8
jihad. *See also* Chechen jihad
Chechen mujahedin, importance of to, 412–19
contributions of Chechnya and Afghanistan to, 217–20
effectiveness of, decline in, 345–53
escalation of, 129–30
global jihad, escalation of, 291–94, 322–23
global jihad, launch of, 199
global jihad, objective of, 312
global jihad, threat from, 331–33
in Iraq, 356–58
against Israel, 317–18
as obligatory duty of Muslims, 181, 354
oil war as jihadist objective, 369
in Palestine, 355–56, 358–59
regional jihad, escalation of to, 357–59, 390
suppression efforts of Russian intelligence, 380–85
against U.S., 113, 151–52, 219–20, 292, 294
against U.S. in Iraq, 209, 305–7, 317–18, 410–12
"The War Can Be Stopped Only by War" (Stomakhin), 317–18

jihad *(cont'd)*
"Working Strategy Lasting Until 2020," 309–12
Zawahiri's management of, 95, 129
Jihad (Yandarbiyev), 363
Jordan, 196–97

Kabardino-Balkariya
jamaats in, 300, 340
mujahedin fighting in, 339–40, 344, 349
Nalchik operations, 281, 340–44, 345
Kadyrov, Akhmad, 271–72, 273, 379–80
Kadyrov, Ramzan, 378–79, 380, 384, 389, 390–92
Karachay-Cherkessia, 340, 344–45
Kaspiysk, Dagestan, 62–63
Katrina, Hurricane, 321–22
al-Khattab, Ibn. *See* Ibn al-Khattab
Khodov, Vladimir Anatolievich, 285–86
Knights Under the Prophet's Banner (Zawahiri), 217–20
Kosovo, 115–16
Kunduz seige, 205–6
Kunta Haji Tariqat, 14, 379

Lebanon
Chechen mujahedin operations in, 220–22, 413
fighting in, 359
Israeli-Lebanese border confrontation, 372–73, 375, 376, 384–85
Russian embassy, attack on, 153–55
Lebed, Aleksandr, 54–55, 57, 102, 104
Lezghins, 27–28
Libya, 31
Livo, 207–9
Lone Wolf terrorist raid, 51

Magas (Ali Musaevich Taziyev)
Beslan school seige, xv, 284–85
death of, xvii
G-8 Summit attack plan, xv
terrorist strikes, role in, 285
Magomedov, Ibragim, 227–30
Makhachkala, Dagestan, 93–94, 97, 98
martyrdom operations
battalion dedicated to Russian cities, 138, 236–37
Black Widows, 257–59, 262–70, 277–78, 308–9, 411–12
in Caucasus, 265–66
in Chechnya, 261–63
fatwa on, 179–83, 280, 308–9
by HAMAS, 308–9
motivation for martyrs, 268–70
of mujahedin, 39
passenger jet bombings, 274, 277–79
in Russia, 259–61, 263–65, 266–68
training for martyr-bombers, 196–97, 261
Maskhadov, Aslan
anti-Semitism of, 100
assassination plan for, 113, 301–2
cease-fire agreement with Russia, 54–55, 296–98, 301
Chechnya as base for campaign against Russia, 64–65
Dubrovka Theater seige, 248
election of, 68
eulogies to, 302, 303, 304
government organized by, 68–69, 73–74
on impact of Moscow bombing plan, 243
Khattab assassination, role in, 228–29
military buildup under, 123–24
negotiations with Russia by, 66, 68–69, 119, 131, 188, 296–98, 301
power and authority, ascent to, 21, 54
power and authority, challenges to, 95–100, 106–7, 119, 123, 298–99
power and authority, loss of, 114, 228–29, 234–35, 295–96
power-sharing agreement, 232
state sponsored terrorism, alignment with, 108–11
Transcaucasian energy company agreement, 84
Massud, Ahmad Shah, 201–2
Muhammad, Omar Bakri, 145–46, 147, 150, 152
mujahedin. *See also* Chechen mujahedin
anti-Semitism of, 190–91
approval of by al Qaeda, 52
characteristics of force divisions, 346–47
effectiveness of, 161–63
infighting and animosity within, 225–27
influx of into Chechnya, 52–53, 59–61, 139–40, 184–86
Islamicization of Chechen struggle, 39

jihad call by Tagayev, 33–34
Martyrdom operations, 39
Nalchik operations, 340–44, 345
power and influence of, 67
recruitment of, 344–45
training of, 34, 52, 59–61, 77, 90–91, 140–41, 159, 174, 203–4
Mujahedin Shura Council, xiii–xiv
Muridism, 50
Muslim population
achievements in Russia by, 33
false Muslims, 408–9
jihad call by Tagayev, 32–34

Nalchik, Kabardino-Balkariya, 281, 340–44, 345
Namangani, Juma, 207–8
Naqshbandiya Sufi brotherhoods, 14, 410–12
National Liberation Army (UCK), 114–15
Northern Alliance, 199, 201–2, 205–6
nuclear terrorism, 49, 102–5

oil and petroleum resources
Azerbaijan pipeline, 174–76, 373–74, 393, 409
Azerbaijan pipeline, sabotage operations against, 369, 372, 374, 375
Baku-Tblisi-Ceyhan (BTC) pipeline, 373–74
German interest in, 16
Grozny-area refineries, 20
Mafiya threat to, 23
oil war strategy of Iran, 365–66, 368–76
Operation Pike, 15–16
pipeline negotiations, 70, 71
raids and sabotage operations against, 53
Transcaucasian energy company, 84
U.S. interest in, 174–75
U.S.-led block to control, 373–74
oligarchs, 22, 167–68, 175
Our Struggle, or The Imam's Rebel Army (Tagayev), 32–34, 87

pacification policy, 12
Pakistan
as advocate for Islamist-Jihadists, 28
Chechen mujahedin in, 323–31, 413–18
drug trade of, 35
evacuation airlift operation, 206–7
intelligence service, 35
jihad, call for support for, 147
sheltering of bin Laden by, 207
support for Chechen jihad, 35, 44
training of mujahedin in, 37–38, 60, 61, 159, 203–4
weapons agreement with Dudayev, 29–30
Palestine
Chechenization of, 3
Intifadah, 190–98, 279
jihad front in, 355–56, 358–59
training for martyr-bombers, 195–97
Pankisi Gorge training camp, 236, 237–39, 240, 253–56
Parkuyev, Abdulla, 120, 121–23
Passport and Visa Service, Ministry of Shariah National Security, 130
Penza hostage incident, 63
Pike Operation, 15–16
Putin, Vladimir, 164–66, 249–50, 298, 301

Qadiriya Sufi Tariqat, 14, 379
al Qaeda
approval of mujahedin by, 52
assistance in the G-8 Summit attack plan, x
European networks, expansion of, 221–22
expansion through Chechenization, 3
goal of in Iraq, 317
nuclear suitcase bombs for, 102–5
weapons for G-8 strike, xv–xvii

Raduyev, Salman
ascent to power by, 54
Lone Wolf terrorist raid, 51
near-miss attack on, 70
Penza hostage incident, 63
terrorism campaign of, 67, 69–70
war stance of, 56–57, 89
WMD strategy, 141
Rahman, Omar Abdel, 146
Ramadan, 149, 188, 292
Great Ramadan Offensive, 316–22
Red Cross incident, 63–64, 66–67
Russia
achievements of Muslims in, 33
anti-Islam policy of, 120

Russia *(cont'd)*
Beslan school seige, ix, xv, 274, 280–90, 291, 292–93
bombing attack plan for, 241–45
Caucasus, importance of, 392–93
Caucasus jihadist front, containment of, 2
cease-fire agreement with Chechnya, 41–42, 46–47, 54–58
Chechenization, toll of, 5–6
Chechnya as base for campaign against, 64–65
Chechnya attack, plans for, 114
Chechnya raids by, 47–50, 186–87
computer seizure, 140
Dagestan raids by, 51
Dubrovka Theater seige, ix, 242–43, 245–53
embassy attack in Lebanon, 153–55
ferry seizure in Russia, 52
harassment campaign of, 76
heroic vision of Muslim revolt, 19
Iranian-Russian relationship, 359–60
Israeli-Lebanese border confrontation, 384–85
kidnapping and execution of diplomats, xii–xv
Kosovo, Russian troops in, 115–16
lessons from Russia's Caucasus experience and, 386–88
martyrdom operations in, 259–61, 263–65, 266–68
negotiations with Chechnya, 66, 71–73, 84–86, 119, 131–32, 187–88, 301
nuclear terrorism incident in, 49
pacification policy of, 12
passenger jet bombings, 274, 277–78, 290
passenger jet hijacking, 193–94
raids and sabotage operations against, 53–54, 56
Second Chechen War, 137–38, 155–57
security of, 315–16
Shamil and Caucasus wars, 7–13, 17–19
terrorism against civilians in, 135–38, 139, 163–64, 232
terrorist strikes against, 44–47, 62–64, 66–67, 91–94, 97–99, 110–11, 117, 143–44, 231–32, 274–75
war with Turkey, 9–11
Western support for, 314–15
withdrawal of troops from Chechnya, 55, 297
Russian intelligence services
Basayev and, 35, 36
breakthroughs and crackdown on Chechen jihad, 223–28, 234–35
Chechen security forces and counterinsurgency efforts, 378–83, 386, 393, 401–2
Cossacks' intelligence gathering, 10–11
infiltration of Islamist-Jihadist movement by, xvi
jihadist movement suppression efforts, 380–85
weapons shipment for G-8 strike, xvi–xix

The Sabres of Paradise (Blanch), 7–8
Sadulayev, Abdul-Khalim, 303–4, 317, 348, 350–51
Sadval, 28
Salman Pak training camp, 239–40
Saudi Arabia, 68, 144–45
SEAL team ambush, 327–31
"Security in Exchange for Independence" (Udugov), 313–16
Seif, Muhammad bin-Abdallah al-, 280, 308
Shaba Farm bombing, 195
Shali, Chechnya, 30
Shamil, Imam, 7–13, 15, 17–19, 122
Shamil, Said, 15–17
Shamil Operation, 17
Sharia practice and enforcement, 62, 70, 73, 74, 106–7, 313–14
shoe bombs, 279
Siddiq, Muhammad, 319–20
Stomakhin, Boris, 317–18
Sudan, 38, 100–101
suicide bombings. *See* martyrdom operations
al-Suwailim, Samir bin Salakh. *See* Ibn al-Khattab
Switzerland, 333

Tagayev, Magomed (Muhammad)
as Basayev spokesman, 134
Holy War, or How to Become Immortal, 117–19

Our Struggle, or The Imam's Rebel Army, 32–34, 87
strategy recommendations, 294
Tajikistan, 48
Taliban
Chechen mujahedin fighting with, 416–18
collapse of, 408
control of Afghanistan by, 199–203
Kunar Province operation, responsibility claim for, 329–30
negotiations with, 109–10, 201–2
popular support for, 326, 330
sheltering of bin Laden by, 148
Taziyev, Ali Musaevich. *See* Magas (Ali Musaevich Taziyev)
terrorism. *See also* bombs and bombings; hostage taking; martyrdom operations
claims of responsibility for, 183
gang activities, 128
Lone Wolf terrorist raid, 51
nuclear terrorism, 49, 102–5
responsibility claims, 290–93
against Russia, 44–47, 62–64, 66–67, 91–94, 97–99, 110–11, 117, 143–44, 231–32, 274–75
Russia, bombing attack plan, 241–45
Russian aircraft, hijacking of, 193–94
against Russian civilians, 135–38, 139, 163–64, 232
Russia to blame for, 65, 97, 107
Tora Bora, battle for, 210–13
Transcaucasian energy company, 84
Turkey
air base, interest in, 84
biological weapons attack, 238–39
manipulation of Muslim revolt, 18–19
pan-Turkism, 4–5, 174
Russian ferry seizure in, 52
training of mujahedin in, 61, 174
U.S.-led block to control oil and gas resources, 373–74
war with Russia, 9–11
weapons purchase by, 30

Udugov, Movladi
Chechen forces in Kosovo, 115
Congress of the Islamic Nation, 89–90
diplomatic relationship with Afghanistan, 100
Dubrovka Theater seige, 248
government role of, 317
independence negotiations with Russia, 72
jihad, escalation of, 302–3
role in Basayev's controlled jihad, 234
Russia to blame for terrorism, 65, 97, 107
"Security in Exchange for Independence," 313–16
Sharia practice and enforcement, 74
on strength of Chechen jihad, xviii
terrorist strikes against Russia, threat of, 85, 164
Umarov, Dokka
on effectiveness of mujahedin, 348–49, 398–99
G-8 Summit attack plan, xvii
power and authority of, 234, 351, 377–78
support for jihad, need for, 352, 353
U.S. energy interests, 374
United States. *See also* war on terrorism
Anaconda Operation, 214–17
attacks on Afghanistan and Sudan, 100–101
Azerbaijan-U.S. relationship, 366–67, 388, 390
Baghdad summit on Iraq, 372
Caucasian common market, 83–84
Chechenization, failure to understand, 5–6, 387
eviction from Caucasus region, 110
global megatrends, disregard for, 2
Great Ramadan Offensive, 316–17
intelligence operations of, 206–7
intervention in Caucasus, 71
jihad, call for support for, 146
jihad against, 113, 151–52, 219–20, 292, 294
jihad against in Iraq, 209, 305–7, 317–18, 410–12
Katrina as punishment from Allah, 321–22
Kunar Province operation, 326–31
Kunduz seige, 206
lessons from Russia's Caucasus experience for, 386–88

United States *(cont'd)*
oil and petroleum resources and, 174–75, 373–74
pan-Turkism, political ramifications of, 5
security of, 315–16
Tora Bora, battle for, 210–13
Uzbekistan, 112, 164–66
Uzbek mujahedin, 205–6, 325, 326

Walid, Abu-al-. *See* Abu-al-Walid
"The War Can Be Stopped Only by War" (Stomakhin), 317–18
war on terrorism
global war within Islam and, 1, 6
as inevitable collision of megatrends, 217
lessons from Russia's Caucasus experience and, 386–88
starting point of, 1
weapons of mass destruction (WMD) strategy, 141–43, 293
Whirlwind Operation, 259–60, 264, 265
"Working Strategy Lasting Until 2020," 309–12
World Islamic Front, 149, 153
World War I, 14–15
World War II, 15–17

Yandarbiyev, Zelimkhan
assassination of, 249
cease-fire agreement with Russia, 56
Dubrovka Theater seige, 248
on Islamic Chechnya, 64
Jihad, 363
role in Basayev's controlled jihad, 234
support request from, 159
uniting Muslims against Russia, 122, 176
Yelstin, Boris, 22, 24, 41–42, 131
Yugoslavia, 114–15

Zarqawi, Abu-Musab al-. *See* Abu-Musab al-Zarqawi
Zawahiri, Ayman al-
arrest of, 58–59
devotion to bin Laden and his organization, 158–59
evacuation from Afghanistan, 207
inspection of networks by, 58, 59, 275–76, 306
jihad, support for, 146
jihad themes, videotapes of, 319, 320–21, 354–57, 358–59
Knights Under the Prophet's Banner, 217–20
management of jihad by, 95, 129
on nuclear suitcase bombs, 104–5
oil war as jihadist objective, 369
Pakistan, importance of, 324
recommendation for Chechnya by, 66
Working Strategy, support for, 310, 311–12